196
W9-CEU-137

international®
AIR POWER
REVIEW

629.13
INT
v.7

AIRtime Publishing
United States of America • United Kingdom

Butler Area Public
218 North McK
Butler, P

international® AIR POWER REVIEW

Published quarterly by AIRtime Publishing Inc.
US office: 120 East Avenue, Norwalk, CT 06851
UK office: CAB International Centre, Nosworthy Way,
Wallingford, Oxfordshire, OX10 8DE

© 2003 AIRtime Publishing Inc.
F-15E cutaway © Mike Badrocke/Aviagraphica
P-47 side profiles © Osprey Publishing Ltd.
Photos and other illustrations are the copyright
of their respective owners

Softbound Edition ISSN 1473-9917 / ISBN 1-880588-48-X
Hardcover Deluxe Casebound Edition ISBN 1-880588-49-8

Publisher
Mel Williams

Editor
David Donald   e-mail: airpower@btinternet.com

Assistant Editors
John Heathcott, Daniel J. March

Sub Editor
Karen Leverington

US Desk
Tom Kaminski

Russia/CIS Desk
Piotr Butowski, Zaur Eylanbekov   e-mail: zaur@airtimepublishing.com

Europe and Rest of World Desk
John Fricker, Jon Lake

Correspondents
Argentina: Jorge Felix Nuñez Padin
Australia: Nigel Pittaway
Belgium: Dirk Lamarque
Brazil: Claudio Lucchesi
Bulgaria: Alexander Mladenov
Canada: Jeff Rankin-Lowe
France: Henri-Pierre Grolleau
Greece: Konstantinos Dimitropoulos
India: Pushpindar Singh
Israel: Shlomo Aloni
Italy: Luigino Caliaro
Japan: Yoshitomo Aoki
Netherlands: Tieme Festner
Romania: Danut Vlad
Spain: Salvador Mafé Huertas
USA: Rick Burgess, Brad Elward, Mark Farmer (North Pacific region),
Peter Mersky, Bill Sweetman

Artists
Mike Badrocke, Valeriy Bulba, Chris Davey, Zaur Eylanbekov,
Mark Rolfe, Mark Styling, Tom Tullis, Iain Wyllie, Vasiliy Zolotov

Designer
Zaur Eylanbekov

Controller
Linda DeAngelis

Origination by Universal Graphics, Singapore
Printed in Singapore by KHL Printing

All rights reserved. No part of this publication may be copied, reproduced,
stored electronically or transmitted in any manner or in any form whatsoever
without the written permission of the publishers and copyright holders.

International Air Power Review is published quarterly in two editions
(Softbound and Deluxe Casebound) and is available by subscription or as
single volumes. Please see details opposite.

Subscriptions & Back Volumes

**Readers in the USA, Canada, Central/South America and the rest
of the world (except UK and Europe) please write to:**
AIRtime Publishing, P.O. Box 5074, Westport, CT 06881, USA
Tel (203) 838-7979 • Fax (203) 838-7344
Toll free 1 800 359-3003
e-mail: airpower@airtimepublishing.com

Readers in the UK & Europe please write to:
AIRtime Publishing, RAFBFE, P.O. Box 1940,
RAF Fairford, Gloucestershire GL7 4NA, England
Tel +44 (0)1285 713456 • Fax +44 (0)1285 713999

**One-year subscription rates (4 quarterly volumes),
inclusive of shipping & handling/postage and packing:**
Softbound Edition
USA $59.95, UK £48, Europe EUR 88, Canada Cdn $99,
Rest of World US $79 (surface) or US $99 (air)
Deluxe Casebound Edition
USA $79.95, UK £68, Europe EUR 120, Canada Cdn $132,
Rest of World US $99 (surface) or US $119 (air)

**Two-year subscription rates (8 quarterly volumes),
inclusive of shipping & handling/postage and packing:**
Softbound Edition
USA $112, UK £92, Europe EUR 169, Canada Cdn $187,
Rest of World US $148 (surface) or US $188 (air)
Deluxe Casebound Edition
USA $149, UK £130, Europe EUR 232, Canada Cdn $246,
Rest of World US $187 (surface) or US $227 (air)

Single-volume/Back Volume Rates by Mail:
Softbound Edition
US $16, UK £10.95, Europe EUR 18.50, Cdn $25.50 (plus s&h/p&p)
Deluxe Casebound Edition
US $20, UK £13.50, Europe EUR 22, Cdn $31 (plus s&h/p&p)

All prices are subject to change without notice.
Canadian residents please add GST. Connecticut residents please add sales tax.

**Shipping and handling (postage and packing) rates
for back volume/non-subscription orders are as follows:**

	USA	UK	Europe	Canada	ROW (surface)	ROW (air)
1 item	$4.50	£4	EUR 8	Cdn $7.50	US $8	US $16
2 items	$6.50	£6	EUR 11.50	Cdn $11	US $12	US $27
3 items	$8.50	£8	EUR 14.50	Cdn $14	US $16	US $36
4 items	$10	£10	EUR 17.50	Cdn $16.50	US $19	US $46
5 items	$11.50	£12	EUR 20.50	Cdn $19	US $23	US $52
6 or more	$13	£13	EUR 23.50	Cdn $21.50	US $25	US $59

John Heathcott

It is with deep sadness that we report the passing away of John Heathcott
after a long illness. He had fought the illness and borne his discomfort
with typical stoicism, and in the same calm yet resolute manner
with which he had approached life.

We had worked alongside John for many years, both at Aerospace
Publishing (for whom he was, among many other things, editor of Wings
of Fame) and at AIRtime Publishing (with whom he was responsible for
the historic subjects in International Air Power Review and other
publications). His vast knowledge of historic aviation and unstinting
devotion to getting the job done right were crucial to the excellence
of the publications. More importantly, 'Kiwi' was a great friend to us —
both in and out of the office.

When John came to Aerospace Publishing from his previous employment
in the RAF Museum research department at Hendon, his abundant
enthusiasm for aviation was readily apparent, and was to grow ever
greater over the coming years. This was matched by a love of all
things to do with railways, and his knowledge and passion for that
subject equalled those for aviation.

However, what came first in John's life was his family, and it is to
them — here in the UK and 'back home' in New Zealand — that our
most heartfelt sympathies go out. We shall miss him greatly.

Volume Seven
Winter 2002

CONTENTS

Acknowledgments
We wish to thank the following for their kind help with the preparation of this issue:
Kevin Smith/Pilatus
COL Marty Post, LT COL Bernard Krueger, MAJs Jon Hackett, Kevin Hudson, Mike Huff, John Ostrowski, Tim Patrick, John Peck, James Reed, CAPTs Tanya Murnock, Scott Trail, LTs Kevin Hyde, Jeremy Yamada, SGT Eric Cantu, and the many others that helped with the WTI article.
Grumman History Center
Anatoliy Artemyev, Sergey Skrynnikov

MAJOR FEATURES PLANNED FOR VOLUME EIGHT
Focus Aircraft: Su-30 family: second-generation Sukhois, **Warplane Classic:** De Havilland Mosquito: fighters and reconnaissance variants, **Technical Briefing:** OH-58 Kiowa Warrior in Bosnia, **Photo-feature:** Lithuania, **Air Power Analysis:** Yugoslavia, **Air Combat:** The rescue of Mussolini, **Type Analysis:** Beriev Be-10 'Mallow'

PROGRAMME UPDATE

Lockheed Martin F/A-22

Having flown its 2,000th hour on 7 June, the USAF-led F/A-22 Combined Test Force at Edwards AFB continues to support ongoing development work, including expansion of the missile launch envelope. Raptor 4004 was sent to Eglin AFB for testing in the McKinley Climatic Laboratory. Operations in temperatures ranging from -65°F (-54°C) and 120°F (49°C) were examined, as well as operations in rain and snow. The trials ended in mid-August.

On 17 September 2002 the Raptor was rechristened F/A-22 to reflect its attack capabilities, which are provided by the JDAM. Other highlights of the 2002 programme included the first supersonic AIM-9 Sidewinder and AIM-120 AMRAAM firings.

In the second half of 2002 the Raptor programme began gearing up for the Dedicated Initial Operational Test and Evaluation (DIOT&E) phase, scheduled to begin in the summer of 2003. DIOT&E will demonstrate the Raptor's lethality, survivability and reliability in a simulated operational environment, and will be handled by the Air Force Operational Test and Evaluation Center (AFOTEC).

DIOT&E will involve five aircraft, comprising four primary aircraft (4008, 4009, 4010 and 4011) and one back-up (4007). To prepare them for the trials all require modifications to bring them to a roughly common standard, as close as possible to the initial production configuration. Raptor 4008 was delivered to Lockheed Martin's Palmdale works on 2 July,

Above: Raptor 4003 launches an AIM-9 Sidewinder for the first time while flying supersonically.

Right: For maximum combat persistence at the expense of stealthiness, the F/A-22 can carry weapons (or fuel tanks) on wing pylons, seen here undergoing flight test for the first time. As well as the 'compressed carriage' AIM-120C AMRAAM, the F/A-22 can carry 'regular' AIM-120As on the wing pylons.

emerging again on 8 October in full DIOT&E configuration. Raptors 4010 and 4011 followed in late 2002. Meanwhile, 4007 was modified at Edwards, while 4009 was modified at Marietta.

Raptors 4010 and 4011 are the first of eight

Production Representative Test Vehicles (PRTV), and the first F/A-22s to be built with production funding. 4010 was officially handed over to the USAF at Marietta on 23 October, before being sent for DIOT&E upgrade.

PROJECT DEVELOPMENT

India

More LCA progress

Confirmation was announced in August of limited series production authorisation for eight indigenous Light Combat Aircraft (LCA) at HAL's Bangalore facility, following an MoU signed between India's Aeronautical Development Agency (ADA), as design authority, and Hindustan Aeronautics Ltd, as manufacturers. All initial LCAs will be powered by GE's 80.1-kN (18,000-lb) F404-F2J3 after-burning turbofan.

ADA's LCA development is still at an early stage, although the second technology demonstrator aircraft (TD2) made its fifth flight, of only 26 minutes, on 17 August, piloted by Sqn Ldr Sunith Krishna. The two LCA technology demonstrators are being supplemented by five prototype aircraft, the first of which (LCA-PV1), was recently rolled-out at Bangalore. It was expected to join LCA TD1 and TD2 in flight-development by the year-end.

South Korea

T-50 begins flight development

After delays of about eight weeks because of software and wiring problems, the T-50 Golden Eagle Mach 1.4

advanced jet-trainer made a successful 39-minute first flight at KAI's Sachon facility on 20 August. The T-50 is being developed jointly by Korea Aerospace Industries and Lockheed Martin Aeronautics, with RoKAF support.

The prototype was rolled out in September 2001, more than 100 days ahead of schedule, and will be joined by three other aircraft to complete flight certification. The second aircraft was completing its ground-checks last summer, to start flying from November. Two structural ground test T-50s have also been built, on which static fatigue trials began last January.

In addition to 50 T-50s and 44 A-50 lead-in fighter/attack versions with multi-mode radar and armament, scheduled for RoKAF procurement, KAI and Lockheed Martin are planning joint international marketing. Initial production, in which Lockheed Martin has a 20 percent share, is planned for autumn 2003, with RoKAF deliveries to follow from 2005.

United Kingdom

Nimrod MRA.Mk 4 rolled-out

Roll-out from the Woodford factory of the first production Nimrod MRA.Mk 4 aircraft (PA1) on 16 August was a major milestone in BAE System's trou-

bled £2.2 billion fixed-price remanu-facturing programme. Intensive ground-testing then started, to achieve a first flight before the year-end. Subsequently, however, BAE admitted that further delays of up to a year were likely before its initial flight, now expected in late 2003. These were due mainly to alignment problems encountered with fitting new wings to the original largely hand-built fuselage-keels. Structural strengthening was also required of the new wing centre-section.

Pressure on production schedules was already being applied at Woodford, for revised delivery dates for the 18-aircraft programme, including the first three for development, specified between 2004-07. Original MRA.Mk 4 contractual targets included a first flight date of February 2000, initial deliveries of the 21 aircraft then being planned from December 2001, and delivery completion by June 2006. After delays costing BAE some £300 million, a revised May 1999 contract set a PA1 first flight date in early 2002, and service entry by 2005.

In February 2002, the MoD deleted three aircraft from the production programme, to save about £360 million ($550 million) in support costs, while maintaining the required opera-tional capabilities from the 18 MRA.Mk 4s' improved productivity. An initial acceptance level to an interim agreed standard was then planned, with the

first contracted MRA.4 delivery in August 2004. Full capabilities were expected with the seventh aircraft delivery and service entry with No. 120 Squadron at Kinloss, contracted for March 2005. The RAF should receive all 18 MRA.4s by 2008, to replace its last Nimrod MR.Mk 2s.

United States

Updated F-16 radar being tested

The USAF's 416th Flight Test Squadron (FLTS) has begun flight-testing a new version of the AN/APG-68 radar that equips the F-16C/D at Edwards AFB, California. The Northrop Grumman AN/APG-68(V)9 radar, offers a 30 percent increase in detection range, along with significant improvements in resolution, growth potential and supportability. It is better equipped to operate in dense electromagnetic envi-ronments and resist jamming. The radar, which will be installed in advanced Block 50/52 variants of the Fighting Falcon, also features improved air-to-air and air-to-ground capabilities. Development testing of the radar began in July 2001 on Northrop Grumman's modified BAC 1-11, which serves as a flying test bed, at its facilities at Baltimore Washington International Airport in Maryland. This testing is scheduled to continue through 2003, while integra-tion testing continues at Edwards. Initial flight testing is being performed

On 21 August Raptor 4003 fired the first supersonic AMRAAM. The aircraft was travelling at Mach 1.2 at an altitude of 12,000 ft (3658 m) over the Pacific test range. Launching the AMRAAM at high supersonic speeds is one of the keys to the F/A-22's mission – launching missiles at high speeds gives them much greater range and shorter fly-out times.

Bell-Boeing V-22 Osprey

MV-22B BuNo. 164942 (EMD prod. no. 10), which is currently the sole flight test aircraft, completed 15 hours of flight testing on 5 July 2002 and subsequently entered into a 35-hour phased maintenance inspection cycle at NAS Patuxent River, Maryland. At the conclusion of the inspection it returned to flight status, completed at least one full conversion between vertical flight to forward flight, reached 250 kt (463 km/h) in forward flight, and climbed to 11,600 ft (3536 m). A second aircraft joined the flight test programme in September 2002 (BuNo. 164940/EMD prod. no. 8). Two CV-22Bs are undergoing testing at the Air Force Flight Test Center at Edwards AFB, California, where BuNo. 164941 (EMD prod. no. 9) is currently supporting testing of the suite of integrated radio frequency countermeasures (SIRFC). The aircraft was recently suspended in the facility's anechoic chamber where it will undergo three months of testing that will verify the interoperability of its electronic systems. BuNo. 164939 (EMD prod. no. 7), the second CV-22B, returned to flight status in the summer.

The Bell-Boeing Joint Program Office recently received a $1.502 billion modification to an existing contract that covers the procurement of nine Fiscal Year (FY) 01 and nine FY02 MV-22B low-rate initial production (LRIP) aircraft for the Marine Corps, along with two FY02 CV-22B production representative test vehicle (PRTV) aircraft for the USAF. MV-22B delivery will be completed by September 2005 and the CV-22Bs no later than October 2005.

The first two PRTV F/A-22s flew in late 2002, comprising Raptor 4011 (left) and 4010 (below). The latter was the first to be handed over to the USAF, and is seen in full USAF camouflage and markings. It wears the 'OT' tailcode for AFOTEC's 53rd Wing, which will perform DIOT&E from Edwards and Nellis. 4011 is seen prior to painting, but with its tailcode applied.

---

on a modified Block 50 F-16C owned by the Hellenic Air Force. Northrop Grumman delivered the first production (V)9 radar in April 2002 and the system will be installed in the first Block 52+ F-16C, which is scheduled for delivery to Greece in October.

Advanced Hawkeye
Northrop Grumman will begin testing a new radar system for the E-2C Hawkeye being developed as part of the Radar Modernization Program (RMP) aboard a rotodome-equipped NC-130H operated by VX-20 at NAS Patuxent River, Maryland. Flight testing of a radar transmitter, receiver, antenna and rotary coupler began in early fall 2002 and continued for a period of six months. During the test programme the NC-130H will not be equipped with the E-2C workstation and data will be transmitted to a ground station . Under development by Lockheed Martin, the new radar is a solid-state, electronically-steered UHF design. Its antenna allows for continuous 360° coverage but with the added ability to perform electronic steering for critical target detection and tracking. Although the new antenna will not need to rotate it will be installed in the aircraft's existing rotodome. The RMP-equipped Advanced Hawkeye will also have theatre missile defence capabilities, multi-sensor integration, a tactical cockpit giving the co-pilot the capabil-

ity to function as a fourth mission system operator, a new communications suite, new generators, improved identification friend-or-foe system, and an updated mission computer and software. Under current plans the first prototype will be delivered in 2007, followed by the production RMP/Advanced Hawkeye in 2008.

USCG Deep Water contract
On 25 June 2002 the US Coast Guard awarded Integrated Coast Guard Systems (ICGS) an $11 billion contract to modernise the ships, aircraft, command and control, and logistics systems that make up its 'Deepwater' forces. ICGS is a partnership that comprises Northrop Grumman and Lockheed Martin. As part of the programme, the Coast Guard will acquire 35 fixed-wing aircraft that will replace the HC-130s and HU-25s currently in service, 34 helicopters, 76 fixed- and rotary-wing unmanned aerial vehicles (UAV) and modernisation for 93 helicopters.

Airborne laser aircraft flies
The first YAL-1A Airborne Laser (ABL) aircraft flew for the first time on

The first prototype KAI T-50 takes off from Sachon on 20 August 2002 for the type's first flight. Lockheed Martin is providing technical assistance, as well as developing and supplying avionics, digital flight control system and wings.

19 July 2001 from Boeing's modification facility at McConnell AFB, in Wichita, Kansas. The extensively modified 747-400 took off at 3:30 pm (local time) on its inaugural flight and flew a 90-minute flight plan that verified the aircraft's aerodynamic performance and system operation. Subsequent testing performed at Wichita will include complete systems functional checks and flight tests to verify aerodynamic performance, and surveillance system checkout. Following flight-worthiness tests, the ABL aircraft relocated in late 2002 to Edwards AFB, California, where its tracking and high-energy laser system will be installed. The ABL system will use a chemical laser aboard the aircraft to shoot down missiles in their boost phase of flight.

X-47A UCAV completes first taxi
The Northrop Grumman X-47A Pegasus experimental unmanned combat air vehicle (UCAV) completed its initial low-speed taxi test at the Naval Air Warfare Center Weapons Division, China Lake, California on 19 July 2002. The test, which comprised five segments, was conducted autonomously after initiation by a ground controller. The test validated start and stop taxi commands and allowed the vehicle to move down the runway in increments of increasing length from 20 to 300 ft (6 to 91 m). The UAV will be used to demonstrate aerodynamic qualities suitable for autonomous flight operations from an aircraft-carrier as part of the company's naval unmanned combat air vehicle (UCAV-N) programme.

UPGRADES AND MODIFICATIONS

Canada

Modifications for Sea Kings
The Canadian Air Force has begun equipping 12 of its 29 CH-124 Sea King maritime helicopters with a defensive electronics suite that will protect the aircraft against missile attacks. The modifications, which include warning sensors, radar warning receivers and a chaff system used to confuse inbound missiles, will be completed by the end of this summer. The Canadian Air Force had been trying to equip the Sea King with a defensive suite since 1991 when five CH-124s were equipped with an interim system during the Gulf War. A prototype for the new defensive system has been undergoing evaluation since 1999. Throughout the interim period Canadian Sea Kings have operated in Somalia, Bosnia and the Gulf with no defensive systems.

Egypt

Boeing to update Chinooks
Boeing's Military Aircraft and Missile Systems Group in Pennsylvania has received a $132 million contract covering the remanufacture of six CH-47C helicopters to the CH-47D configuration for the Egyptian air Force. The modifications will be completed by 31 August 2004.

Peru

Upgrade funding sought
Increases of nearly 40 percent were being sought over the $650 million 2002 national defence budget for the coming year, for essential equipment procurement, upgrade and maintenance programmes. Current year air force allocations have already been increased by over $12 million for essential maintenance, to return four of the FAP's long-grounded 17 Antonov An-32Bs and all five Lockheed L-100-20 turboprop transports to service.

Upgrade or refurbishment are also planned for the FAP's 10 Dassault Mirage 2000Ps and 18 MiG-29s, together with the upgrade of two Mi-8Ts to Mi-8MTV-1 configuration. Procurement of up to eight new Kazan-built Mi-17MD assault helicopters is being considered as an alternative to restoring some of the

FAP's 14 existing Mi-17s, of which fewer than six are currently airworthy. Peru's army is similarly placed with its 23 Mi-17s.

Among increased deliveries of Bell Huey II modernisation kits in Latin America, totalling over 85 to mid-2002, Peru's Department of State received eight kits in June, for installation in the remaining FAP Bell UH-1Hs.

Russia

VVS Il-76s to be re-engined
Ilyushin is to re-engine and upgrade "a small number" of the 200 Russian air forces (VVS) Il-76M military transports, to meet ICAO noise, emission and air traffic management standards, for international operation. Main changes involve replacement of the original 117.7-kN (26,455-lb) Shvetsov D-30KP-2 engines by 143.8-kN (35,200-lb) Perm/P&W PS-90A-76 turbofans, costing about $12 million per aircraft. Ilyushin AK General Director Viktor Livanov said that the redesignated Il-76MD-90 should complete certification by mid-2003.

The upgraded Il-76MD-90s will be generally similar to the stretched PS-90-powered Il-76MFs ordered for the VVS, which was due to receive its first two of these aircraft from the Tashkent Chkalov Aircraft Plant in Uzbekistan earlier this year. More VVS Il-76MF orders are planned when funds permit. The three Il-76-based A-50Eh long-range AWACS aircraft now being modified by Beriev in Russia, for installation by IAI of Elta Phalcon phased-array radar from a $1.05 billion Indian contract, will also have PS-90A-76 engines.

United Arab Emirates

AH-64A Longbow upgrades
Congressional approval was received earlier this year for a $1.5 billion upgrade of 30 Abu Dhabi air force Boeing/MDH AH-64A attack helicopters to AH-64D standards, with mast-mounted Longbow fire-control radar and associated systems. Also included are 240 AGM-114L Longbow RF Hellfire laser-guided autonomous fragmentation missiles and 49 AGM-114M anti-ship versions.

The US has also supplied price and availability briefings to Saudi Arabia to upgrade its 12 AH-64As to AH-64Ds.

Tornados on test – seen at the Decimomannu weapons test facility/range in Sardinia are the first Mid-Life Update aircraft for the AMI (above), under test with Alenia, and an RSV aircraft (below) carrying two instrumented MBDA Storm Shadow rounds.

Saudi Arabia is additionally interested in acquiring another 12 Longbow Apaches.

United Kingdom

Harrier GR.Mk 7A flies
Flight development of the first RAF BAE Harrier GR.Mk 7 with an uprated Mk 107 turbofan started at Warton on 20 September. The successful initial 45-minute sortie followed the December 1999 MoD contract to replace the GR.Mk 7's original 95.6-kN (21,500-lb) thrust Pegasus Mk 105s with Mk 107 engines, similar to the Pegasus 11-61s in the US Marine Corps AV-8B Plus Harrier IIs.

The uprated, lower-maintenance Pegasus Mk 107 develops an extra 13.34-kN (3,000-lb) thrust in ambient temperatures above 30°C. This allows improved hover and weapons bring-back performance at maximum operating weights of 15422 kg (34,000 lb) instead of 14515 kg (32,000 lb), even in hot and high conditions.

Rolls-Royce delivered the first Pegasus 107, remanufactured from a Mk 105 powerplant, from an order for 40, originally with options for 86 more, on 30 November 2000. BAE Systems Customer Solutions & Support (CSS) division also received an MoD contract for modification kits, including new composite rear-fuselage sections, for installation in 40 redesignated Harrier GR.Mk 7As, in a combined £150 million RR/BAE programme.

These will maintain a force of 30 fully-upgraded aircraft, flown by RAF and RN pilots, in three squadrons in Joint Force Harrier (JFH), for which the first 20 GR.Mk 7As will be completed by April 2004. The remain-

der will receive Mk 107 turbofan installations during the concurrent Harrier GR.Mk 9 weapons systems upgrade programme from a bridging contract with BAE last April, to GR.Mk 9A standards. BAE was then still negotiating a full GR.Mk 9 avionics upgrade contract, costing at least £540 million.

By 2007, JFH will effectively become a single-type unit, although with a mixture of 70 or more Harrier GR.Mk 7/7A and GR.Mk 9/9A variants. It will also operate nine of 12 two-seat Harrier T.Mk 10 V/STOL conversion trainers being upgraded with GR.Mk 9 avionics, while retaining their original Pegasus Mk 105 powerplants, to become T.Mk 12s in an additional squadron.

United States

Customs Service Orions
Lockheed Martin was recently awarded a $27 million contract to upgrade four P-3B Airborne Early Warning (AEW) aircraft operated by the US Customs Service (USCS) Air and Marine Interdiction Division (AMID). The programme will provide the aircraft, which are equipped with the AN/APS-145 radar, with a common glass cockpit that features new flight management system, digital autopilot, digital engine instruments, tactical display, updated mission system and upgraded radar – along with a common communications and navigation system update. Modification work began at the contractor's facility in Greenville, South Carolina, in July 2002 and it will be completed by September 2003. The USCS operates eight P-3B AEW aircraft along with eight 'slick' P-3A/Bs that are equipped with a variety of surveillance systems.

Hawkeye modification
Northrop Grumman has received a $15.5 million contract covering the performance of standard depot-level maintenance (SDLM) and modification of one Group II E-2C to the latest

The Bulgarian air force has converted two of its four Mi-17PP electronic warfare helicopters into firefighters. All EW equipment is stripped out and a single 3-tonne water tank and dispenser is installed in the cabin. The tank is built by TEREM's Georgy Benkovsky facility.

This VS-38 S-3B carries an AGM-84H SLAM-ER missile after the weapon was cleared for service in September.

Right: Testing has begun of a Boeing E-6B Mercury with an EFIS cockpit and updated communications systems.

Hawkeye 2000 configuration. The aircraft will be returned to the Navy by February 2004.

AWACS test aircraft upgraded

Known as Test System 3 (TS-3) JE-3C serial 73-1674 – which serves as the test aircraft for the USAF airborne warning and control system (AWACS) fleet – recently underwent a lengthy and extensive nine-month modification and refurbishment effort at Boeing's Modification Center in Wichita, Kansas. The aircraft has been the principal AWACS test asset since its delivery in 1973 and has been the proving vehicle for virtually all AWACS modifications. Beginning in June 2001 the aircraft was provided with internal and external improvements to strengthen the airframe structure and increase power and cooling capacities that will support multi-sensor command, control intelligence, surveillance and reconnaissance (C²ISR) systems integration. As part of the airframe structural upgrades, Boeing fitted larger cheek fairings, which will house reconfigurable antennas to support future technology experiments. In addition, the aft section and tail received modifications that will support the evaluation of trailing wire technologies. Following the modifications, the aircraft returned to Boeing Field in Seattle, Washington, where it is normally based.

Updated E-6B flies

Flight-testing of the first E-6B Mercury equipped with 'glass' cockpit displays began on 1 August 2002 at Boeing's Maintenance and Modification Center in Wichita, Kansas. The cockpit displays are developed from those used on the latest generation Boeing 737 commercial airliners. During the initial flight the aircraft, which serves as a communication link between national command authorities and strategic land- and sea-based nuclear forces missiles, was put through a series of basic manoeuvres, including acceleration, deceleration and banked turns to ensure the instruments, displays and flight management systems operated properly. Flight-testing will be accomplished by a joint Boeing/Navy test team over the next three months. In addition to the instrumentation, the aircraft's communications suite has been upgraded to provide enhanced capabilities and the flight test programme includes verification of the new battle management, command and control communications mission equipment. The first of 15 E-6Bs is scheduled to enter a modifi-

cation programme in late 2002 and the last modified aircraft will be delivered by 2005.

MH-47E modification awarded

The US Special Operations Command has awarded Boeing a contract to convert two additional CH-47D helicopters to the MH-47E special operations aircraft (SOA) configuration. The new aircraft will replace two Chinooks lost recently in Enduring Freedom.

C-5 AMP modifications

The first C-5B aircraft to be inducted into the C-5 Avionics Modernization Program (AMP) arrived at Lockheed Martin's facility in Marietta, Georgia, and installation of the associated hardware and software began on 13 June 2002. The aircraft, serial number 85-0004, is assigned to the 60th Air Mobility Wing at Travis AFB, California. A C-5A serves as the second prototype and modification of that aircraft began in the summer of 2002. First flight of the completed C-5B is scheduled for February 2003. The two prototypes are part of the Development Integration and Test (DIT) phase of the Engineering and Manufacturing Development (EMD) contract. The EMD phase is expected to last 23 months. Under the DIT phase the aircraft's analogue flight and engine instruments are being replaced by seven 6-in (152-mm) by 8-in (203-mm) flat panel liquid crystal displays. Navigation equipment is being updated through the installation of a dual imbedded, 12-channel Global Positioning System/Inertial

Navigation System. A new communications suite includes an updated SATCOM and HF datalink, an enhanced ground proximity warning system (EGPWS) and traffic alert and collision avoidance system (TCAS).

C-130 LAIRCM

The USAF has awarded Northrop Grumman a contract to develop an infrared countermeasures system for its fleet of C-130 transport aircraft. The two-year, $12.9 million contract covers the engineering and manufacturing development of the large aircraft infrared countermeasures (LAIRCM) system, which will provide the Hercules with protection against heat-seeking missiles, and includes production options for the installation of the system in seven aircraft.

Ukrainian upgrade for L-39C

During the Aviasvit XXI exposition in Kiev (14–18 September 2002) an upgraded demonstrator for the L-39C Albatros trainer was shown. This upgrade is the result of co-operation by four companies: the Odesaviaremservis military repair plant in Odessa; ZMBK Progress engine design bureau at Zaporozhye, the Motor Sich engine production plant at Zaporozhye, and the IAI Lahav Division. The upgrade is aimed principally at the Ukrainian air force, although there are no orders at present. Ukraine presently has 329 L-39C trainers, with a further 293 aircraft in storage.

The modernised aircraft has been equipped with the Progress/Zaporozhye AI-25TLSh engine (Sh for *shturmovoi*, attack); the engine differs from the basic AI-25TL by having an additional 'combat' operating range with increased thrust and reduced throttle response time, improving the dynamic characteristics of the aircraft. The new operational range of the engine will be used mainly for employing the L-39 in the light attack role. Further plans provide for the modernisation of the engine control system and for increasing the engine's service life. Three AI-25TLSh engines have already been produced. With upgraded engine fitted, the demonstrator flew for the first time at Odessa on 13 June 2002, and had

The cockpit of the Ukrainian/Israeli upgraded L-39C is vastly different from before. A single large MFD is fitted, with an Upfront Controller and head-up display. HOTAS controls are installed. The cockpit's modernity is mirrored in the new navigation system.

made 16 flights at the time of writing, at which point recorded data was being analysed. Total running time for the 5,000 AI-25TL engines used in 37 countries amounts to around 6 million hours.

However, the most important alteration is the almost completely new avionics suite developed by IAI's Lahav Division. It makes the cockpit of L-39C trainer similar to those of contemporary combat aircraft and enables combat training to be undertaken (a 'virtual' radar imitates air-to-air combat on the monitor screen); the whole avionics suite is interconnected via Mil Std 1553 databuses. The cockpits are each equipped with a multi-function liquid crystal display, while the front cockpit has been equipped with a head-up display (HUD). A considerably updated flight-navigation system includes inertial navigation system, gyro-stabilised platform and air data computer. Hands on throttle and stick (HOTAS) principles have been implemented. The aircraft with upgraded avionics flew at Odessa on 30 August 2002.

The upgrade is planned to be undertaken as aircraft undergo a life extension rework (to 30 years), allowing the L-39C to remain in operation until a new-generation training aircraft enters service. Odesaviaremservis plant in Odessa, Ukraine, which is to perform the L-39C upgrades, also repairs MiG-21s, MiG-23s, MiG-27s and L-39s, as well as their engines. The plant also converts An-24 and An-26 transports into passenger carriers. The company also offers modernisation of MiG-21s with the Sura helmet-mounted target designation system combined with R-60 and R-73 air-to-air missiles.

Piotr Butowski

New radars for Ka-27/28 upgrade

Two radars, competing for the modernisation of the Ka-27 'Helix-A' shipborne ASW helicopter (and its Ka-28 export version) were shown at the hydroaviation exposition at Gelendzhik, Russia, at the beginning of September 2002. The new radar will replace the present Initsiativa-2KM. Both the Russian Navy and India, which operates 14 Ka-28 helicopters, have ordered a Ka-27/28 upgrade, and it is expected that China, which has operated the Ka-28 since 1999, will also be interested in the new radar. The radar is intended for detecting surface and air targets, for guiding missiles, and for navigational purposes.

Moscow-based Phazotron-NIIR is offering the Kopyo-A (or Alba), which is a version of the Kopyo (spear) fighter radar used in the upgrade of MiG-21bis fighters for India. Commonality with the production radar is a great advantage of Kopyo-A. Its individual modules, including the processing system and algorithms, have already been tested. Four Kopyo-A radars are now under construction; one of them is already undergoing bench tests and will be installed in the helicopter at the end of 2002. Phazotron considers the 360° mechanical scanning antenna made of glass-reinforced plastic to be a great advantage because of its light weight, low price and greater strength by comparison with a metal antenna. The Kopyo-A is capable of detecting an air target (with RCS of 5 m²) from a distance of 70 km (43 miles), a small surface object (RCS 1 m²) from 30 km (19 miles), a motor boat (RCS 150 m²) from 130 km (81 miles) and a large naval ship from the distance of the horizon. The radar operates with a 3-cm wavelength (X-band) and weighs 100 kg (220 lb).

Kopyo-A's rival is the Lira (lyre) radar being developed by the NIIS (Nauchno-Issledovatelsky Institut Sistemotekhniki) institute, a member of the Leninets holding group in St Petersburg. The Lira – shown at Gelendzhik under its export name of Strizh (swift) – belongs to the Novella (short story) family of systems

The Ka-27 'Helix' is the Russian navy's standard shipborne ASW helicopter, and in its export Ka-28 form also serves with India and China. Upgraded aircraft will feature new radar, uprated engines and improved defences.

(export name Sea Dragon). The biggest of these systems, the 1SD (Sea Dragon 1), is intended for the new large maritime patrol aircraft. The slightly smaller 2SD is now installed in upgraded Il-38 'Mays'; the third radar, 3SD, is the Strizh intended for small patrol aircraft and helicopters, while the smallest 4SD can be installed in light aircraft and pilotless vehicles. As well as the radar, the system can include various electro-optical, radio sonobuoy, dipping sonar, electronic support measures and magnetometric subsystems.

Strizh (Lira, 3SD) can detect a small surface object (RCS 1 m²) from a distance of 35 km (22 miles), a motor boat from a distance of 150 km (93 miles) and an air target from 90-100 km (56-62 miles).

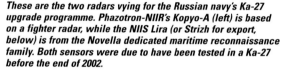

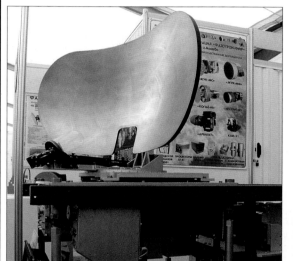

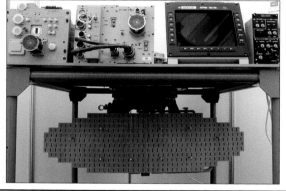

These are the two radars vying for the Russian navy's Ka-27 upgrade programme. Phazotron-NIIR's Kopyo-A (left) is based on a fighter radar, while the NIIS Lira (or Strizh for export, below) is from the Novella dedicated maritime reconnaissance family. Both sensors were due to have been tested in a Ka-27 before the end of 2002.

The radar is equipped with a slot-type antenna and it weighs 100 kg (220 lb). NIIS considers a high peak power 30 kW and antenna gain of 32 dB as the main advantages of the radar. Another advantage of Strizh (Lira) is commonality with the Novella family, already in service in the upgraded Il-38. The Strizh prototype has been already installed in a Ka-27 helicopter, and tests will start soon.

Besides the new mission system, the upgraded Ka-27/Ka-28 will also be equipped with a self-protection system including warning receivers, active electronic jammer and flare dispensers. More powerful VK-2500 turboshaft engines and an upgraded rotor system with new blades may be introduced to increase both take-off weight and payload. A Kamov official stated that the upgraded helicopter will undergo flight tests at the end of 2003.

Piotr Butowski

PROCUREMENT AND DELIVERIES

Algeria

'Flankers' supplied?
Recent Moscow reports that the Algerian air force (AQJJ) received 10 Sukhoi Su-27UBK two-seat combat trainers in 2001 were the first indication that this country may have become a 'Flanker' operator. If true, the AQJJ had presumably also received some single-seat Su-27s, although deliveries had not previously been reported.

Apart from the first three KnAAPO-built prototypes, most Su-27UB production has been undertaken by the Irkutsk Aviation Industrial Association (IAPO) factory in the Sukhoi group. Su-27UB production was reportedly completed by IAPO in November, with the last of the type, designated Su-27UBKs, arriving in China. IAPO production then switched to the generally similar but upgraded two-seat Su-30KN, with standard Su-30 provision for inflight refuelling equipment, for long-range air-defence missions.

Austria

Eurofighter reviewed after flood
Civic funding priorities resulting from heavy summer floods in Central Europe, coupled with a political crisis leading to parliamentary dissolution in September, raised serious doubts concerning Austria's planned Eurofighter procurement. Austrian Chancellor Wolfgang Schuessel said in August that it was impossible to achieve the full requirement for 24 air defence Eurofighters, but the government still hoped to fund 18.

Even then, the originally-proposed Eu1.79 billion ($1.76 billion) combat aircraft programme was expected to be challenged in the year-end national elections. Contract negotiations with the EADS-led consortium, involving agreed industrial offsets totalling 200 percent of the contract value, had perforce been suspended. Renewed submissions had then been proposed by the unsuccessful contenders.

Gripen International renewed its bid for the JAS 39C, including an

Eu500 million lease of up to 18 earlier versions from 2003, as did Lockheed Martin and the US government for Block 50/52 F-16C/D or surplus F-16A/B MLU variants. RSK MiG, however, upgraded its original Eu900 million MiG-29SMT/UBT submission to a dozen extensively redesigned, advanced and more costly MiG-29M1 digital fly-by-wire MRCA versions, plus six two-seat MiG-29M2s. Instead of industrial offsets, RSK MiG proposed exchanging its MiG-29 package, to help reduce Russia's trade debts.

Brazil

Kfirs considered
Budget economies and presidential elections due on 15 November resulted in further delays in selection of Brazil's $750 million F-X BR requirement for 12-24 new combat aircraft until late 2002. In the late summer, the Gripen International JAS 39C and the MiG-29SMT were dropped from the five submissions, leaving the Dassault Mirage 2000-5 Mk 2BR, Lockheed Martin Block 50/52+ F-16C/D, and Sukhoi Su-35MK/UB with canards and thrust-vectoring in contention.

The F-X BR requirement is to replace the FAB's 18 Mirage IIIEBR/DBR multi-role fighters, which Dassault has also offered to upgrade for only $20 million, for extended service. Discussions have also been held with IAI's Lahav division for lower-cost interim procurement by Brazil of 12 upgraded ex-IDF/AF Kfir C-10s, with digital avionics, including a Rafael helmet-mounted sight and Python 4 advanced AAMs.

Chile

FACh procurement reviewed
Regional economic problems and associated national budget cuts, plus internal political dissension between the government and the armed forces, have threatened prospects for several current Chilean military procurement programmes. As the largest of these, a contract for Chile's planned $500 million FMS purchase of six F110-GE-129-powered Block 50 F-16Cs and four two-seat F-16Ds, for which an LOA was signed in February, still awaited finalisation in late 2002.

Funding and technical problems also appear to have ended FACh procurement of 12 used Bell 412 util-

Above: On 25 October 2002 the IDF/AF welcomed the first of its Grob G 109A-I Snunit (swallow) flight screeners. They serve with the Air Force Academy at Hatzerim, and have replaced the PA-18 Super Cub which performed the role for 42 years. The Grobs are being operated by Elbit under a 10-year PFI contract.

Above right: Canada's newest military type is the AgustaWestland CH-149 Cormorant, which entered service with 442 Squadron at Comox in August 2002 in the SAR and training roles. Other Cormorants will serve at Trenton (Ontario), Greenwood (Nova Scotia) and Gander (Newfoundland).

ity and SAR helicopters from US civil sources, after delivery of only four, costing some $10 million, between mid-2000 and April 2001. Apart from their generally poor condition, including corrosion in one aircraft which had to be scrapped, the Bell 412s required extensive modifications to upgrade and standardise their navigation and communications systems.

Czech Republic

Gripen procurement deferred
National flood damage in mid-August, costing some $3 billion for rectification, resulted in suspension on 11 September of the CKr62 billion ($1.953 billion) planned Czech air force (CLPO) procurement programme for 24 new JAS 39C/D Gripens. Evaluations were then reopened for purchase or lease of used combat aircraft, for which F-16A/B MLUs were reportedly favoured from Belgium or the US.

Selection was expected by the year-end. BAE Systems and Gripen International said that Czech industrial offset programmes already started as part of their JAS 39 tenders, would be completed in their own right.

France

Rafale procurement boosted
Increases of 7.5 percent in the 2003 French military budget, announced in September, to a total of Eu31 billion ($30 billion), will include an 11.2 percent growth in equipment allocations, to Eu13.6 billion (over 43 percent of the total), as part of a new Eu98.9 billion six-year defence programme. This included funding for France's long-awaited second aircraft-carrier; AdA's first three Airbus A400Ms; two long-range Elint aircraft to replace AdA's MDC DC-8 Sarigue; and upgrades to 24 ALAT Eurocopter

In a ceremony at Lohegaon Air Force Station, Pune, the Indian Air Force accepted its first batch of 10 Su-30MKIs for service with No. 20 'Lightnings' Squadron. Also at Pune is No. 24 Squadron, which flies the Su-30MK-1.

Cougars and 45 Pumas.

Main aerospace items in the 2003 budget include Eu963 million for 59 more Dassault Rafales, and Eu426 million for 680 MBDA MICA beyond-visual-range AAMs from 1,135 required. Deliveries are also expected in 2003 of 101 MBDA APACHE/SCALP cruise missiles, and the French navy's third (first upgraded) Northrop Grumman Hawkeye 2000 AEW aircraft. Funding is also included for upgrades to similar standards of Aéronavale's first two E-2C Hawkeyes.

Allocations of Eu3.12 billion, covering 46 more AdA Rafale B/C versions and 13 naval Rafale Ms, account for 15 percent of total 2003 arms appropriations. They increase total orders to 120, and overall programme funding to Eu4.085 billion, from specified requirements for 234 AdA Rafales, plus 60 for Aéronavale, costing Eu24 billion. As the sole recipient to date, the navy is currently flying 11 Rafales.

Hungary

Later Gripens sought
Changes were being discussed late last year in Hungary's 10-year $490 million lease of 12 single-seat JAS 39As and two twin-seat JAS 39Bs, signed via Gripen International in November 2001, to obtain later equipment. Agreement was reported to amend the Hungarian contract for acquisition of similar numbers of Batch 3

JAS 39C/Ds, despite their higher costs and later delivery dates. Hungary was originally due to receive its JAS 39A/Bs between late 2004-June 2005. Later deliveries would probably require extending Hungary's recently-placed MiG-29 upgrade contract with RSK MiG from 14 to perhaps all 27 of its 'Fulcrums'.

India

Trainer decisions expected
In early August, Defence Minister Fernandes told the Parliamentary Consultative Committee (PCC) that the Price Negotiation Committee, established for IAF advanced jet-trainer (AJT) procurement, had submitted its report. A decision on the AJT programme, for which IAF procurement of 66 BAE Hawk 100s costing $1.38 billion, had been favoured, was then expected from the Union Cabinet.

Later reports, however, indicated that the IAF was still considering Aero Vodochody's lower-cost offer of L-159B advanced trainers, which CAS Air Chief Marshal Krishnaswamy said in October were equally acceptable to BAE Hawks. Possible IAF Hawk procurement was also discussed briefly on 12 October by UK premier Tony Blair with his Indian opposite number, who then indicated that a final decision had not been reached.

Sticking point still appeared to concern unit price, quoted in the Indian press as $12-14 million for the L-159B as against up to $21 million for the Hawk. In recent negotiations, 24 Hawks were proposed to be purchased directly from BAE Systems, while the remaining 42 would be built jointly under licence by HAL. Offering similar facilities, Aero Vodochody, in

which Boeing has a 35 percent share-holding and supplies the L-159A/B mission system package, is claiming a capability to deliver at least half of the IAF's 66 required AJTs by late 2003.

Earlier, Fernandes said that negotiations for Indian acquisition of Russia's uncompleted *Admiral Gorshkov* aircraft-carrier were nearing finalisation. Its delivery, however, would then take 3-4 years, and probably require Indian expenditure of $450-500 million for its completion and modernisation.

The Defence Minister informed PCC members that despite its age, INS *Viraat*, India's only current aircraft-carrier, still had a residual life of 10-12 years, during which two more similar vessels may be needed. In July, approval was given for local construction of a 37,500-tonne indigenous air defence carrier, accommodating up to 18 Sea Harriers and possibly MiG-29Ks, plus 10 Advanced Light Helicopters and Kamov Ka-31s. Estimated basic cost for India's Air Defence Ship (ADS), probably with foreign design inputs, although built from January 2004 by the state-owned Cochin Shipyard, is around $800 million, for December 2009 commissioning.

Indonesia

Mil helicopter procurement
Following earlier reports of deferment of earlier planned Indonesian navy procurement via Rosoboronexport of Mil Mi-17s and Mi-2 helicopters, the TNI-AL has gone ahead with a similar contract with Russia's Ulan Ude Aircraft Plant (UUAP). Costing over $12 million, including TNI-AL air and ground crew training, this involves the supply of eight refurbished Mi-8s and two new Mi-171s, the latter basically

Having previously served as a VIP transport with the 86th Wing's 75th Airlift Squadron in Germany, this C-20A has been reassigned to NASA Dryden for flight research duties.

civil-configured, with uprated 1864-kW (2,500-shp) Klimov VK-2500 engines, a rear-loading ramp, and improved avionics.

Israel

Apache Longbows for IAF

The US Army Aviation and Missile Command recently awarded McDonnell Douglas Helicopters, a wholly owned subsidiary of Boeing, a $51.8 million modification to an existing contract covering the purchase of eight AH-64D Longbow Apache attack helicopters for the Israeli Air Force. Rather than remanufacturing earlier AH-64A models the Israeli aircraft will be newly constructed.

Italy

Boeing 767 tanker confirmed

Following late 2001 Italian air force (AMI) selection of four tanker/transport versions of the Boeing 767-200ER to replace its four Boeing 707-328Bs, Italy's Defence Administration authorised Boeing and Alenia Aeronautica in July to proceed with the $720 million joint design, development, production and logistics support programme. As launch customer for the B767TT, powered by GE CF6-80C2 turbofans, (as in Japan's similar aircraft), the AMI expects initial deliveries in early 2005.

After Boeing Wichita prototype modifications, further B767 tanker conversions will be undertaken in Italy by Alenia Aerospazio and its Aeronavali subsidiary. Apart from three-point hose and drogue underwing and rear-fuselage air refuelling systems, the AMI 767s will have an alternative refuelling boom, to operate with its 30 refurbished Block 15 F-16As and four Block 10 F-16B OCUs. Their delivery is due next year, on a 10-year $777 million USAF lease.

Italy is also discussing with Boeing possible acquisition of the company's AEW&C B737, to enhance its air defence system, and the AMI has received funding to acquire a third Airbus A319 Corporate Jet (ACJ), and third Dassault Falcon 900EX. These will replace two Gulfstream G.IIIs, also operated by 31° Stormo from Rome-Ciampino since 1985, by the year-end.

Jordan

More F-16s planned

Having operated 12 ex-USAF F-16As and four two-seat F-16Bs on a five-year lease with No. 2 Squadron at El Azraq, the RJAF is planning a similar follow-on deal to establish a fighter weapons school. RJAF C-in-C, Maj. Gen. Prince Faisal bin al-Hussein, said that this would mean acquiring another six F-16As and four F-16Bs for operational conversion and air combat training. They would be available to other F-16 operators in the Middle East and elsewhere, to supplement similar RJAF facilities currently offered with some of its 42/13 Northrop F-5E/Fs for lead-in fighter training.

In July, the RJAF received the first of 16 Slingsby T.67 basic trainers, costing an estimated $12 million and reportedly funded by Oman, to replace its 16 BAE Bulldog 125/As, delivered in 1974-81.

Kuwait

HIDAS selected for AH-64Ds

Sixteen new Boeing MDH AH-64D day/night attack helicopters being ordered by the Kuwait air force from a new $2.1 billion FMS contract for Longbow Apache procurement, are being equipped with BAE Systems' advanced programmable helicopter integrated defensive aids systems (HIDAS). Most US Army Apaches are currently limited to basic DAS, with little more than a radar warning receiver and chaff/flare dispensers.

Developed for the UK Army Air Corp's 67 AgustaWestland Longbow WAH-64s, BAE Systems' HIDAS supplements the former Lockheed Martin Sanders (now BAE) standard APR-48 radar frequency interferometer with a BAE/LMS ultra-violet and infra-red AN/AAR-57 common missile warning unit. These are integrated with BAE Systems Sky Guardian 2000 RWR, with centralised control, and Type 1223 laser-warning system, and further auto-linked with Thales/Vinten 455 chaff and two-row flare dispensers. Provision is also made for an IR jammer, such as the BAE ALQ-144, or the next-generation Northrop Grumman AN/AAQ-24 Nemesis directed IR-counter-measures (DIRCM).

Apart from the AH-64Ds and four spare T700-GE-701C turboshafts, Kuwait's revised FMS proposal includes eight Lockheed Martin/ Northrop Grumman AN/APG-78 Longbow millimetric-wave fire-control radars; Lockheed Martin's Arrowhead modernised target acquisition/designation system; plus 96 Boeing/Lockheed Martin AGM-114L3 and 288 AGM-114K3 Hellfire ATMs.

Nepal

UK funds more Mi-17s

Recent acquisitions by the Royal Nepalese Army Air Service of three Mil Mi-17 'Hip' transport helicopters, via Kyrgyzstan, were being boosted earlier this year from military aid funding provided by the UK. At £3 million ($4.6 million), this represented almost half the UK's total aid package of £6.5 million, and included two more Mi-17s, for delivery by October, increasing RNAAS 'Hip' totals to five. The RNAAS also recently received two Chetaks from India.

Nigeria

Russian helicopters delivered

Russia recently delivered three Mil

Boeing's 'Bird of Prey'

On 18 October 2002 Boeing revealed its hitherto unknown 'Bird of Prey' stealth technology demonstrator, which flew 36 times between the autumn of 1996 and 1999. A product of the Boeing (formerly McDonnell Douglas) Phantom Works, the 'Bird of Prey' shows certain similarities with the unmanned X-36 technology demonstrator, and with the current X-45 UCAV, especially in the the low-observable intake design. For initial tests the aircraft had a ventral fin fitted but this was later removed. The aircraft was constructed from 1992 at Palmdale, California.

The 'Bird of Prey' – seen left at its public unveiling and above during an early test flight with the ventral fin still fitted – was a stealth/construction technique demonstrator which contributed data to several Boeing projects, including the X-32 JSF contender. Power was provided by a Pratt & Whitney JT15D engine, and weight was about 7,500 lb (3400 kg). It was unpressurised, had only basic instrumentation, and was restricted to a speed of about 260 kt (480 km/h; 300 mph). Three pilots flew the aircraft, including Joe Feelock.

Israel's Night Owls – Yanshuf 3 in service

On 8 August 2002 at Hatzerim air base, the Israel Defence Force/Air Force (IDF/AF) celebrated the service introduction of the Sikorsky S-70A-55/UH-60L Yanshuf (Night Owl) 3 assault helicopters with the 'Desert Birds' Squadron. The first five Yanshuf 3 helicopters were delivered to Hatzerim in a USAF Lockheed C-5 on 5 August and deliveries of the 24 helicopters will be completed by January 2003, when the fifth and last shipment will arrive in Israel.

Israel forwarded a request to the US to purchase the helicopters in September 2000 and this was announced by the US Department of Defense on 27 September 2000. The request was speedily approved by December 2000, so that Israel would enjoy the reduced cost of the US five-year H-60 purchase plan that was to be concluded by 31 January 2001. According to Sikorsky, the IDF/AF delivered about 300 technical drawings to cover requested modifications and about 20 non-standard systems were installed in the Yanshuf 3. These included a secure communication system, a 'black box' and an IAI Elta EW system, yet deliveries commenced according to schedule within 18 months of contract signature.

Shlomo Aloni

Mi-34C light helicopters to Nigeria, where they joined five similar aircraft that had been delivered previously. The aircraft will be used for training and other duties.

Oman

More Mushshaks delivered
Deliveries of five more MFI-17 two-seat primary trainer/light ground-attack aircraft from the Pakistan Aeronautical Complex factory at Kamra from 1 August, increased Royal Air Force of Oman totals to eight. Others have recently been sold to Iran (25) and Syria (6) from over 300 built to date, mostly for Pakistan's armed forces.

Philippines

C-130K procurement delayed
Defence funding problems and other equipment priorities had delayed contract finalisation by November for the Philippine air force's planned $43 million procurement of four surplus RAF Lockheed C-130K tactical transports. This has also delayed associated Lockheed Martin plans, in conjunction with Asian Aerospace, for joint finance of a new aircraft overhaul and maintenance centre at the ex-US Clark AFB, near Manila. Initially, this would take over technical support of the PhilAF's 13 C-130B/H and L-100 Hercules, some of which are currently in storage, from AIROD in Malaysia.

PhilAF funding priority is currently being given to buying three (originally six) maritime patrol aircraft for some $170 million, for which a suitably-equipped twin-turboprop Bombardier DHC-8 is reportedly favoured. Lockheed Martin has also proposed one or more quick-change maritime patrol mission-systems modules for 'roll-on' installation in PhilAF C-130s.

Poland

Deadline for new aircraft bids
BAE/SAAB's Gripen International was one of several respondents to recent Polish government Requests for Proposal (RFP) for up to 48 new fighters. US concern was reported about meeting the 12 November deadline for responses, based on Block 50/52+ F-16C/D submissions, with AIM-9X, AIM-120 and JDAM. This was due to Congressional delays in approving the accompanying $3.8 billion 15-year FMS programme loan to Poland, finally cleared on 10 October.

Similar credits are also being offered by Gripen International, and by Dassault for its Mirage 2000-5EPL (Mk 2) bid, with 85 percent French government backing. Poland wants a minimum 100 percent industrial offsets returns on programme costs. Selection was due by the year-end.

Russia

VVS backs both trainers
Planned procurement of both Yak-130 and MiG-AT was announced in October by VVS C-in-C Col Gen. Vladimir Mikhailov. He said that only 10 percent of Yak-130 development costs since 1995 was from state sources. These will now fund the first 10 of up to 120 planned Motor Sich AI-222-powered Sokol-built Yak-130s.

Development and pre-series production of an initial 15 MiG-ATs had been 40 percent funded by a French government loan of Eu58 million ($57 million), via SNECMA for Larzac 04-R20 powerplants and Thales for digital avionics. RSK MiG had provided some $80 million, to fund the first five MiG-ATs to complete development and certification for prospective export, plus 10 for service trials. Hitherto, the VVS has been barred from accepting any aircraft incorporating non-CIS engines, equipment or systems. Their replacement in the MiG-AT, however, would greatly increase programme costs and complications.

South Korea

Surplus P-3Bs requested
The Republic of Korea has requested nine excess P-3B Orions as part of a foreign military sale (FMS) package. The aircraft will be used to patrol South Korea's economic exclusion zone, and conduct anti-submarine and anti-surface warfare as well as support search and rescue operations. The FMS package includes 16 T56-A-14 and 20 T56-A-10W engines, refurbishment and modification of engines, propellers and landing gear; the installation of secure communications systems, missile warning systems, countermeasures dispensing systems, aircraft cockpit enhancements, acoustic receiver and processor

Although old, the AGM-65 Maverick is still widely used for precision attacks. Here one is launched by an F-16C Block 30 of the 522nd FS 'Fireballs', 27th Fighter Wing at Cannon AFB, New Mexico.

system, missile warning and missile countermeasures dispensing systems; data management system; spares; and support equipment. The total cost of the programme is estimated to be $66 million.

Spain

Spanish army helicopter order
Competition is intensifying between Boeing and Eurocopter for an Eu1.3bn ($1.27 billion) Spanish Army Aviation (FAET) requirement to replace 28 MBB BO 105ATH attack helicopters, operated alongside some 30 BO 105GSH reconnaissance versions since 1981. Procurement is planned of 20-25 new helicopters, for which FAET has been evaluating Boeing's AH-64D Longbow Apache, and the smaller and cheaper Eurocopter Tiger since mid-2001.

Boeing is reportedly offering standard production AH-64Ds to Spain, but may also offer an interim lease of surplus AH-64As, as in the Netherlands. Both Boeing and Eurocopter have co-operation agreements with Spanish industry for offset contracts. As part of the EADS group, which also includes CASA, Eurocopter already has a Spanish subsidiary, at Cuatro Vientos airfield, Madrid.

An early decision on Spain's attack helicopter requirement was not expected, however, because of on-going defence procurement funding problems. Spain's 2002 military budget totalled only Eu6 billion ($5.89 billion).

Sweden

SAF receives first JAS 39C
Formal delivery took place on 6 September to Sweden's Defence Materiel Administration (FMV), of the Swedish air force's first upgraded JAS 39C Gripen (s/n 39.208). Apart

from inflight refuelling equipment and on-board oxygen generation system (OBOGS), the JAS 39Cs incorporate colour multi-function cockpit displays, new communications and computer software, and strengthened wings for advanced-weapons pylons.

The last 20 of 110 Flygvapnet Batch 2 aircraft are being completed as JAS 39Cs, to NATO-compatible Export Baseline standards. These will also feature in Flygvapnet's third and last batch of 64 JAS 39C/Ds, to complete Swedish orders for 204 Gripens, including 28 two-seat versions. Some may now be transferred to export customers, but the remainder will replace all current SAAB Viggen combat/reconnaissance aircraft by 2007.

A109 helicopter deliveries begin
Acceptance by the Swedish Defence Materiel Administration took place at AgustaWestland's Vergiate factory in Italy on 10 September of the first two of 20 A109 Power training/utility helicopters ordered in June 2001. Designated Hkp 15s, the A109s will replace three older types currently operated by Sweden's Tri-Service Helicopter Wing, for additional ASW, SAR and medevac roles, including some from ships.

Thailand

Third RTAF F-16 squadron
Formal acceptance was made on 17 August by Royal Thai Air Force C-in-C Air Chief Marshal Thares Punsri, at Korat Air Base, of the first five of 16 refurbished ex-USAF Lockheed Block 15 F-16A/Bs, ordered from a $133 million mid-2000 Peace Naresuan IV FMS contract. Five more F-16s were expected in October, and the remaining six early next year.

Withdrawn from Aerospace

Maintenance and Regeneration Center storage in Tucson, Arizona, the 15 F-16As and one F-16B two-seat combat trainer will equip a third RTAF squadron, to supplement 26 new-built F-16As and 10 F-16Bs. These were delivered between 1988-1996, and have flown over 55,000 flight hours, without loss.

Lockheed Martin modification kits for Falcon Up structural upgrades, and some new equipment and systems, were installed in the RTAF's latest F-16A/Bs by the USAF, prior to delivery. Pratt & Whitney upgraded their powerplants to F100-PW-220E configuration, for commonality with the RTAF's other F-16s.

Tunisia

C-130B delivered
The Ogden Air Logistic Center (OG-ALC) at Hill AFB, Utah recently delivered a refurbished C-130B airlift aircraft to the Tunisian Air Force. Built in 1958, the Hercules had been in storage at the Aerospace Maintenance and Regeneration Center (AMARC) at Davis Monthan AFB, Arizona between 1993 and July 2001 when it was flown to Hill for overhaul.

United Kingdom

STOVL retained in JSF selection
In the first of two long-awaited key decisions announced on 30 September concerning Britain's planned two new aircraft-carriers (CVFs), Procurement Minister Lord Bach said that the short take-off and vertical landing (STOVL) version of the Lockheed Martin F-35 Joint Strike Fighter had been selected to meet the UK's Future Joint Combat Aircraft (FJCA) requirement.

Procurement of up to 150 F-35Bs, he said, similar to those developed for the US Marine Corps, is planned to replace current BAE Harrier FA.Mk 2s and GR.Mk 7/9s of Britain's Joint Harrier Force by 2015.

In a programme worth up to £10 billion ($15.5 billion), the F-35Bs will enter UK service with the first of the RN's two new aircraft-carriers (CVFs) in 2012. A key factor in the UK's decision was that the USMC's F-35B will be the first JSF version to enter service, thereby also meeting Britain's time-scale.

Additional Bell 412s for DHFS
As a result of an $11 million contract between Bell Helicopter and Bristow Helicopters Ltd, the United Kingdom's tri-service Defence Helicopter Flying School (DHFS) at RAF Shawbury will be obtaining two additional Bell 412 training helicopters. The Bell 412 is known as the Griffon HT.Mk 1 in UK service and nine are currently in service with the tri-service school. Bristol is one of three partners in FBS Inc., which administers the DHFS.

United States

USAF orders Globemaster IIIs
On 15 August 2002 Boeing and the USAF completed negotiations on a $9.7 billion contract covering the purchase of 60 additional C-17A Globemaster III transports. This multi-year contract will bring the USAF's inventory of C-17As to 180. To date Boeing has delivered 89 of the airlifters.

Army orders Jet Rangers
The US Army Aviation and Missile Command recently awarded Bell

Helicopter a $23.9 million modification to an existing contract for covering the purchase of 15 TH-67A+ JetRanger training helicopters for the US Army. The helicopters will be assigned to the US Army Aviation Center at Fort Rucker, Alabama, in support of the service's Flight School XXI programme.

First SHARP pod accepted
On 24 June 2002 the US Navy took delivery of the first of five engineering and manufacturing development (EMD) examples of the Shared Reconnaissance Pod (SHARP) from Raytheon. The pod will support developmental testing and integration on the F/A-18E/F Super Hornet at the Naval Air Warfare Center Aircraft Division at NAS Patuxent River, Maryland. The pod, which is designed to provide a high- and medium-altitude tactical reconnaissance capability, will make its initial deployment with the F/A-18Fs assigned to VFA-41 aboard the USS *Nimitz* (CVN 68) in mid-2003. The system will also be compatible with the F/A-18C and D Hornet variants. The SHARP, which is carried on the F/A-18 centreline station, is the size of a 330-US gal (1249-litre) fuel tank and is equipped with sensors manufactured by Recon Optical Inc.

First production ATFLIR
Raytheon delivered the first low-rate initial production (LRIP) AN/ASQ-228 Advanced Targeting FLIR (ATFLIR) pod to the US Navy on 21 May 2002. The delivery marks the beginning of a multi-year plan to purchase 574 pods for the US Navy and Marine Corps F/A-18C/D and E/F strike aircraft. Raytheon is building 15 LRIP 1 pods under a sub-contract from Boeing. The

ATFLIR achieved early operational capability with VFA-115, the Navy's first F/A-18E squadron in June 2002, which deployed with three refurbished engineering development model pods. It combines the functions of the AN/AAS-38 targeting pod with those of the AN/AAR-50 NAVFLIR and the LDT/Strike Camera currently operational on the F/A-18.

ATFLIR comprises a targeting FLIR, a CCD television camera, and a high-power laser that has been demonstrated at altitudes of up to 50,000 ft (15240 m). In addition, a navigation FLIR is installed in the adapter that connects the pod to the aircraft. The navigation FLIR is currently carried in a separate pod on a second weapon station. Raytheon delivered eight pods during the development phase, which included more than 500 developmental test and operational test flights. An operational evaluation (OPEVAL) of the ATFLIR is scheduled for October 2002 and it will achieve initial operating capability (IOC) during 2003. The Navy recently issued Boeing a $95.4 modification to an existing contract covering a second low-rate initial production (LRIP) purchase of 28 AN/ASQ-228s. The contract comprises the production of 24 ATFLIRs for the F/A-18E/F and four ATFLIR pods and two pod adapters for the F/A-18C/D.

Powered gliders for Academy
The USAF Academy has taken delivery of the first examples of the TG-14A powered glider for the school's airmanship programmes. Based upon the Grupo Aeromot AMT-200S Super Ximango, the new trainers are built in Brazil and powered by a Rotax 912A engine that develops 81 hp (60.4 kW). A total of 15 is on order.

Republic of China Air Force on display

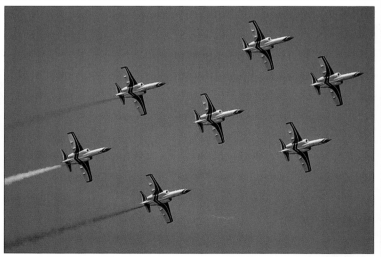

Held at Ching Chuan Kang AB, Taiwan's Air Force Day celebrations included the first public showing of the Pathfinder/Sharpshooter (downgraded LANTIRN) pods on an F-16A of the 21st Fighter Squadron/4th Group (above). Below is an F-16B wearing the low-viz markings of the Hualien-based 5th Group. The 1st Group at Tainan has also applied unit markings, as shown on this F-CK-1B (below right). Displaying at the show were the 'Thunder Tigers' demonstration team, which flies the AIDC AT-3 (right).

AIR ARM REVIEW

Afghanistan

Russia aids air force revival
Military aid from Moscow to help rebuild the fragmented Afghan National Army is initially expected to include a few transport and support aircraft, including helicopters, plus vehicles and ground equipment, in a $35-40 million package. Currently, the Afghan air force, commanded by General Mohammad Dauran, has no effective combat element, beyond about three remaining Aero L-39C jet-trainers.

Two Antonov An-26s and one An-32 have reportedly been returned to service, together with about five Mil Mi-35 attack helicopters, plus two Mi-8s and five newer Mi-17s for transport roles. Russia promised deliveries by the year-end of up to 10 aircraft, including an Antonov An-12, plus Mi-8 and Mi-24 helicopters.

Australia

First Hawk LIFT graduates
The first eight RAAF students to complete fast-jet training purely on BAE Hawk Mk 127 lead-in-fighter trainers graduated from the F/A-18 operational conversion course in June. Deliveries of 33 Hawks began to the RAAF in October 2000, from an $A850 million ($466.6 million) 1997 contract.

Eighteen Hawks are operated by 76 Squadron at RAAF Williamtown, New South Wales. Fourteen are based with 79 Squadron at RAAF Pearce, Western Australia, and the remaining Hawk is on loan to BAE Systems Australia for development work. The RAAF is currently completing its third Initial Fighter Course on the Hawk 127, for economic F/A-18 conversions.

Canada

Cormorant enters service
Concluding nine months of testing and training, the CH-149 Cormorant officially entered service with the Canadian Air Force's 19 Wing at CFB Comox, British Columbia, on 26 July 2002, when the aircraft began holding 'search and rescue stand-by'. Serial 149902 had the distinction of carrying out the Cormorant's initial rescue on 28 July 2002, when an injured sailor was lifted from a 328-ft (100-m)

freighter in the Hecate Strait 125 miles (200 km) off northern Vancouver Island. Five CH-149s are operated by 442 Transport & Rescue Squadron at Comox, however, two of the aircraft are used primarily for training. Although the Cormorant initially shared rescue duties with the Labrador at 19 Wing, by mid-August the Labradors had been removed from service and transferred to other units. The first three helicopters slated for SAR duty on Canada's East Coast arrived at CFS Gander, Newfoundland, on 8 July 2002. Serials 149903, 149906 and 149908 will replace the CH-113 Labradors currently assigned to 103 Search and Rescue Squadron, which is part of 9 Wing. Canada purchased 15 examples of the CH-149 from EH Industries at a cost of $512 million, and the CH-149 will replace Canada's 12 Labradors by the end of 2003.

More T-6As arrive
Two additional Raytheon T-6A (CT-156) advanced turboprop trainers were due for delivery in late 2002 for the NATO Flying Training in Canada (NFTC) organisation. These were ordered from an $11.6 million contract to meet additional NFTC commitments, and increase T-6A totals in Canadian service to 26. †he most recent NFTC customer is Hungary, which signed a 17-year contract in March 2002, for initial training of two instructors, and seven students a year thereafter.

Greece

AT-6Bs for light ground-attack
An armed version of Raytheon's T-6A Texan II development of the Pilatus PC-9 advanced turboprop trainer has completed flight- and weapons-testing at Eglin AFB, Florida, for the Hellenic air force. Twenty of 45 Texan IIs ordered by the HAF will be fitted with hard points for various automatic weapons, bombs and rockets, and will be designated AT-6B.

India

MiG-23/25s to be retired
Plans were revealed in July to retire the IAF's ageing MiG-23BN/MF variable-geometry fighters and few remaining Mach 2.8 MiG-25R/U strategic high-altitude reconnaissance

aircraft. This would involve imminent withdrawal of about 123 aircraft, mostly MiG-23s, followed by several hundred earlier MiG-21s.

The IAF operates three squadrons of ground-attack MiG-23BNs, and one squadron of MiG-23MF interceptors. It also flies 16 MiG-23BNs on electronic warfare roles, plus six two-seat MiG-23UBs. Eight MiG-25R/Us remain in the IAF inventory, but few are currently airworthy. Their reconnaissance roles have now been taken over by military satellites and UAVs.

MiG-23BN ground-attack roles are being taken over by IAF MiG-27M 'Flogger-Js', of which 30 of about 130 in service will be upgraded by HAL, together with 50 of 80 or more deep-penetration SEPECAT/HAL Jaguars. The IAF's Jaguar force, plus its Dassault Mirage 2000s, will receive over 100 Rafael Python 4 high off-boresight close-combat IR AAMs and Derby beyond-visual-range radar-guided AAMs, from negotiations with Israel in September.

New Zealand

Support units redeployed
Following recent disbandment of the RNZAF's Air Combat Force, and retirement of its 14 MDC A-4K and five TA-4K Skyhawks from their Ohakea base, their accommodation has been taken over by two flying units transferred from Whenupai, near Auckland. Ohakea now houses the 14 Bell UH-1Hs of No. 3 Battlefield Support Squadron and the three Beech B200 King Airs of No. 42 Multi-Engined Conversion Squadron.

Apart from RNZAF HQ, the Auckland base still accommodates New Zealand's main military airlift force, including the two Boeing 727-220C general transports of No. 40

To celebrate 30 years of Belgian C-130 operations, CH-02 was painted up in special markings by 20 Sm/Esc. Twelve aircraft were delivered in 1972/73, of which one was lost. The aircraft have been updated with defensive countermeasures.

Squadron and five Lockheed C-130Hs of No. 5 Squadron. Auckland is also home base of the RNZN's four leased Kaman SH-2F ASW SeaSprite helicopters, soon to be replaced by five upgraded SH-2G(NZ)s.

Replacements are being sought for the Boeing 727, plus missions systems, communications, and defensive aids upgrades for the RNZAF's Lockheed C-130s and P-3Ks. The government is currently committed to capital injections of up to $NZ1 billion ($481 million) for the NZ Defence Forces over the next decade, although procurement projects may exceed $NZ3 billion.

Pakistan

Former USAF C-130Es requested
On 16 July, the Defense Security Co-operation Agency requested that the US Congress approve the possible sale of six surplus C-130E transports and associated equipment to Pakistan at a cost of $75 million. The Government of Pakistan requested six aircraft with engines, one C-130E for cannibalisation with engines, and upgrade of T56-A-15 engines. Pakistan needs these aircraft to support a current and long-term airlift and to support the US with Operation Enduring Freedom.

Portugal

EC 635 helicopter cancellation
Compensation was being sought by the Lisbon government from

Tornados figured prominently in 2002's crop of special schemes. The most eye-catching were these ECR aircraft: 155º Gruppo/50º Stormo's 60th anniversary aircraft (above) and JBG 32's 'firebird' (right).

Left: 'The Southern Cobra Squadron' of the IDF/AF, which flies the Bell AH-1 Zefa (viper), has adopted a snake decoration for its aircraft, similar to that worn by its sister unit, 'The Northern Cobra Squadron'.

Above: Ex-Israeli A-4N Skyhawks are operated by BAE Systems Flight Systems (previously Tracor) on target-towing duties under contract to the Luftwaffe. This dart-equipped aircraft taxis at Decimomannu for a gunnery sortie.

Eurocopter in September, following cancellation in early August of an Eu34 million ($33.6 million) Portuguese army aviation launch order for nine EC 635TI seven-seat armed utility and reconnaissance helicopters. Ordered in October 1999, the EC 635s were due for delivery between June 2001 and March 2002, to equip a newly-formed army aviation element at Tancos air base.

Some Portuguese army air and ground personnel had already started training with Eurocopter, but no EC 635s had been received by early August. Portugal's order cancellation followed failure to reach agreement with Eurocopter on weapons system integration terms, after concurrence on service entry deferment. Lisbon's National Defence Ministry then sought up to Eu6.5 million ($6.43 million) in compensation on August 9.

Germany was the first military customer for the civil-configured EC 135 version of the EC 635, its army having received 15 and the Border Guard nine from September 2000.

Russia

Mil Mi-26 loss

A Chechen rebel shoulder-launched SAM reportedly caused the crash of a VVS Mil Mi-26 heavy-lift helicopter on 19 August, killing 115 of 147 people on board. With a normal maximum capacity of 86 passengers, the Mi-26 was approaching the air base at Khankala from Mozdok at about 200 m (650 ft), when one of its two ZMKB Progress D-136 turboshaft engines caught fire. An impact fire followed a crash landing, from which only 27 passengers and five crew members escaped.

Escorting Mi-8 helicopter crews reported seeing a missile trail, and a discarded SA-7 Strela 2M SAM launcher was later found in the vicinity. Russian Army Aviation C-in-C Col Gen. Vitaly Popov was suspended, pending investigation into the crash, in which serious overloading was a contributory factor. Over 300 Mi-26s have been built to date by Rostvertol, mostly for Russian military service.

United States

Harriers deploy with Litening II

AV-8Bs operated by VMA-542 recently became the first Harrier unit to fly combat missions over Afghanistan equipped with laser-guided munitions and the AN/AAS-28 Litening II targeting pod. The aircraft are attached to Marine Medium Helicopter 261 (Reinforced) [HMM-261(R)] as part of the 22nd Marine Expeditionary Unit (Special Operations Capable) [MEU(SOC)], which was deployed aboard the USS *Wasp* (LHD 1) amphibious ready group (ARG). The pod, which entered the Marine Corps inventory in August 2000, incorporates a charged coupled device (CCD-TV) camera used for video reconnaissance, and a forward-looking infrared (FLIR) and laser spot tracker/range finder sensors that are used for precision targeting. Following integration and operational testing the pod was released for fleet service in January 2002.

JPATS achieves IOC

Air Education and Training Command recently announced that the Joint Primary Aircraft Training System (JPATS) has achieved initial operational capability. JPATS is comprised of a Ground Based Training System (GBTS) and the Raytheon T-6A Texan II, which is the system's flying platform. The components of JPATS are all now operational with the 479th Flying Training Group at Moody AFB, Georgia, where two student classes have already completed the T-6A portion of Specialized Undergraduate Pilot Training (SUPT). The group began operating at full T-6A student pilot production capacity in July 2002 and will train approximately 250 students per year.

C-37 replaces Stratolifter

The 65th Airlift Squadron placed a new C-37A Gulfstream V in service at Hickam AFB, Hawaii on 1 May 2002. Serial 01-0065, which will serve as a CINC Support aircraft for the commanders of the US Pacific Command (USPACOM) and the Pacific Air Forces (PACAF), replaced a C-135E that was later transferred to the 412th Flight Test Squadron (FLTS) at Edwards AFB, California, on 26 July 2002. The C-135E, serial 57-2589, will replace the squadron's C-135C, known as the 'Speckled Trout', as a communications test bed and executive transport. The 65th AS is assigned to the 15th Air Base Wing (ABW) at Hickam, while the 412th FLTS is part of the 412th Test Wing (TW) at Edwards.

Longbow units convert

Following a ceremony held at Desiderio Army Airfield, Camp Humphreys, on 5 June 2002, the 3rd Squadron, 6th Cavalry Brigade (3-6 CAV) departed the Republic of Korea for Fort Hood, Texas. The unit's three troops will transition from the AH-64A to the AH-64D with the 21st Cavalry Brigade. The unit's AH-64As were shipped to the Boeing facility in Mesa, Arizona, where they will undergo conversion to Longbow configuration. The unit will return to Korea in the summer of 2003 with its modified helicopters.

The US Army's 6th Squadron, 6th Cavalry Brigade (Attack) (6-6 CAV) has been certified combat-ready after converting from the AH-64A Apache to the AH-64D Apache Longbow and completing extensive training with the 21st Cavalry Regiment at Fort Hood, Texas. The battalion will return to its home base at Illesheim Army Airfield, Germany, this summer as the first Longbow Apache unit to be stationed in Europe. The unit is one of two regiments that comprise the 11th Aviation Group (Attack).

B-1B consolidation continues

More than a year after the USAF announced plans to consolidate its B-1B fleet at just two locations, the first steps have taken place. The plan calls for the B-1 fleet to be reduced by more than 30 aircraft and consolidating the remaining bombers at Ellsworth AFB, South Dakota, and Dyess AFB, Texas. This involved the transfer of seven B-1Bs from Mountain Home AFB, Idaho, to Ellsworth and aircraft that had been assigned to the Kansas and Georgia Air National Guard units were relocated to Ellsworth and Dyess. As part of the moves all B-1B training has been consolidated at Dyess, while Detachment 1 of the USAF Weapons School and Detachment 2 of the 53rd Test and Evaluation Group have been relocated from Ellsworth. Although neither unit has aircraft assigned, the detachments respectively train B-1 instructors and perform operational tests and evaluations.

Dyess accepted 12 Air National Guard aircraft, increasing the number of assigned aircraft to 52. However, 12 of these will be placed in storage at the Aerospace Maintenance and Regeneration Center at Davis Monthan AFB, Arizona, later this year. Eight

Based at Hurlburt Field, Florida, the 6th Special Operations Squadron is part of the 16th SOW. It operates this An-32B to 'familiarise special forces troops with foreign aircraft types'. The An-32B is registered to Skylink Aviation of Cheyenne, Wyoming, and wears the appropriate registration N6505. The 6th SOS also flies the CASA 212 which, after many years' service without an MDS number assigned, has recently been designated C-41A.

further examples will be permanently retired and placed on static display. The money saved by reducing the fleet size will be invested into the defensive systems and weapons-modernisation efforts, including the integration of the Wind Corrected Munitions Dispenser (WCMD), Joint Stand-off Weapon (JSOW) and the Joint Air-to-Surface Stand-off Missile (JASSM) as part of the Block E upgrade. The consolidation will be completed by 1 October 2003.

The last three B-1Bs assigned to the Kansas Air National Guard's 184th Bomb Wing/127th Bomb Squadron departed McConnell AFB shortly before 12 pm on 4 August 2002, enroute to Ellsworth and Dyess. The final aircraft included serials 85-0069, 85-0081, and 86-0136.

The 184th Bomb Wing had already begun its transition from the B-1B to the KC-135R tanker. Seven Stratotankers had been transferred to the wing from the 22nd Air Refueling Squadron (ARS) at Mountain Home AFB, Idaho. The 127th Bomb Squadron will be redesignated as the 127th ARS on 1 October 2002. The 184th will accordingly be redesignated an Air Refueling Wing on the same day and will transfer to the control of Air Mobility Command.

In preparation for transitioning to the E-8C Joint STARS, the Georgia Air National Guard's 128th Bomb Squadron made one last flyby on June 2002 when the 116th Bomb Wing held 'fini flight' festivities. The units will take control of the E-8C fleet currently operated by the 93rd Air Control Wing and will itself be redesignated the 116th ACW. The 128th BS will accordingly become the 128th Airborne Command & Control Squadron (ACCS).

New JSTARS training squadron

The 330th Combat Training Squadron (CTS) was activated as part of the 93rd Air Control Wing (ACW)/Operations Group (OG) at Robins AFB, Georgia, on 13 August 2002. The unit will take over the E-8C training mission that had been conducted by the 93rd Training Squadron (TRS). The unit will operate a pair of TE-8As that had been assigned to the 93rd TRS.

US Army revives fleet flight test

On 31 May 2002 the US Army Aviation Technical Test Center (ATTC) at Fort Rucker, Alabama, resumed its 'Lead the Fleet' test programme. The programme, which was initiated in

1986 but halted in 1995 by funding constraints, is used to determine the long-term performance of aircraft, systems and hardware using a variety of test scenarios. 'Lead the Fleet' was reactivated because of an increase in the number of alerts regarding potential safety problems in recent years. These alerts have often led to groundings or special inspections, maintenance procedures and reporting.

Located at Cairns Army Airfield, ATTC – which serves as the aircraft developmental test branch of the Army Test and Evaluation Command – is tasked with testing aircraft, aviation systems and associated aviation support equipment from development to fielding and throughout operational service. The test aircraft will carry extra weight to approximate the effects of normal ammunition combat loads as well as internal and slung loads, and will accumulate flight hours at about twice the rate of aircraft in field units. Testing aircraft at accelerated flight-hours and in specific mission profiles has revealed aircraft system and hardware failures in the past. Four helicopter models are initially included in the programme, comprising the AH-64A Apache, the UH-60A/L Blackhawk and the CH-47D Chinook. In the near future, the centre will also begin testing the AH-64D Longbow Apache and the OH-58D Kiowa Warrior. The service believes that the flight test data will improve aircraft readiness through the early identification of problems.

New rescue units

Preparations are underway at Davis Monthan AFB, Arizona, for the establishment of a pair of new combat search and rescue squadrons, which will be assigned to the 355th Wing. Officially designated as Rescue Squadrons (RQS), the two units will eventually operate a total of 12 HH-60G helicopters and 10 HC-130 long-range rescue and tanker aircraft. The units will be activated during 2003.

New Osprey test unit activated

Detachment 2 of the 18th Flight Test Squadron was officially activated as the USAF component of the V-22 Multiservice Operational Test Team (MOTT) on 31 July 2002 at NAS Patuxent River, Maryland. The MOTT, which is currently comprised of the 11 USAF personnel assigned to Detachment 2, an Air Force Operational Test Evaluation Center

During the type's first cruise, an F/A-18E Super Hornet of VFA-115 lands aboard USS Abraham Lincoln during work-up in the Pacific. In November 2002 the unit notched up the first combat action for the Super Hornet, being used for precision strikes against air defence sites in Iraq.

Test Director, and 18 Marine Corps personnel, will begin operational assessments and evaluations of the Osprey next summer. The unit will initially assess the first MV-22B Block A upgrade low-rate initial production aircraft. These aircraft have a redesigned nacelle and software upgrades. The MOTT will operate and maintain the aircraft in an operationally representative environment and will eventually conduct the second phase of the V-22's operational evaluation using MV-22B Block A production aircraft from various ship and shore locations. The 18th Flight Test Squadron is based at Hurlburt Field, Florida, and functions as the operational agency for the Air Force Special Operations Command.

Airlift squadron reactivated

The 16th Airlift Squadron (AS) was reactivated as part of the 437th Airlift Wing (AW) at Charleston AFB, South Carolina, on 26 July 2002. The squadron will eventually be assigned 12 C-17A Globemaster III transports. The squadron had been deactivated in September 2000 following the retirement of the last C-141Bs assigned to the wing. The squadron is scheduled to fly its first mission on 1 October 2002.

Re-engined Rivet Joint deploys

Air Combat Command has begun deploying re-engined RC-135W Rivet Joint reconnaissance aircraft, and the first of these intelligence platforms deployed to Prince Sultan Air Base, Saudi Arabia, where it was assigned to the 763rd Expeditionary Reconnaissance Squadron. Serial 63-9792 is equipped with four CFM International F108-CF-100 turbofan engines that provide the aircraft with 16,000 lb (71.2 kN) of additional thrust

over the earlier TF33 engines. The USAF is equipping the entire fleet of 15 Rivet Joints with the newer engines, which also power the KC-135R/T tanker. Additional modification programmes include updating the RC-135's cockpit with digital avionics, while the mission electronic systems are also being improved. The USAF's entire fleet of Rivet Joints is permanently assigned to the 55th Wing at Offutt AFB, Nebraska. Boeing recently received a contract to provide five additional re-engine kits for the RC-135 fleet.

Marines transfer helicopters

The US Navy's Combat Support Squadron Five (HC-5) at Andersen AFB, Guam, recently took delivery of two low-time HH-46Ds that had been assigned to the search and rescue (SAR) unit at MCAS Iwakuni, Japan. The transfer was made possible when the Japan Air Self Defence Force assumed the responsibility for SAR duties at Iwakuni.

Although HC-5 is undergoing conversion to the MH-60S, until that aircraft receives approval to conduct vertical replenishment (VERTREP) missions the squadron will continue to use the Sea Knight for these duties. Whereas HC-5's aircraft have accumulated an average of 13,000 flight hours, the former USMC examples have clocked up only 7,000 hours. The Navy and Boeing are currently working to extend the Sea Knight's service life to 14,500 hours.

OPERATIONS AND DEPLOYMENTS

United States

Carriers deploy

The USS *George Washington* (CVN 73) carrier battle group (CVBG) deployed from Norfolk, Virginia, on 20 June 2002 and relieved the USS *John F. Kennedy* (CV 67) on 19 July 2002. The *Kennedy* CVBG had been conducting combat operations over Afghanistan from the Arabian Sea. The USS *Abraham Lincoln* (CVN 72) and its

CVBG deployed from Everett, Washington, on 20 July and, following operations in the Western Pacific, was also scheduled to support Operation Enduring Freedom from the Arabian Sea. Carrier Air Wings (CVW) Seventeen and Fourteen were embarked aboard the respective ships.

USAF reorganises AEF

The USAF has announced that it will retain the current Air Expeditionary

Force (AEF) deployment structure and alignment. Under this structure, 10 AEFs are utilised in five pairs with 90-day temporary duty (TDY) periods within a 15-month total cycle length. The service has, however, decided to realign the 'on-call' or '911' air expeditionary wings (AEW). The wings had been structured around the 4th Fighter Wing at Seymour Johnson AFB, North Carolina, and the 366th Wing at Mountain Home AFB, Idaho. The realignment will see the resources of those AEWs assigned throughout the 10 existing AEFs by August 2002 and

tasked accordingly when AEF Cycle 4 begins in June 2003.

Global Hawk milestone

Although the system is still undergoing test and has not yet formally entered the USAF inventory, Northrop Grumman's RQ-4A Global Hawk marked a major milestone on 15 June 2002 when the unmanned air vehicle flew its 1,000th combat flight hour in support of Operation Enduring Freedom. The high-altitude, long-endurance system has been deployed since the autumn of 2001.

Boeing X-45 UCAV

Autonomous unmanned combat vehicles become reality

With Predators and Global Hawks in regular service, it is now natural to talk about full-size aircraft that are remotely operated by a 'pilot' who is sitting in a trailer with a joystick and a video screen. Now it's time to get used to aircraft where the pilot's role is simply to insert a disk and go have lunch while the aircraft takes off and flies its entire mission profile – autonomously.

During the winter of 2001-2002, the public became aware of the value of unmanned, remotely-piloted aircraft in the support of combat operations. Both the General Atomics RQ-1 Predator and the Ryan/Northrop Grumman RQ-4 Global Hawk were used by the US Air Force for reconnaissance operations over Afghanistan, and the Predator was field-modified to carry AGM-114 air attack missiles.

Simultaneously with the Afghanistan deployments, the Air Force was preparing for the 22 May 2002 first flight test of the Boeing X-45A. This new generation of unmanned vehicle is distinguished from its predecessors in two important ways. First, it is designed to fly its mission autonomously and, second, the X-45A is the prototype of the first generation of unmanned aircraft designed from the outset for combat.

The United States armed services began experiments with unpiloted aircraft in the 1960s, but it would be the early 1990s before developments in control technology finally reached a level of sophistication that made such vehicles practical. These aircraft would be known by the acronym UAV, which stood for Unmanned Aerial Vehicle. During the Clinton Administration there was a brief flirtation with political correctness in which the nomenclature was temporarily rewritten as Uninhabited Aerial Vehicle. However, it must have been pointed out that aircraft are never actually inhabited, and the UAVs are now once again known as Unmanned Aerial Vehicles. Apparently, no-one ever suggested the term uncrewed. Now, with the Boeing X-45A UCAV, the word 'Combat' is inserted in the acronym for the first time. Even the proof-of-concept prototypes are designed with a pair of bomb bays.

Essentially, America's first UCAV is a three part programme. Step one (Spiral 0) is the test and evaluation of a pair of X-45A prototype aircraft, which are two-thirds scale models of the eventual operational UCAV. Second, there will be a full-scale X-45B test aircraft (Spiral 1), and finally an operational UCAV (Spiral 2). In 2002, this aircraft was being tentatively discussed under the attack designation 'A-45'.

The operational UCAV is being groomed to take human beings out of the Suppression of Enemy Air Defences (SEAD) mission, which is typically the first and most dangerous of an air campaign. Before bombers and attack aircraft can strike other targets, they must first remove the threat posed by enemy anti-aircraft defences. With UCAVs, such high-risk work can be done without imperiling human pilots. This will make the decision to launch high-risk attacks easier to make. As Major Rob Vanderberry of Air Combat Command put it, "These aircraft will allow Air Force leaders to breathe easier when making a combat decision. What UCAV lets us do is attack a target without the concern of losing a pilot, or having someone become a prisoner of war."

Taking the pilot out also reduces the size and radar signature of the aircraft, and it reduces the acquisition and maintenance cost. The operational UCAV is estimated to have a price tag of about $10 million, about a third of the cost of a 'next-generation' tactical combat aircraft. Boeing estimates that the UCAVs would cost up to 65 per cent less to produce than future manned fighter aircraft, and up to 75 per cent less to operate and maintain than current aircraft.

X-45 statistics

The X-45A is 26 ft 5 in (8.05 m) long, with a wingspan of 33 ft 8 in (10.26 m). It has no vertical tail surfaces, so 'height' is a misnomer in listing specifications. The fuselage is 3 ft 7 in (1.09 m) thick. By contrast, the X-45B will be about 32 ft (9.75 m) long and 4 ft (1.2 m) thick, with a wingspan of 47 ft (14.32 m). According to the plan, the operational UCAV will have the same dimensions as the X-45B. The X-45A weighs approximately 8,000 lb (3629 kg) empty, and carries a 3,000-lb (1360-kg) payload in two weapons bays. For flight test purposes, one bay is fitted with an instrumentation pallet so that crews can have easy access to evaluate test results between flights.

Both X-45As were constructed at the Boeing (formerly McDonnell) facility at Lambert Field near St Louis, Missouri, and shipped to Edwards AFB in California aboard a C-17A Globemaster III. The operational UCAVs will also be deployed to forward areas of action the same way. The UCAVs are designed to be broken down and packed in such a way that the Globemaster can carry up to six at a time.

Left: X-45A no. 1 is seen during the UCAV's first flight from Edwards AFB on 22 May 2002. The follow-on X-45B will be a larger machine of similar configuration, and is intended to be representative of the operational aircraft. While hardware development continues, commanders and politicians debate how a UCAV might be employed, and how far to develop automated vehicle technology. At present the SEAD role is seen as the primary task.

Boeing's UCAV concept is designed to be easily and rapidly transportable to wherever it is needed. The Mission Control System (left) is truck-mounted and has two operator stations (primary to left and back-up to right). The UCAV itself is stored and shipped in a purpose-designed crate (above).

The X-45A was unveiled to the media on 11 July 2002. Present were both prototypes (no. 1 in blue trim and no. 2 in red), together with Boeing's Canadair CL-30 (CT-133) chase aircraft. Lending scale to the X-45A at right is Lt Col Michael Leahy, USAF UCAV programme manager.

After rolling out at Lambert Field on 27 September 2000, the first X-45A went out to California for equipment installation and for flight test, which began 20 months later. The second X-45A was on hand by the time that the first made its 14-minute debut flight in May 2002, and the two aircraft were seen together publicly for the first time on 11 July 2002. The two gloss white aircraft are distinguished from one another visually by the first aircraft being painted with blue trim, and the second being trimmed in red. The X-45As are expected to be joined by the first of two matt grey X-45Bs sometime during 2004.

Test and evaluation of the X-45A is a joint project of Boeing's Phantom Works component as well as the Defense Advanced Research Projects Agency (DARPA) and the Air Force Research Laboratory (AFRL), which is located at Edwards AFB. The aircraft are actually housed at the National Aeronautics & Space Administration's Dryden Flight Test Center, which is also situated on the 300,000-acre (121410-hectare) base.

During the early days of the UCAV technology demonstration, the programme manager on the government side was Air Force Lieutenant Colonel Michael Leahy, PhD. He summarised the importance of the UCAV concept by observing that it would "exploit real-time on-board and off-board sensors for quick detection, identification and location of fixed, relocatable and mobile targets. The system's secure communications and advanced cognitive decision aids will provide ground-based, human operators with situational awareness and positive air vehicle control necessary to authorise munitions release."

Of course, a key part of the UCAV's ability to do its job is the incorporation of low-observable 'stealth' technology, and such considerations have been critical in the design of the X-45A and its operational progeny. There is no vertical tail and, although the X-45A aircraft both have nose probes for testing purposes, the X-45B and the 'A-45' will be 'clean.' As David Lanman, deputy chief of UCAV advanced technology demonstration at AFRL, puts it: "UCAV systems [must be] as low-observable as possible, to achieve their intended missions. . . If you had a radar antenna or refuelling boom projecting into the air from the surface of a UCAV, that would significantly cut down on its mandated stealthiness."

According to Leahy, the UCAV programme objective is "to have all the testing completed by 2005. Then we would have the necessary information needed to field these aircraft by 2010 if the Air Force decides to use them."

Ready to ship

In the field, the 'A-45s' could be stored in ready-to-ship containers for years until they were needed. At that point, the UCAVs could be deployed in their delivery truck-sized containers along with their mobile Mission Control Station, which is about the size of an average recreational vehicle. Though the UCAV is designed to operate autonomously and is not remotely-controlled, a command and control link would be maintained between the aircraft and a crew of two in the Mission Control Station.

When the need arises, the container-packed UCAVs would be flown to a forward location within an 800-mile (1290-km) radius of the intended target. Crews would then unpack and assemble them, a task which is intended to take just a few hours. Meanwhile, the Mission Control Station would be made ready and the strike mission would be launched.

At the moment, the Air Force envisions that the UCAV will be deployed with free-fall gravity bombs, as well as GPS-guided GBU-31 and GBU-32 Joint Direct Attack Munitions (JDAM). If the retrofit of Predators with Hellfires in 2002 is any indication, the UCAV will almost certainly also have guided missiles in the bag of tricks that it carries into enemy territory.

Bill Yenne

Pilatus PC-21

New-generation trainer for the 21st century

The PC-7 Proof of Concept vehicle demonstrated salient features of the PC-21, including short-span wing, new cockpit display architecture and PT6A-68 engine.

Since the flight of the first production PC-7 in 1978, Pilatus has established itself as the world leader in turboprop-powered trainers. The 'big brother' PC-9 was introduced in 1984, and the success of the Pilatus family was underlined by the adoption of the PC-9 Mk II (as the Raytheon T-6A Texan II) for the joint USAF/US Navy basic trainer requirement. However, the T-6A deal left Pilatus itself in an unusual situation with regard to future competitions. Under the terms of the agreement, Raytheon was free to market its T-6A in direct competition with offerings from Pilatus (in much the same way as the Boeing T-45 Goshawk competes with the BAE Systems Hawk). In certain key markets, Pilatus could foresee that the greater political 'clout' of the US company would work against the Swiss-built products and, although the PC-9 is considerably cheaper than the T-6A, that it would be difficult to compete on cost alone with the US aircraft. While further improvement of the PC-9 may have reaped some benefits, it was felt that the development of a new and far more advanced aircraft was a more effective solution.

Accordingly, on 11 November 1998 the Pilatus board gave the go-ahead for a privately-funded venture, worth SFr 200 million. In an ironic twist, funding for the project largely came from revenue generated by the T-6A programme. Work began in January 1999 on what was a 'clean sheet' design. It had three principal goals: to achieve better aerodynamics and performance to enable it to simulate modern jet fighters more closely; to address a wide range of through-life cost, maintenance and reliability issues to make it highly attractive in the marketplace; and to feature a better mission system than any other trainer – including jet-powered rivals. Development proceeded in consultation with several potential customers.

On 1 May 2002 the new trainer design, designated PC-21, was rolled out at the Stans facility. While bearing a family resemblance to its fore-runners, the PC-21 is a dramatically different aircraft, with exceptionally clean lines and a very short (8.77-m/28-ft 9-in), slightly swept wing. The 1,600-shp (1194-kW) Pratt & Whitney Canada PT6A-68B engine and short wing had already been tested on a company-owned PC-7 Mk II, which had been modified for Proof of Concept work and to establish just how far the aerodynamics could safely be pushed. This vehicle, in fact, had an even shorter wing, at just 8.40 m (27 ft 7 in) span. It was also used to test new avionics architecture and cockpit displays, and had first flown in this new fit on 17 December 2000.

As well as its low aspect ratio wing, the PC-21 introduces spoilers for the first time in a turboprop trainer, features which are to be found on many operational jets. Combined with the ailerons they confer roll-rates similar to those found in jet fighters. An automatic yaw compensation system provides jet-like handling, while a power management system allows the engine power profile to be tailored to the phase of training. Take-off and landing characteristics are benign, as befits the aircraft's trainer role, while speed and climb performance are breathtaking. Low-level maximum speed is an impressive 330 kt (611 km/h; 380 mph), which ensures a 300-kt (555-km/h; 345-mph) navigation exercise capability. The aircraft can be dived to Mach 0.8. The small wing is highly loaded, providing an accurate and stable platform for simulated or actual weapons aiming and release. Naturally, the PC-21 is fully aerobatic, and its flexible graphite propeller allows it to be flicked with ease (earlier rigid aluminium props precluded this). A strake kit can be added for spin resistance. *G* limits are +8/-4.

All primary structure is aluminium, for ease of maintenance and strength, although composites are incorporated into secondary structures. Milled and chemically etched parts provide outstanding accuracy and fleet commonality. The cockpit is provided with Martin-Baker's state-of-the-art Mk 16L ejection seats, and the canopy features a fracturing linear shaped charge. The cockpit has been designed 'big' to cater for a wide anthropometric size range, and offers an outstanding view. With the rear seat in the high position a forward lookdown angle of 7° is possible, while in the front the lookdown angle is 11° – sufficient for a simulated 100-ft (30.5-m) CCIP bomb release.

Pilatus has paid great attention to the cost and ease of ownership, and the PC-21 has been developed with private finance initiatives (PFIs) in mind, whereby contractors are looking for known costs across potentially large fleets, and accurately predictable through-life costings. The company predicts the unit cost will be around $6.5 million, or the same as a T-6A.

State-of-the-art Health and Usage Monitoring (HUM) systems are fitted to keep maintenance costs low and reliability high. The PC-21 has 25 per cent greater fuel than its predecessors, allowing fast turn-rounds and back-to-back training missions without refuelling, in turn providing further significant cost savings over a lifetime.

Mission systems

Undeniably impressive from the outside, it is perhaps inside where the PC-21 is the most advanced. The cockpit is a state-of-the-art fully digitised 'glass' cockpit, with three 6 x 8-in active-matrix liquid crystal displays (AMLCDs) in each cockpit, and no analogue instruments. There are two standby AMLCDs in each cockpit. The front cockpit has a HUD, and the rear cockpit can have an AMLCD HUD repeater fitted. Inputs to the system are made by HOTAS, MFD soft keys or by the large UpFront Controller Panel (UFCP). The latter is of fighter size thanks to the innovative positioning of the camera forward of the HUD, rather than the

P01 overflies Stans airfield during its first flight on 1 July 2002, during which the undercarriage remained extended while basic low-speed handling qualities were verified. Early flight tests quickly established the excellent handling at speeds up to 330 kt (611 km/h).

PC-21 specification

Powerplant: one Pratt & Whitney Canada PT6A-68B turboprop rated at 1,600-shp (1194-kW)
Dimensions: wing span 8.77 m (28 ft 9 in); tailplane span 4.00 m (13 ft 1.5 in); length overall 11.19 m (36 ft 8.5 in); height overall 3.91 m (12 ft 10 in); fuselage width 1.00 m (3 ft 3.5 in)
Undercarriage: wheelbase 2.50 m (8 ft 2 in); wheel track 2.74 m (8 ft 11.5 in)
Weights: basic empty 2250 kg (4,960 lb); maximum take-off (utility) 4250 kg (9,370 lb); maximum take-off (aerobatic) 3100 kg (6,834 lb); maximum external load 1150 kg (2,535 lb)
Speed: maximum operating 370 kt (685 km/h; 425 mph) or Mach 0.7; maximum dive speed 420 kt (777 km/h; 483 mph) or Mach 0.8; stall speed less than 80 kt (148 km/h; 92 mph)

Roomy and well-equipped, the PC-21's cockpit provides an outstanding training environment for students destined for fast jets. The displays can be configured in the air to show data applicable for all instructional needs from basic flight training to weapons release simulation and tactical mission planning. The PC-21's high speed also provides realistic low-level navigation training for fast-jet pilots.

usual position aft. The camera itself does not record the actual HUD image, but rather a digital overlay generated by the HUD system, thus ensuring high brightness for the repeater and digital recorders. The cockpit is Gen-3 NVG-compatible.

Controlling the system are three flight safety-critical mission computers. The main computer is made by GD(UK) and the two Primary Flight Display computers are manufactured by Barco. Systems are tied together by Mil Std 1553 and ARINC 429 databuses. The system is of open architecture, allowing processor upgrades and plug-ins to be accepted without major reconfiguration. The cockpit can be made to look and feel like any advanced fighter aircraft, tailored to a particular customer's needs.

What the system provides the operator with is an extremely flexible training platform which can meet a wide variety of requirements. When operating in 'normal' state, the displays in front and rear cockpits are tied together. In the 'split' state the instructor can override the student and can command extra features or simulate emergencies. If a student has made a mistake, the

Roll-control spoilers and automatic yaw control (which negates the effect of propeller torque) confer a jet-like feel on the PC-21. Despite the short-span wing, landing speed remains commendably low for the training role.

instructor can 'decouple' from the front seat and monitor the correct settings, leaving the student to follow his/her erroneous inputs.

Weapons and mission training is also possible through the system, and is where Pilatus can offer the biggest savings to potential customers by switching these traditional jet tasks to the cheaper turboprop platform. The system has a simulated stores management system, and a simulated weapons release function. This can be fully controlled in the air, allowing mid-air 'reloads'. When operating in the 'split' mode, the instructor can use the system to simulate weapon hang-ups. A datalink emulation card is an option to simulate the use of radar, while the system can support FLIR and electro-optical sensors.

Today's advanced combat aircraft feature mission planning systems, through which waypoints, threat locations and other navigational/situational data are pre-loaded into the aircraft before take-off. The PC-21 has a similar function, the mission planning system also acting as a system which records all data – not just the displays – for effective debriefing even after a solo sortie. The open architecture of the system will allow further functions to be incorporated in the future, such as simulator A/A and A/G radar, Direct Voice Input and Rangeless Airborne Intercept Debriefing Systems (RAIDS) for real-time mission monitoring. All of the PC-21's inflight systems are mirrored in the software of ground-based training aids, from simple computer-based aids to full simulators.

As with the PC-7 and PC-9 before it, the PC-21 can be converted for actual weapon delivery if required, and would make a very

effective weapons trainer or light attack/FAC platform with a healthy external stores load. In this case, Pilatus would have to apply for a war material licence, as it is prevented by Swiss law from supplying the aircraft in attack configuration. However, that has not stopped local modifications being made to PC-7s and -9s overseas, and the light attack potential has not been overlooked with the PC-21, either.

Flight programme

Following its rollout in May, the first prototype (P01) made its first flight with Bill Tyndall at the controls on 1 July 2002. A fault-free initial flight test programme – during which the critical flight envelope (up to 4 *g* and 330 KTAS), stalling and engine relights were tested – led to a dispensation from the Swiss authorities for the aircraft to attend the RIAT and Farnborough air shows in the UK after just 10 hours 13 minutes flight time. (The normal minimum for an experimental aircraft to leave the country is 50 hours). For the UK trip the rear seat was installed and telemetry equipment removed from the rear fuselage. (While the test equipment is in place the rear seat is removed for centre of gravity reasons.) On its return to Switzerland, P01 was laid up while the engine was removed for a thorough inspection.

Certification is expected in 2004, by which time the second prototype (P02) should have flown. Owing to the early test success of P01, completion of P02 has been delayed and it is scheduled to fly no sooner than April 2004. Under the present planning it will not be required in the certification process, and will be used primarily as a customer demonstration aircraft and to act as a pre-production machine. It should be noted that the PC-21 has been built on 'hard' tools from the outset, and no major changes are envisaged between P01 and full-production machines.

While Pilatus would agree that only a jet trainer can really be used for instruction in advanced air defence/air combat manoeuvring, the high-performance turboprop trainer with a suitably advanced mission system can be much more cost-effective in most other training roles, from BFT (basic flying training) through to Fighter Lead-in and air-to-surface mission management. That the costs of a PC-21 equate to about one-fifth of those of a jet offering comparable systems capability make the third generation of Swiss turboprop trainers a highly attractive proposition. Future UK training requirements, and equipping the Eurotraining consortium, are high on Pilatus's list of targets.

David Donald

Kamov Ka-31

Shipborne AEW for the Indian Navy

Two Kamov Ka-31 early warning helicopters designed for the Indian Navy are now completing tests on the Kamov airfield at Chkalovskiy, outside Moscow. In August 1999, India ordered four Ka-31s for service aboard the *Viraat* aircraft-carrier and three 'Krivak-III' frigates which are under construction at St Petersburg; in February 2001, five more helicopters were ordered.

On 16 May 2001, the first series helicopter took-off at Kumertau. The production process is subdivided into two stages: the airframes made by Kumertau's KumAPO factory are transferred to the Kamov workshop at Lyubertsy near Moscow, where mission systems are installed. The first two Ka-31s are due for delivery to India on 7 December 2002, with the remainder following in 2003. The Russian armed forces are also interested in the Ka-31 and this helicopter has been included in the state armament programme for 2002-2010.

Design work on the Ka-252RLD (Radio-Lokatsionnogo Dozora, radar surveillance) shipborne early warning helicopter, later redesignated Ka-31, started in 1985, coincident with the start of construction of the first full-size Russian aircraft-carrier – the *Tbilisi* (now *Admiral Kuznetsov*). The Ka-31 was developed from the Ka-27 'Helix' anti-submarine helicopter with almost identical propulsion system,

Ka-31 specification

Rotor diameter	15.9 m (52 ft 2 in) each
Fuselage length	11.25 m (36 ft 11 in)
Maximum height	5.60 m (18 ft 4.5 in)
Wheel base	3.05 m (10 ft)
Main wheel track	3.5 m (11 ft 6 in)
Front wheel track	2.41 m (7 ft 11 in)
Max. take-off weight	12500 kg (27,557 lb)
Maximum speed	250 km/h (155 mph)
Cruising speed	220 km/h (137 mph)
Operating speed	100-120 km/h (62-75 mph)
Operating altitude	up to 3500 m (11,483 ft)
Operational radius	up to 100 km (62 miles)
Operational endurance	2 hours 30 minutes
Range	600 km (373 miles)

transmission, rotors and tail unit. The fuselage was modified for equipment installation, and the undercarriage was all-new.

Bearing side number 208, the first prototype flew for the first time in October 1987, piloted by V. Zhuravlov. This prototype was used for testing the extending radar antenna, retractable landing gear, and for tests of the helicopter's aerodynamics when operating with the rotating antenna of the radar. The prototype was not fitted with the complete range of mission equipment, and several antennas and devices, such as the equipment cooling fan, found on subsequent machines were absent.

In 1990, Ka-31 031 flew for the first time, fitted with the mission system; some time later 032 appeared. In 1992 both helicopters, 031 and 032, were spotted on board *Admiral Kuznetsov* for the first time. In August 1995 031 was shown to the public for the first time during the MAKS'95 show at Zhukovskiy.

Upon completion of the state acceptance tests, a decision on starting Ka-31 series production was taken in 1996. However, the available money was sufficient only for the modification of the Ka-27 production line at the KumAPO plant at Kumertau, Bashkortostan, to meet the needs of Ka-31 production. Actual production of Ka-31 helicopters could only be started after the receipt of an order from India.

Integrating the radar system with other equipment of the helicopter was the main concern of the development programme, as well as maintaining electronic compatibility with ship- and shore-based equipment, and ensuring the stability of the helicopter with a large flat antenna rotating under the fuselage. Thanks to the SAU-37D automatic flight control system, developed by KBPA (Konstruktorskoye Byuro Promyshlennoi Avtomatiki, Industrial Instrument Design Bureau) in Saratov, the rocking of helicopter while the antenna rotates does not exceed 2°, which is equivalent to the condition of flight in turbulent air. The PNK-37DM

flight-navigation system holds the pre-set heading and altitude, follows the selected path, performs the landing approach and finds the exact position of the helicopter without external sources of information. The Ka-31 has a crew of two, comprising pilot and navigator. Cockpit arrangements differs from version to version, but has two to four MFI-10 multi-function displays.

Developed by NIIRT (Nauchno-Issledovatelskiy Institut Radiotekhniki, Radio Scientific-Research Institute) in Nizhnyi Novgorod, the E-801M Oko (eye) radar system is installed in the Ka-31 helicopter. The radar detects and tracks air targets, including small targets flying at very low altitude, and surface sea targets. The 6 x 1 m (20 x 3-ft) rectangular radar antenna weighs 200 kg (440 lb). In cruise position the antenna is folded away horizontally, held flush with the bottom of fuselage. When operating, the antenna is lowered, swivelled into a vertical position and rotated around its axis every 10 seconds. The undercarriage legs are raised in order to avoid signal disturbances. In emergency, the antenna can be raised up manually or jettisoned.

Capable of tracing up to 20 targets simultaneously, the radar unit operates in the 10-cm (4-in) wavelength band. It can detect a fighter aircraft at a distance of 110-115 km (68-71 miles), whereas surface ships can be detected within the limits of the horizon (200 km/124 miles from 3000 m/9,842 ft altitude). After lowering the aerial and selecting the operating mode, which is performed by the navigator, the further operation of the E-801M system is executed automatically; the navigator only supervises the system operation. Data for target co-ordinates, speed, heading and identity, found by the radar, are transmitted via coded radio data link to shipborne or shore-based command posts. The information on targets is not processed on the helicopter, but is transmitted directly to the ship or shore command post; the navigator can only watch the situation on the display.

Power for the Ka-31 comes from two TV3-117VMAR turboshaft engines rated at 1641 kW (2,200 hp) each. A TA-8-Ka auxiliary power unit is installed instead of the Ka-27's AI-9 to provide the additional power supply required by the radar system. Fourteen fuel tanks contain 3060 litres (673 Imp gal).

Although developed initially as a ship-based radar picket, the Ka-31 is being studied for other roles. A new version, designated 23D2, is currently under development and being offered by Kamov as a tactical early warning vehicle for the Army and Air Force, intended to detect low-flying aircraft, helicopters and missiles. It can also be used as an airborne relay for transmitting to the ground reconnaissance data from other helicopters (such as the Ka-60R and Ka-52), or from pilotless vehicles. It would also be able to transmit target designation data to Ka-50 combat helicopters. The NIIRT institute has also produced a land-based version of the Oko system, installed on a vehicle. In this case, the antenna is raised on a strut to a height of 12-16 m (40-52 ft).

Piotr Butowski

The first Ka-31 for the Indian Navy is seen in Russia prior to delivery. The rotating antenna for the Oko radar is visible in the stowed position under the cabin.

Sea King ASaC.Mk 7

Enhanced AEW for the 'Bag'

Thales Sensors has delivered the first pair of upgraded Westland Sea King ASaC.Mk 7s (Airborne Surveillance and Control) to the Royal Navy. The first two examples arrived at 849 NAS, RNAS Culdrose on 17 May 2002, with deliveries of the 11 additional ASaC.Mk7s (eight upgraded AEW.Mk 2As and three converted HAS.Mk 5 anti-submarine warfare variants) continuing through to 2004, at a rate of around one per month to give a final complement of 13. The squadron has the very apt motto of 'Primus Video' ('The first to see').

The US$155 million programme followed a Ministry of Defence (MoD) invitation to tender in early 1995 for the Mission Systems Update (MSU) of the Royal Navy's existing Sea King AEW.Mk 2As. Thales Sensors (formerly Racal Defence Ltd) was selected as prime contractor, with work beginning in February 1997, leading to flight trials in April 2001. Subcontractor Westland installed the new mission equipment in the first two examples at its Weston-super-Mare plant, with remaining modification work being conducted by a Thales-led team at the AMU at RNAS Culdrose.

The upgrade centres around the Thales Searchwater 2000 AEW radar and Cerberus mission system, which radically enhances the capabilities of the 'Airborne Early Warning' Sea Kings. The Radar System Update (RSU) has seen the original AEW.Mk 2A's THORN EMI Searchwater radar being replaced with the Searchwater 2000 AEW pulse-Doppler radar. This extends the helicopter's surveillance range with an improved capability to detect small airborne targets, with greater flexibility in operation. Previously only 40 tracks could be maintained, with operating height of the Sea King having to be carefully calculated to ensure optimum radar performance. The new system permits over 250 target tracks, with the two operators working at the new Human Computer Interface (HCI), jointly designed with the Royal Navy to provide improved situational awareness. The ASaC.Mk 7 also features a Rockwell Collins AN/URC-138(V)1(C) Link-16 JTIDS datalink terminal, AN/APX-113(V) IFF and Northrop Grumman LN-100G GPS/INS ring laser gyro navigation system. The Sea King AEW.Mk 2A received Have Quick II secure speech radios in 1995, and the ASaC.Mk 7

retains them along with the existing MIR-2 Orange Crop Electronic Support Measures (ESM) system for warning of illumination from threat radars.

The Sea King AEW.Mk 2 was always seen as a 'quick fix' AEW platform for the Royal Navy, but the upgrade ASaC.Mk 7 is seen as a quantum leap forward, with huge potential. The loss of ships during the Falklands conflict graphically illustrated the requirement for an embarked AEW capability. A pair of Sea Kings was rapidly configured to carry the externally mounted (bagged) Searchwater radar and set sail on HMS *Illustrious*, arriving in theatre as the conflict came to an end, but 'The Bag' was born and 849 NAS was re-commissioned at RNAS Culdrose to give the fleet a valuable AEW capability.

Today, a typical ASaC.Mk 7 mission involves a three-man crew, with the Observer in the left-hand cockpit seat staying 'up front' for landing and take-off phases and moving 'down the back' to join his radar Observer colleague once 'on task'. Weight savings from a reduced crew and revised systems fit has given the upgraded Sea King an extra hour's endurance, with a four-hour mission now possible. The Searchwater 2000 AEW radar has advanced long-range air-to-air look-up/look-down capa-

There are no major external differences to differentiate the Mk 7 'Bag' from the Mk 2, but it is a far more capable aircraft in terms of both systems and mission capability. Improvements in the latter come courtesy of a major weight-saving process which increases endurance on patrol.

bility, with both overwater and overland performance. The radar Observers will typically build up an annotated radar picture and continuously update air or sea friendly forces via encrypted voice or datalink transmissions. The Mk 7 is Link-16 JTIDS-equipped and is also compatible with Link-11 and the future Link-22. Unlike some other 'AWACS' types, the two-man radar crew does not have the facility to advise 'players' on appropriate tactics, but they serve to give valuable time-critical target information. The potential for the new system is being realised by the very crews that worked with Thales to make the upgrade such a success in the first place.

The Cerberus mission system has been designed to be installed on a wide range of platforms. The Royal Navy's Maritime Future Organic Airborne Surveillance and Control (MASC) programme for AEW support of the Royal Navy's future carriers (CVF) could see the mission suite being moved from the ageing Sea King to another type. Thales is keen to promote the potential for the AgustaWestland Merlin to receive the system, however the Navy reportedly favours an E-2 Hawkeye-based platform.

In the meantime, the only limitation for the new system would appear to lie in the demise of the Sea Harrier FA.Mk 2 in Royal Navy service. The ASaC.Mk 7s will soon be relying on the less-capable JTIDS-equipped Harrier GR.Mk 7/9s with short-range AIM-9 Sidewinder missiles to act upon its information to defend the fleet, not to mention the valued 'eye in the sky' Sea King.

Jamie Hunter

Internally the Sea King ASaC.Mk 7 upgrade differs little from the Mk 2 configuration on the flight deck (above), but has new HCI displays for the two Observers (right). The upgrade has been considered a major success from both operational and financial viewpoints.

Gripen International for South Africa

The SAAF's Advanced Light Fighter Aircraft

Right: The SAAF's Lt Col Mike Edwards 'flies' the Gripen full dome simulator at Såtenäs. South African industry is now a major supplier to the Gripen programme after offset deals have been fulfilled.

With full deliveries due from 2008, South Africa is gearing up to introduce the Saab Gripen as the basis of its air force for the first half of the 21st Century. The choice of Gripen reflects the change in South African defence policy in the 1990s to a defensive posture in a tighter budgetary climate, but with the need to maintain a qualitative superiority over its neighbours.

Back in 1990 the situation was vastly different, and the SAAF had very different requirements. The force was active in southern Angola, where the Cheetah was being increasingly 'outgunned' by opposing fighters, and Namibia was occupied. The SAAF's requirement envisaged 32 heavy fighters in the Eurofighter class and 48 Hawk/AMX light attack aircraft to replace 40 Cheetahs, 21 Mirage F1s and around 200 Impalas. However, in 1991 the South African Defence Force withdrew from Namibia and operations in Angola decreased dramatically. In 1994 the apartheid regime crumbled and Nelson Mandela swept to power.

In 1994/95 a major defence review was undertaken, which placed the armed forces on a far more defensive posture, but at the same time recognising South Africa's dominant position in the southern half of the continent. With the November 1997 retirement of the Mirage F1AZ the SAAF contracted to a combat force of one squadron of Denel Cheetah C/Ds. New aircraft procurement was revised completely, resulting in a call for 28 Advanced Light Fighter Aircraft (ALFA) and 24 Lead-In Fighter Trainers (LIFT). The fighter requirement stated the need for an aircraft capable of hot-and-high operations and the ability to operate from highway strips. Based on threat analyses, it had to be a true multi-role platform capable of defeating what were termed 'medium' fighters, while also offering a range of precision attack and reconnaissance options. In line with a far more restrictive budget than the SAAF had once enjoyed, it had to be highly reliable, easy to maintain and be of low cost – both in terms of acquisition and through-life costs. Furthermore, at least 100 per cent offsets were required to maintain South Africa's growing defence and other industries. Delivery before 2010 was specified due to the expected out-of-service date for the Cheetah.

On 18 November 1998 the results of both competitions were announced: the ALFA requirement would be met by the Saab-BAE Systems (now Gripen International) Gripen (19 single-seaters and nine two-seaters), while the LIFT would be the BAE Systems Hawk. The deal was formally signed in Pretoria on 3 December 1999. In November 1998, Grintek Electronics was awarded the first of several contracts by Saab-BAE Systems to supply avionics items, and in March 1999 Denel was given the contract to supply NATO-compatible weapons pylons for export Gripens. Defence contracts account for about 50 per cent of the offset deal, with the remainder fulfilled by other national industries.

South Africa's Gripens will be to the Export Baseline Standard (EBS), with NATO compatibility, climate adaptation, VOR/ILS, OnBoard Oxygen Generating Systems (OBOGS – to enhance deployability), English language cockpit and inflight refuelling probe as standard. South Africa's aircraft will also feature a Link ZA datalink, RSA-specific comms and IFF, Helmet-Mounted Display (HMD), TACAN and RSA-specific weapons.

Included in the latter is the R-Darter active-radar medium-range missile, in the AMRAAM class. This weapon is due to be test-fired from a Gripen in 2007/08 at the Tactical Fighter Development Centre (TFDC) at Bredasdorp. A short-range air-to-air missile programme – Project Kamas – is under way to provide a fifth-generation SRAAM for the Gripen. RFIs have been issued and a decision is expected around 2005 to give sufficient time for the new weapon to be integrated with the Gripen. Once seen as the logical choice, the local V-3E A-Darter is possibly no longer under active development.

Air-to-surface weapons will include a stand-off weapon, currently under review, and laser-guided bombs. A designation pod is required for the latter but no choice has been made publicly, although Gripen International offers the Rafael/Zeiss Litening as its main export

Gripen '208' is a late Batch 2 Flygvapnet aircraft, and the first to be completed to JAS 39C standard. A key feature of the 'C' and all export aircraft is the retractable inflight refuelling probe housed above the port intake.

option. Similarly, no decision has been announced regarding a reconnaissance pod: the choices are pods developed by Saab and by Thales (Vinten). South Africa's Gripens will feature the EWS 39 electronic warfare suite, as developed for Sweden's JAS 39C/D aircraft. This, ultimately, will include a towed radar decoy.

Deployment of the ALFA and LIFT aircraft is due to begin with the delivery of the first Hawks to TFDC for Integrated Test and Evaluation (IT&E) in 2003. The Hawks will ultimately serve with 85 Central Flying School at Hoedspruit and will have a primary training role (with an emergency attack 'war' role if required). One two-seat Gripen is due to be delivered to the TFDC in 2006, and will be used for IT&E. In 2007/08 the Operational Test and Evaluation (OT&E) phase will begin, including weapons trials. The remaining eight two-seat Gripens will be delivered in early 2008, to allow training to begin with Swedish instructors. A full simulator is also provided under the Gripen International deal. The 19 single-seaters will be delivered in 2010.

In South African service the Gripen is expected to retain its Swedish name, despite a tradition of naming types locally. The operating unit will be No. 2 Squadron at Louis Trichardt Air Base. It is envisaged that all pilots will be qualified in any of the Gripen's roles, but there will be specialist role flight commanders, notably for reconnaissance. The heavy proportion of two-seaters reflects not only the SAAF's training needs, but also the expectation that they will be used in combat operations, providing an on-scene situation command capacity.

David Donald

This enhanced photo shows a two-seat Gripen carrying R-Darter missiles under the wing pylons. This AMRAAM-class missile will be the principal air-to-air weapon of SAAF aircraft. Rafael Python 4 SRAAMs are shown on the wingtip rails – a possible candidate for the South African Project Kamas requirement.

Sikorsky MH-60S Knighthawk

'Sierra' training with the 'Packrats' of HC-3

The US Navy is investing heavily in order to address widespread obsolescence amongst its active-duty and reserve helicopter fleets. The Helicopter Master Plan is designed to modernise and consolidate the Navy's helicopter force through a migration from eight different types to two 'linchpin airframes', both new variants of the generic H-60 – the MH-60S Knighthawk and the MH-60R Strikehawk.

The MH-60S Knighthawk will initially replace the remaining H-46 Sea Knights in the Navy's Combat Support (HC) Squadrons, followed by HH-60Hs in the Reserve HCS squadrons, thereafter the base flight-operated UH-3H and HH-1N helicopters, and finally the HH-60H aircraft in the Fleet (HS) squadrons. A total of 237 new-build MH-60 'Sierras' has been ordered.

The MH-60S is the first US Navy helicopter to be produced by Sikorsky since manufacture of the HH-60H ended in 1996. It is designed with the Helicopter Combat Support role very much in mind, combining the robust load-carrying ability of the Army's Blackhawk with the 'navalised' features of the SH-60B Seahawk. It is based on a standard UH-60L Blackhawk, with the SH-60's T700-GE-401C engines, rotor system and dynamic components including the automatic rotor blade folding system, folding tail pylon, improved durability gearbox, rotor brake and automatic flight control computer. The new Knighthawk also features a larger cabin, double-door configuration, an integrated cargo-handling system, a 9,000-lb (4082-kg) capacity cargo hook and provision for external pylons.

Notable visual differences between the original Navy SH-60 Seahawk and the new MH-60S Knighthawk include a tail wheel which sits further aft, permitting more aggressive approaches to confined landing zones, and engine exhaust venting which effectively reduces the helicopter's heat signature. The cabin is redesigned to make it cargo- and passenger-compatible, with opposing main cabin sliding doors to facilitate rapid troop ingress and egress.

Knighthawk takes shape

Sikorsky first demonstrated the MH-60S concept in 1995 with a modified UH-60L Black Hawk. The success of this venture, coupled with the urgency of the programme, led to detailed design work commencing in October 1996. The prototype YCH-60S made its maiden flight on 6 October 1997 with the new design proving highly successful in VERTREP (vertical replenishment) duties.

Lockheed Martin secured a US$61 million contract in the third quarter of 1998 for development of a common 'glass cockpit' prototype for both the MH-60S and MH-60R. The new cockpit is centred on Lockheed Martin-developed computer systems, but using commercial PowerPC processors, with data presented to pilots via an electronic flight instrument display and multi-function mission display. The cockpit incorporates four 8 x 10-in (20 x 25-cm) Multi-Function Displays, two operator key sets and digital communications suite as well as Litton (now Northrop Grumman) integrated INS/GPS, mass memory unit, mission and flight management computers

MH-60R Strikehawk

The MH-60R will replace the existing fleet of SH-60B and SH-60F helicopters operated by the active-duty HS/HSL community. MH-60R is the new designation for the original SH-60R upgrade (also known as LAMPS Block II), which combines SH-60B Seahawk capabilities with the dipping sonar of the SH-60F.

The original US Navy Helicopter Master Plan called for all 170 SH-60Bs and 18 SH-60Fs to be re-worked to MH-60R configuration by 2011. However, development delays and cost concerns prompted the Navy to restructure the plan in 2001 and to opt for a predominantly new-build programme. Consequently, out of a total requirement of 243 aircraft, two will be MH-60R prototypes, four will be SH-60Bs remanufactured as MH-60R test airframes, and three will be Low-Rate Initial Production (LRIP) remanufactured aircraft; the remainder will be all-new production MH-60Rs. The 'Romeo' is expected to enter the fleet in the first quarter of 2005.

The MH-60R brings a significant upgrade in capability, including the Multi-Mode Radar, AQS-22 airborne low-frequency dipping sonar, electronic support measures, forward-looking infrared radar, datalink, acoustics and Integrated Self Defence package.

and applicable operational software for both versions.

A decision to proceed with low-rate initial production (LRIP) was made in 1998, initially for six examples. The first production Knighthawk made its maiden flight from Sikorsky's Stratford facility on 27 January 2000. Following company trials the test Knighthawks moved to NAS Patuxent River, Maryland in May 2000 to begin testing and operational evaluation with the Rotary Wing Aircraft Test Squadron (RWATS). A total of four examples was used in the trials, which culminated in the Operational Evaluation (OPEVAL). Initially planned for early 2001, the OPEVAL slipped to November of that year following avionics integration issues.

In May 2002 it was reported that the Knighthawk had failed its OPEVAL with the type being officially graded 'not effective' for the VERTREP role. The helicopter had apparently failed to meet a two-hour fuel endurance requirement, as well as suffering computer shortcomings. However, according to H-60 multi-mission helicopter programme manager Captain Bill Shannon, the MH-60S had erroneously been evaluated against an obsolete Operational Requirements Document (ORD). With these issues ironed out and new computer software coming on line, the MH-60S was cleared to proceed towards Initial Operational Clearance (IOC) in September 2002.

After initial 'release to service' trials had been completed the first production Knighthawk was delivered to HC-3 'Packrats' at NAS North Island on 19 December 2000. The 'Packrats' are

Left: Both pilots of the MH-60S are provided with large MFDs for the display of aircraft system, flight and tactical data. A handful of traditional dial-type instruments are retained in the central dashboard as emergency back-ups. Communications and navigation equipment controls are located in the central console between the pilots.

Below: An HC-3 'Packrats' MH-60S sits on the North Island ramp next to the type it is replacing: the CH-46D. Training for the latter type was performed by HC-3 until September 2002.

HC-3's CAG-bird wears full-colour markings with an eagle on the fin. The squadron badge is worn on the cabin sides. All prospective MH-60S crew will pass through the FRS, although the syllabus will be altered according to their future assigned roles.

Features of the MH-60S include heat-diffusing exhaust shields (left) and a heavy-duty winch for SAR ops (above). A central hook gives the Knighthawk its main cargo-carrying capability.

the first Fleet Replacement Squadron (FRS) for the MH-60S and commenced training new aircrew in October 2001, with the first 'Sierra' students graduating in April 2002.

Lieutenant Commander Mario Mifsud, Executive Officer at HC-3, confirms that things are progressing well with the Navy's newest helicopter: "Having, for a time, combined the roles of FRS for the HH-46D and the new MH-60S, we finally retired our last Sea Knight in June 2002. Instruction of any remaining aircrew assigned to the 'Phrog' will continue to be undertaken by HC-3 but we will be utilising aircraft from HC-11. We are projected to complete H-46 training in September 2002."

The major effort at HC-3 is now centred on the new 'Sierra' and the squadron already has nine of the new Knighthawks, and expect this to increase to 10 in the near future. It is anticipated that the squadron may operate up to 16 aircraft once Knighthawk deliveries to the Fleet gather pace.

The training syllabus for the Block 1 version of the MH-60S closely follows the initial core role that the aircraft will fulfil in the Fleet in replacing the venerable H-46 Sea Knight – providing carrier battle group logistical support through VERTREP, remote site logistics missions and Search and Rescue for the Large Deck Amphibious Assault Ships.

According to Lt Cdr Mifsud "the backbone of the transition of the Fleet's [HC] squadrons to the Knighthawk are the CAT 2 students. These are pilots who are other type-qualified (i.e. H-3/H-46) and who will go on to be Instructors on the new aircraft or senior aircrew in the Fleet squadrons due to receive the 'S'."

HC-3 administers a number of programmes in addition to providing trained pilots and aircrew to the Fleet. As the Model Manager for the Naval Air Training and Operating Procedures Standardisation (NATOPS) and the NVD (Night Vision Device) programme for the MH-60S, HC-3 will evaluate the NATOPS and NVD programmes of each squadron annually.

By September 2002 HC-3 instructors had graduated MH-60S crews for operational fleet squadrons such as HC-6 'Chargers' at NAS Norfolk, which is already using the aircraft in Fleet service.

Full-rate approved

In early September 2002 the US Navy granted Milestone III Full-Rate Production for the MH-60S, giving the 'green light' to the purchase of the 237 examples required through to 2010, subject to annual budget allocations.

Multi-Mission Helicopter Programme Manager Captain Bill Shannon remarked; "The MH-60S has gone from concept, through devel-

opment, into production and fleet introduction faster than any contemporary programme. Specifically, the programme went from Milestone II to Fleet IOC (Initial Operational Capability) in four years. Adding to the significance of the shortened timeframe is that the development effort was completed for $1.4 million less than the objective cost of $72 million."

The future

The primary missions of the MH-60S Knighthawk will include day and night VERTREP, day and night amphibious SAR, vertical onboard delivery and airhead operations. Secondary missions of the MH-60S will include Combat Search and Rescue (CSAR), Special Warfare Support (SWS), recovery of torpedoes, drones, unmanned aerial vehicles and unmanned undersea vehicles, non-combatant evacuation operations, aeromedical evacuations, humanitarian assistance, executive transport and disaster relief.

The CSAR/SWS version of the MH-60S will have additional mission equipment installed that will provide capabilities for CSAR and SWS in both the active carrier-based Helicopter Antisubmarine Squadrons (HS) and in the Reserve Helicopter Combat Support (Special) (HCS) Squadrons.

Sikorsky is also in the design phase for the Airborne Mine Countermeasures (AMCM) derivative that will add an operator's station to the cabin, additional internal fuel stores, and towing capability to the aircraft. The AMCM-capable MH-60S will provide the Carrier Battle Groups (CVBG) and Amphibious Ready Groups (ARG) with organic airborne mine countermeasures (OAMCM) capability. Initial Operational Capability is planned for calendar year 2005.

Jamie Hunter and Richard Collens

Agusta MH-68 'Shark'

HITRON-10's gunships

In November 2000 the US Coast Guard took delivery of its first example of the Agusta A109E Power helicopter. Designated MH-68A and known unofficially as 'Sharks', the helicopters are assigned to Helicopter Interdiction Tactical Squadron Ten (HITRON-10) at Cecil Field Airport near Jacksonville, Florida, and support the service's Airborne Use of Force (AUF) mission. The 'Shark' is the only aircraft in Coast Guard service not intended for multiple missions. In fact, HITRON is tasked as a maritime airborne law enforcement unit. Its primary mission is to interdict long-range, fast-moving vessels that are attempting to smuggle illegal drugs into the United States.

In September 1999 the Coast Guard revealed the existence of a proof-of-concept demonstration known as Operation New Frontier. As part of this demonstration, the service leased two examples of the MD Helicopters MD-900 Explorer, which were known unofficially as the MH-90 Enforcer. The aircraft, which were later exchanged for two similar examples, were deployed aboard a number of Coast Guard cutters in an attempt to interdict the so-called 'go-fasts' before they could deliver their illegal cargo. The successful demonstration ended in March 2000 and, after conducting an open competition, the service announced it had selected the A-109E and would lease eight examples for the next phase of the AUF programme.

The MH-68A is flown by a three-member crew, comprising two pilots and an aviation gunner. Cockpit equipment includes a three-axis autopilot flight director, a moving-map display, NVG-compatible LCD glass displays and a colour weather radar. Power comes from

Mounted in the port cabin doorway, the M240 machine-gun has a wide field of fire. Its principal role is for firing warning shots. This aircraft is in a standard AUF interception fit, with only one flotation bag on each side and no rescue hoist fitted.

two 732-shp (546-kw) Pratt & Whitney Canada PW206C turboshaft engines, controlled by a full-authority digital engine control (FADEC) system. The rotor system does not fold but the blades can be removed if shipboard stowage is required. The rotor system features a titanium main head and elastomeric bearings and is equipped with a brake system. The blade grips and rotor blades are constructed from composite materials.

MH-68As are fitted with equipment not typically found on other Coast Guard helicopters (FLIR/LLTV, Nitesun spotlight, siren, loudspeaker and a blue strobe light). Although a rescue hoist is available, it is seldom installed or used. Unique to the MH-68A is the 7.62-mm M240 machine-gun, which is installed in a flexible mount on the port side of the aircraft. In addition, a shoulder-fired 0.50-in Robar RC-50 precision rifle, equipped with an electronic laser sight, is carried. Although manufactured in Italy, final assembly of the aircraft was accomplished by Agusta USA in Philadelphia, Pennsylvania, which is also responsible for providing life cycle support.

HITRON deploys aviation detachments aboard the Coast Guard's high- and medium-endurance cutters. Because of the limited space aboard the cutters, two aviation detachments are normally deployed to a pair of cutters, with each vessel supporting one helicopter, two pilots, two aviation gunners and a maintenance technician.

A pair of MH-68As flies a port security mission over St John's River near HITRON-10's base at Cecil Field, Jacksonville. Note the FLIR/LLTV and Nitesun turrets under the nose. The aircraft have hoists fitted and two flotation bags on each side.

Once alerted to a suspected smuggler, the two crews work together to stop the vessel so that it can be boarded for inspection by personnel from a surface vessel. Initially they will attempt to contact the suspect vessel via radio or the aircraft's loudspeaker, while emphasising the law enforcement mission by using the blue light and siren. Failure to stop will result in warning shots being fired across the vessel's bow from the M240. If necessary, the Robar is used to disable the vessel's engine(s).

Initially based at the Coast Guard Aviation Training Center at Bates Field, Mobile, Alabama, HITRON relocated to Cecil Field Airport in April 2000, from where all training and maintenance is conducted. HITRON operates eight MH-68As, the last of which was delivered in August 2001, and is allocated 32 pilots, 30 aviation gunners, 10 maintenance personnel and two maintenance supervisors.

With the current two-year lease nearing an end the service recently released a request for proposals (RFP) covering the purchase of a permanent fleet of helicopters for HITRON. Although a number of suppliers were expected to bid for the contract, the winner has not yet been announced.

Tom Kaminski

Su-25 'Frogfoot' family
Developments and upgrades

Ugly, sturdy and hard-hitting, the Su-25 emerged in the 1990s as arguably Russia's most important combat aircraft. This situation was created by a variety of factors, most notably the collapse of funding for the armed forces, which severely restricted the procurement and operation of more sophisticated systems. At the same time, the Russian air force rapidly switched from its Cold War posture to one of internal policing, faced with growing unrest around the fringes of the former Soviet Union.

Left: It is uncertain how many Su-25UBs were built, although the number is probably around 60/65. This example, posing with an Mi-17 built at the same plant, wears the bear badge of the Ulan Ude factory which built all the two-seaters.

Right: The hump-backed Su-25T programme greatly increased the aircraft's abilities, but it foundered with the break-up of the Soviet Union. This is one of the service test batch, seen over the Volga during a test flight from Akhtubinsk.

Below: In its original guise the Su-25 was designed to fly short-range close air support missions, and it remains a powerful weapon in this role, despite its austere avionics fit. With five hardpoints available under each wing, it can carry an impressive rocket and bomb load.

Small and robust, the Sukhoi Su-25 'Frogfoot' can now justifiably claim to occupy a prominent position among the generation of combat aircraft that was fielded *en masse* in the 1980s and still forms the backbone of the Russian air arm. Although lacking the streamlined silhouette and stunning air show performance of both the MiG-29 and Su-27 air superiority fighters, the somewhat ugly and often undervalued 'Frogfoot' has gradually but indisputably emerged as the most useful combat aircraft of its generation in Russian air force service.

In the 1990s, the ubiquitous jet attack aircraft saw active use in numerous local conflicts that erupted over the territory of what is now known as the Commonwealth of Independent States (CIS), and on at least six occasions was used in anger by foreign operators.

Recent developments of the Su-25 core programme, especially the two newly-promoted upgrade packages – Russian and Israeli/Georgian ones – as well as the latest series of export successes between 1998 and 2001, have refuted to some degree the conclusion of some analysts who maintained that the first-generation Su-25 is effectively dead in its single-seat form. In fact, the basic 'Frogfoot-A' has proved yet again that – like its American counterpart, the A-10 – the type can be considered a fairly successful and long-lasting programme.

Luckless Su-25T/TM

In the early and mid-1990s, the Su-25T (design bureau internal designation T8M) was built as a specialised anti-tank aircraft for the Central European war theatre. Equipped with

the highly-automated SUV-25T Voskhod (Rise) navigation/attack suite, the aircraft's prime anti-tank weapon was the then-new 9M120 Vikhr tube-launched laser-guided anti-tank missile. A total of 16 Vikhr missiles, able to penetrate armour up 900 mm (35 in) thick, could be carried on two eight-round underwing launchers, in addition to all 1980s-vintage Soviet laser- and TV-guided missiles and bombs. The T8M-1 prototype was flown for the first time in September 1984, and was soon followed by two more prototypes and eight Tbilisi-built pre-production aircraft to be used in the flight test programme; another airframe was used for ground static testing. The protracted development programme, especially the trials of the complex nav/attack suite, held back the type's introduction into regular squadron service prior to the collapse of the Soviet Union.

In short, the very promising Su-25T had the bad luck to approach the completion of its flight test programme at the wrong time, i.e., just as the Soviet Union and its mighty aviation industry and air force suddenly collapsed in 1991. To make matters worse, it originally had been planned to build the new aircraft in a factory outside Russia, which was politically unsuitable. And finally, some analysts argue that there was no urgent need for such a highly-specialised, rather expensive and somewhat over-sophisticated new combat aircraft to enter service with the violently downsized and poorly financed VVS (Voenno-Vozdushnye Sily – Russian air force); in fact, only six pre-production Su-25Ts are known to have been taken on strength by the VVS in the early 1990s.

Not surprisingly, in the early and mid-1990s, the Su-25T programme reported disappointingly slow progress. The first machine of the 12-aircraft operational trials batch to be built in the Tbilisi, Georgia-based factory (formerly known as Factory No. 31 and now as Tbilisi Aerospace Manufacturing – TAM) was flight-tested in June 1990. A total of eight aircraft was used in a flight test programme that comprised over 3,000 flights and no fewer than 40 Vikhr live launches. The type's state trials were reported to have been completed in September 1993.

In the early 1990s, in the wake of the dissolution of the USSR, some 18 fully- or partially-completed Su-25T airframes of the pre-production batch were left in the Tbilisi-based factory, along with about three dozen basic Su-25s in various states of assembly.

The new type received its baptism of fire during the second Chechen war. Between late 1999 and mid-2000, two Su-25Ts drawn from the Lipetsk-based 4th TsPLSiBP (Aircrew Conversion and Combat Training Centre) were deployed to Mozdok in nearby Dagestan and were used to destroy a number of high-value targets in Chechnya, such as a satellite communications facility, radio relay station, the fortified house of well-known Chechen field commander Shamil Bassaev, a hangar housing defence equipment and, curiously, an An-2 light transport on the ground, suspected to be ferrying weapons from nearby Georgia. Kh-29L (AS-14 'Kedge') and Kh-25ML (AS-10 'Kegler') laser-guided missiles were the main types of weapons employed in Chechnya, but KAB-500 laser-guided and ODAB-500 1,102-lb (500-kg) free-fall fuel-air bombs were occasionally used to destroy underground shelters and weapons storage facilities.

As well as the Su-25UB/UBK combat trainer, the Ulan Ude factory built 12 hooked Su-25UTG carrier trainers for the AVMF. For some time it was thought that a carrier-capable single-seater would provide an anti-ship element in the AVMF's carrier air wing, using Kh-31 and Kh-35 missiles.

In the late 1990s, reports surfaced that the Georgian air arm had ordered as many as 50 Su-25Ts. These can be considered entirely erroneous, as that cash-strapped air arm cannot currently afford even a few more basic Su-25s.

'Frogfoot' for Ethiopia

Two Su-25Ts (TKs), drawn from the Lipetsk-based Aircrew Conversion and Combat Training Centre, and two Su-25UBKs (drawn from VVS regular units) were delivered to the Ethiopian air force in March 2000. The Su-25UBKs were refurbished by the VVS-controlled 121st ARZ (Overhaul Facility) based at Kubinka, and the Su-25Ts were prepared for

Lessons from Afghanistan saw defences on Su-25s added and improved. The Su-25T introduced a redesigned tail area with a new cylindrical fairing housing ECM and IR jammers, and two 32-round ASO-2V chaff/flare launchers.

An Su-25T is seen with a wide array of weapons, from standard FAB bombs in the foreground to Kh-29L missiles on the inboard pylons. The aircraft also carries the five-round B-13L 122-mm rocket pod and KAB-500 laser-guided bombs.

export by Lipetsk's own technical service, rather than passing through costly and protracted overhaul. One of the Su-25UBKs was reported to have been written off in a landing accident in April or May 2000, probably at Debre Zeit air base near the Ethiopian capital of Addis Ababa.

Piloted by Ethiopian aircrew, the three surviving aircraft took part in the closing stage of the war against Eritrea, which ended on 10 June 2000. One Kh-29T TV-guided and two Kh-29L laser-guided missiles were reportedly used, in addition to a large number of S-24 240-mm (9.5-in) rockets. A total of 17 combat sorties was flown in three weeks of fighting as the aircraft operated from Debre Zeit and Mercele airfields.

This instance of successful Su-25T export after a decade of fruitless marketing efforts was eventually revealed by the Sukhoi Shturmoviks company president and designer general, Vladimir P. Babak. In an article written for the Russian defence magazine *Voyennyi Parad* (*Military Parade*) in June 2000, he alluded that Su-25TK pilots from a hot-climate export operator (not identified by Babak but believed to be the Ethiopian air arm) had mastered the employment of laser-guided missiles and successfully used them in combat.

In 2000, six to 10 ex-Russian Su-25s (including a pair of second-hand two-seaters) were acquired by Eritrea, and probably used in anger in the closing stage of the war against Ethiopia.

In the early 1990s, Russian aircraft industry authorities had decided to transfer Su-25T production to the Ulan Ude-based factory (formerly known as Factory No. 99 and now as Ulan Ude Aviation Plant – UUAP), which is located in the Russian Republic of Buryatiya near Lake Baikal. This factory had previously been intended to act as the sole manufacturer

of the two-seat Su-25UB conversion and continuation trainer and Su-25UTG carrier-based trainer (about 50 and 12, respectively, built until the late 1990s). From the start, the Su-25T's production rate at UUAP was extraordinarily slow due to the lack of proper funding; announcements appearing throughout the mid-1990s indicated that the VVS was no longer interested in supporting the Su-25T project. Supporters of this view still maintain that the anti-tank role can be handed over entirely to army aviation, which began to acquire the new Kamov Ka-50 attack helicopter, also armed with the Vikhr anti-tank missile.

Attempts to sell or lease to Bulgaria and Slovakia a small number of Su-25Ts – up to eight, probably the aircraft taken from the uncompleted pre-production batch at TAM – also proved unsuccessful.

Eventually, the Su-25T programme was salvaged, thanks to funding provided by the Sukhoi Shturmoviks company (the authority responsible for the Su-25's design, upgrades, after-sales support and marketing) and some of the subcontractors.

All-weather Su-25TM

It is of note that the development programme for a further upgraded 'Super Frogfoot' commenced as early as 1986. It was a night-capable, all-weather Su-25T derivative, designated Su-25TM (bureau designation T8M). This variant was to be equipped with pod-mounted Kinzhal (Dagger) millimetric wave radar and/or the Khod (Motion) FLIR. However, radar and FLIR trials revealed that the systems had lower-than-expected performance and reliability and, eventually, the Phazotron-NIIR Kopyo (Spear) 3-cm radar was chosen. New equipment also included the two-pod Omul ECM system, and the T8TM was able to use a wide variety of new precision-guided air-to-surface and air-to-air weapons.

Two Su-25Ts were upgraded in the early 1990s to be used as Su-25TM development platforms. These were the T8M-1, redesignated T8TM-1; and the T8M-4, redesignated T8TM-2. The former made its maiden flight in its new guise on 4 February 1991, and undertook testing and evaluation of the Khod FLIR and the new EW system. The T8TM-2 was used in development trials of the Kinzhal system, and

Between 1991 and 1994 the Khod imaging infra-red pod was tested by the Su-25T. This aircraft is shown with a Kh-31P anti-radar missile, along with Vikhrs.

Another Su-25T pod is the Merkuriy LLLTV, which allows the pilot to see a bridge from a distance of up to 8 km (5 miles), or a tank from 3 km (1.9 miles).

On the inboard pylon of this Su-25T is the Kh-29T missile, the TV-guided version. This weapon is easily distinguished from the laser-guided version by having a much larger seeker head.

The twin-barrelled 30-mm GSh-302 cannon of the Su-25T is mounted externally in the NPPU-8M installation, with 200 rounds. The standard Su-25 has a similar weapon installed internally in the VPU-17A system, with 250 rounds. Su-25Ts have Pastel RWR and other EW antennas in the wingtip fairings.

later for the enhanced Shkval-M development, as well as for the new SUO-39 weapons control system. In late 1994, the T8TM-2 was fitted with a dummy Kopyo pod on the centreline hard-point, and in August 1995 it was displayed at the MAKS-1995 exhibition.

The first 'Super Frogfoot' (T8TM-3) to be assembled at UUAP, serialled 'Bort 20', took to the air on 15 August 1995, followed by the second (T8TM-4) on 15 March 1998. The former was used in aerodynamic trials with the Kopyo pod, whereas the latter featured the full-standard SUO-39 fire control system and was used for trials of the radar and the new two-pod MSP EW system. The T8TM-4 was displayed for the first time at MAKS-99 wearing the serial 'White 20'.

A total of seven Su-25TM airframes was laid down on UUAP's production line, but only five have been delivered to the VVS, and the current programme status is unclear. According to often over-optimistic Sukhoi Shturmoviks press releases, the Su-25TM's state trials were to have been completed by the end of 2001, but, given the perpetual delays and scarce funding, it would be realistic to predict that the new type will be cleared for regular VVS service no earlier than the end of 2003.

It was reported that UUAP is capable of building up to 12 Su-25TMs annually, should proper programme funding materialise. A plan for the procurement of as many as 24 Su-25TMs to equip a number of regional rapid-reaction aircraft groups (using a mixture of Su-25TMs and Su-25s) was briefly announced in 1999, but it is now considered impossible for such a plan to be pursued, due to lack of funding.

Yet again, however, this does not mark the end of the Su-25TM programme. A number of reports in late 2000 and early 2001 mentioned that the assembly of aircraft numbers four and five has been completed, thanks to funds generated by the Ethiopian sale.

The Su-25T's export price is now quoted as being between US$8.5 and $10 million per unit, depending on the weapons, spare parts, training and support equipment package; the more sophisticated Su-25TM is being offered for around US$12 million, including the pod-mounted Kopyo-25 radar. Newly-built Su-25s are now being offered by TAM for up to US$6 million, with refurbished ex-VVS aircraft going for US$3 to $5 million.

VVS upgrade efforts

Considerable delays in the Su-25T/TM programme had left the Russian air force in a rather uncomfortable situation by the early 1990s, as the service began to experience a notable shortage of cost-effective precision strike aircraft. The basic Su-25 was originally designed as an affordable single-role aircraft, being optimised for clear-weather co-ordinated close air support of advancing ground forces. It was intended to operate only up to 54 nm (100 km) behind the forward edge of battle. In the event of large-scale conventional war in Central Europe or the Far East in the 1980s, the jet Shturmovik's main task would have been to complement more sophisticated types such as the Su-17M4 and MiG-27M/K.

Left: The Su-25T was based on the airframe of the Su-25UB, utilising the space of the second seat to house additional avionics for the SUV-25T Voskhod nav/attack suite. The space was also used to increase fuel capacity from the UB's 2725 kg (6,008 lb) to 3840 kg (8,466 lb) – more than the standard single-seater.

Su-25 'Frogfoot', 4th Air Army

Su-25s from the 4th Air Army bore the brunt of attack operations during both Chechen wars. During wartime missions the Su-25s were mainly used to deliver a variety of unguided bombs and rockets, as well as undertaking strafing attacks. Two Su-25Ts were deployed in an operational evaluation programme which involved several launches of Kh-25 and Kh-29 laser-guided weapons.

The Su-25TM built on the advances made by the Su-25T, and added true night/all-weather capability. This came courtesy of a podded Kopyo radar carried on the centreline and FLIR. After trials with two converted Tbilisi-built Su-25Ts, the first Ulan Ude-built Su-25TM flew in August 1995, but progress with the programme has been slow, and only around five are believed to have been completed. The first aircraft is shown here (left) with the wing-mounted Omul ECM pod. Another aircraft (below) carries R-73, Vikhr, Kh-29T TV-guided missile and a Kh-58 anti-radiation missile.

Both Su-25T and TM have a wider nose housing the Shkval-M system (similar to that used by the Ka-50), comprising a high-resolution TV, Vikhr laser guidance system and the Prichal laser rangefinder/designator.

However, the situation changed entirely in the mid- and late 1990s: following the end of the Cold War, such a front-line fleet composition was considered prohibitively expensive and unnecessarily complex. Moreover, the subsequent sharp reductions in the Russian defence budget led to some abrupt air force inventory rationalisation moves.

In the early and mid-1990s, the entire Su-17M3/M4, MiG-23MLD and MiG-27M/D/K fleets (a total, in round numbers, of nearly 1,500 aircraft) were withdrawn from use. Significant reductions in the number of active Su-24Ms and MiG-29s also took place. And finally, all development and test programmes for new-generation multi-role tactical fighters and dedicated strike aircraft were either notably slowed or scrapped entirely.

In 1990, the mighty Soviet air force had an inventory of 385 Su-25s based in East Germany and the European part of the Soviet Union, with 40 more operated in the Far East; no fewer than 80 examples were in service with naval aviation. As many as 60 per cent of the 'Frogfoots' (between 280 and 300 airframes, including up to eight serviceable Su-25Ts) were retained in Russian air force and naval aviation service following the Soviet Union's collapse. By the late 1990s, the Russian air arm had an active inventory of around 200 to 220 'Frogfoots'; Belarus inherited 99 and Ukraine about 80; and a dozen or slightly more Su-25s went to other CIS republics such as Armenia and Azerbaijan. In the mid-1990s, a small number of ex-VVS Su-25s had been redistributed to the air arms of CIS republics such as Turkmenistan (at least five aircraft noted in regular service in 2000) and Uzbekistan.

Meanwhile, the role of the Russian air force's tactical fighter fleet changed considerably as emphasis shifted to the capability of participating in local rather than in global armed conflicts; inevitably, this led to the definition of a number of new requirements. The Su-25 fleet has escaped reductions on as large as scale as those suffered by the MiG-29 and Su-24M fleets, and only a small number of airframes produced prior to 1985 have been either grounded for use as spare part sources or offered for export.

Eventually, the faithful 'Frogfoot' emerged as the most cost-effective Russian air force aircraft to cope with the challenges of the low-intensity wars that erupted in various corners of the former Soviet empire.

'Frogfoot' in action

Major conflicts in which the VVS Su-25 fleet was used in anger in the post-Soviet era were those between Georgia and Abkhasia in 1993 and 1994, and also in 1994 during the Russian offensive against Moslem fundamentalists near the Tajikistan-Afghanistan border. Very soon afterwards, the 'Frogfoot' saw extensive use in the bloody Chechen wars as VVS Su-25s flew more than 6,000 missions in both campaigns there. Three aircraft were lost to enemy air defences in the first Chechen war (1993-1994), and another five or six were lost either to rebel AAA or in collisions with mountains in the second campaign (1999-2001). During the latter war, the 'Frogfoot' fleet was used mainly in the 'bomb-truck'/'rocket battery' role, leaving the complex guided weapons and bad weather missions to the Su-24Ms and two Su-25Ts.

As expected, the simple and robust 'Frogfoot' lived up to VVS expectations in these encounters, demonstrating yet again fairly good reliability and mission availability, given the primitive Afghanistan-like operating conditions and scant field maintenance. This is particularly true for the airframe and powerplant.

The type's mission avionics are no longer considered modern, as it has unimpressive capability and low precision for some complex types of missions such as would be required in future low-intensity conflicts. Ten years ago these taskings would have been assigned to the now-defunct Su-17M4 and MiG-27M/K fleets. Obviously, most – if not all – of these missions are within the Su-25T/TM's capability, but this new, potent, rather expensive and still-

This Kopyo-equipped Su-25TM carries the Kh-35 (AS-20 'Kayak') anti-ship missile. Note also the downward-firing chaff/flare launcher fitted to the lower fuselage below the engine intake.

One of two Su-25SMs initially upgraded by the 121st ARZ at Kubinka was displayed at the MAKS exhibition at Zhukovskiy in 2001. The new variant did not begin flight-testing until March 2002, and conversion of further examples has proceeded slowly.

unproven type will not enter mass squadron service for at least three to five years (and it is still unclear if the new type will ever enter squadron service, due to the competition posed by a number of higher-priority programmes). Thus, it was decided in early 1998 to upgrade part of the existing VVS 'Frogfoot' inventory.

Initially, it was intended that the majority of the existing VVS 'Frogfoots' would undergo such an upgrade, but eventually the percentage of aircraft to be upgraded fell to the more modest 40 per cent; in numerical terms, this accounts for about 80 aircraft, which should be sufficient to equip five or six first-line squadrons. However, some senior Sukhoi officials interviewed at the ILA 2002 exhibition in Berlin maintained that the VVS has expressed a desire to upgrade most, and perhaps all, of its currently active Su-25s and Su-25UBs.

Su-25SM/UBM – Russian upgrade

Preliminary design work on the 'Frogfoot' upgrade started at Sukhoi Shturmoviks in mid-1998, and the concept was announced publicly later that year. A highly sophisticated – and hence rather expensive – upgrade standard was initially offered for both the single- and twin-seat VVS Su-25s. This comprised the Pantera (Panther) nav/attack suite, nose-mounted Kopyo-25SM radar and pod-mounted electronic warfare suite, as well as a number of measures for improving the type's survivability and maintainability, and extending airframe and engine service life, thus reducing life-cycle costs by 30 per cent. Introduction of radar-absorbent coatings and a new type of paint, in order to reduce the probability of radar and visual detection, were also on offer. Range extension would be possible from additional external tanks on twin-carriage pylons. Most of the components were already proven, or were planned for use in the Su-25TM.

The first Su-25SM conversion is seen on display with laser-guided Kh-29L (foreground) and Kh-25ML missiles. The SM has a single LCD display in the cockpit on to which imagery from EO-guided weapons can be displayed, along with navigation and flight data.

By early 2000, VVS and Sukhoi Shturmoviks officials announced that they would go ahead with a far cheaper and less extensive upgrade standard for the single-seaters, priced at between US$500,000 and $1 million per unit. This comprises a downgraded version of the Pantera nav/attack suite, and perhaps a new EW suite for some of the aircraft. The upgraded 'Frogfoot' was redesignated Su-25SM (*stroyevoy modernizirovannnyi* – Line Upgrade, indicating that all work is to be carried out at the VVS's own maintenance facilities).

Depending on funding available beyond 2002, it would be possible for a small number of Su-25UBs to be upgraded to the definitive standard that included the Kopyo-25SM radar. These aircraft would be designated Su-25UBM and would possess genuine anti-ship and beyond visual range (BVR) air-to-air capability. The former capability would come from the integration of the Kh-31A (AS-17 'Krypton') and Kh-35 (AS-20 'Kayak') anti-ship missiles, while BVR air-to-air capability would be possible via the active radar-homing R-77 (AA-12 'Adder') air-to-air missile. In order to minimise the necessary airframe structural re-work, the Kopyo-25SM would be pod-mounted under the fuselage (as on the Su-25TM) instead of in the nose, as originally planed.

However, the Russian air arm currently has a limited number of Su-25UBs – no more than two dozen are on hand – most, if not all, of which are being heavily utilised in conversion

This model shows the initial Su-25SM configuration, with Kopyo-25SM radar in a neat nose radome. This proposal was dropped as being too expensive and the current Su-25SM upgrade has no radar.

A nose-mounted radar was also part of the original Su-25UBM proposal. The Kopyo-25SM radar may be retained in the revised concept, but only as a pod-mounted unit. Note the provision of radar-guided AAMs, including the R-77, and Kh-58 ARM.

and advanced training roles; small numbers have seen action in Chechnya. The acquisition of 10 to 15 newly-built Su-25UBs, to be upgraded on the production line or later to the Su-25UBM standard, is a possibility.

The Su-25SM's Pantera nav/attack suite is built around a new digital computer, borrowed from the Su-25TM. All the analog components of the existing KN-23-1 nav/attack suite, which have poor reliability and are maintenance-intensive, will be replaced by mostly digital ones. Aircraft navigation accuracy will be improved tenfold thanks to the A-737 satellite navigation receiver integrated into the existing ICV gyro reference platform (the receiver can use signals from satellites of both the Russian GLONASS and the US NAVSTAR Global Positioning System). This will bring the Su-25SM closer in line with the Su-25T/TM's remarkable navigation accuracy of about 0.2 per cent deviation from track on a typical route. Navigation (non-visual) bombing in bad weather and at night against fixed targets that have a precisely known position, using data from the upgraded navigation system, is also becoming possible.

TV and laser weapons

The Su-25SM's guided weapons suite was enriched with the Kh-29T TV-guided air-to-surface missile and the KAB-500Kr TV-guided bomb (both 1980s-vintage, but still considered to be very effective lock-on before launch weapons). The upgraded aircraft is to be capable of dropping and firing several types of weapons in a single attack pass, and also to use the highly agile R-73 air-to-air missile, an extremely effective weapon for slow-speed

Two views show the Scorpion demonstrator's cockpit. New to the Su-25 is a head-up display – earlier aircraft had a simple sight. The HUD, upfront controller, two large MFDs and HOTAS provide a more modern feel to the cockpit although some old instruments are retained. Cockpit instrumentation is a mixture of Russian and English.

target intercepts over the battlefield, especially when integrated with a helmet-mounted sight.

Some of the existing nav/attack system components that have demonstrated good performance and reliability – such as the Klyon-PS laser rangefinder/target marker and the ASP-17BTz-8 electro-optical sight – were retained.

According to reports from the Sukhoi Shturmoviks company, the Su-25SM's cockpit had undergone a redesign in order to improve the ergonomics, thus reducing pilot workload during complex attack missions. One LCD colour display was installed on the instrument panel (two LCDs would probably be employed in the front and rear cockpits of the Kopyo-25SM-equipped two-seat aircraft, in addition to a new HUD), and possibly a Russian interpretation of the HOTAS controls concept was adopted, at least partially, by the Su-25SM/UBM. There is (or will be added at a later stage) provision for the integration of new-generation Russian FLIR/laser designation pods, such as the UOMZ Sapsan, to enable day/night laser-guided bomb employment. An ECM system comprising the two-pod Omul jammer covering the frequency band between

7 and 10 GHz, which can operate in co-operation with the Pastel radar warning/emitter locator system, is probably also being planned to equip some of the upgraded Su-25SMs.

The majority of the VVS Su-25s originally slated to serve until about 2010 are now thought to be under-utilised, falling well below the half-way point of their 2,500-hour limit. Airframe life extension to a total of 40 years and 4,000 hours is being offered as part of the Su-25SM upgrade package (in such a case, TBO is 2,000 hours); it may also be adopted for the rest of the VVS Su-25s, and is known to have been offered to foreign operators.

The Su-25SM is set to retain the tried Gavrilov R-95Sh turbojets. Although often described as old-tech and rather uneconomical, the 1960s-vintage R-95 is still readily available in large numbers. Remarkably reliable and fairly cheap to maintain, it features a total life of 1,500 hours, with overhauls required every 500 hours. Engine life extensions up to 2,000 or even 2,500 hours are options.

Rework begins

In March 2001, the first two VVS Su-25s were sent for upgrade to the VVS 121st ARZ (Overhaul Facility) at Kubinka near Moscow. One of these, serialled 'Bort 33', was displayed on the static line at the MAKS-01 exhibition in August that year, in what was reported as being an upgraded configuration (cockpit photos were prohibited), although it still awaited the commencement of flight testing. By early March, reliable information was still lacking about whether the Su-25SM had begun its flight test programme, leading to the conclusion that the programme still suffered from chronic lack of funding. Eventually, the first upgraded

Elbit's Scorpion demonstrator, resplendent in Georgian Air Force markings, has made high-profile appearances at the Paris and Farnborough air shows, including impressive flying displays by Yehuda Shafir, Elbit's chief test pilot.

Although it lacks radar, the Su-25KM is a far more capable upgrade package than the Su-25SM, due to its modern avionics architecture which allows the integration of modern precision weapons and man/machine interface, including helmet-mounted sight.

Su-25SM made its first test flight in its new guise on 5 March 2002. Tests were expected to have been completed by August 2002, and production upgrades started in spring 2002 (a total of five Su-25s, including one two-seater, was reported to have been upgraded by the Kubinka-based 121st Overhaul Facility).

Su-25KM Scorpion

The most radical Su-25 upgrade package yet announced is that offered by Elbit Systems, an acknowledged leader in ex-Soviet platform upgrades that has established successful co-operation with the original manufacturer TAM. The joint venture's aim is to produce a 'digitalised' 'Frogfoot' derivative; the Elbit/TAM upgrade package, known as Scorpion, was developed in late 2000 and early 2001. The technology demonstrator aircraft, wearing Georgian air force insignia and the serial '01', was flown for the first time by Yehuda Shafir on 18 April 2001. He later successfully demonstrated the Scorpion, known also as the Su-25KM (KM standing for *kommercheskiy*, Modernised), at the Le Bourget air show in June that year. In 2002, the Su-25KM was demonstrated at the Farnborough air show.

For Elbit Systems, the Su-25KM is yet another Russian combat platform to be upgraded using proven and fairly modern digital avionics, a process that considerably improves the type's overall combat capability. Main improvements are to the nav/attack system and man-machine interface/situational awareness, with the introduction of all-weather day/night capability and advanced debriefing facilities.

The Su-25KM's new avionics suite, featuring an all-new 'glass' cockpit, was developed around Elbit's Multi-Role Modular Computer (MMRC) controlling dual Mil-Std 1553 data-buses – one responsible for navigation, the other for the weapons delivery system. The 'glass' cockpit is dominated by an El-Op Head-Up Display (HUD), up-front control panel, two 6 x 8-in (15 x 20-cm) LCD colour displays, and solid-state backup instruments. Precise navigation capability is provided by a laser-gyro INS/GPS hydride system. NATO- and ICAO-compatible navaids – VOR/ILS and DME (TACAN optional) – are added, as are two new radios (one UHF and one VHF).

Elbit's Display And Sight Helmet (DASH) is offered, significantly improving the pilot's situation awareness. The DASH is reported to have been integrated into both the R-73 air-to-air missile and the new navigation system, for marking points of interest on the ground during strike or reconnaissance missions. Complete mission pre-plan capability is also included in the core package. Some of the original Russian-made instruments are retained, such as the SPO-15LM radar warning receiver (RWR), RV-15 radar altimeter, AoA indicator, engine RPM and fuel meter. New weapons that the Su-25KM can employ include the above-mentioned R-73 air-to-air missile (also manufactured by TAM), as well as Elbit's infra-red-guided Opher and laser-guided Lizard bombs.

The Scorpion package is thought to be more advanced and flexible, to some degree, than the competing Su-25SM export derivative. More importantly, unlike the Elbit-upgraded MiG-21 Lancer and MiG-29 Sniper, the Scorpion is being strongly backed by the original manufacturer, which has legal rights to the type's design and in-service documentation, and possesses the required expertise and experience to introduce major changes to the aircraft's structure and avionics suite, as well as to offer logistic support and depot-level maintenance.

An aggressive marketing campaign, initiated by Elbit in 2001 and continued in 2002, is aimed at a number of current European and Third World 'Frogfoot' operators. However, at

Weapon options introduced by the Su-25KM Scorpion include the R-73 air-to-air missile (carried on the centre pylon) to augment the existing R-60 (outboard), and the Elbit Opher infra-red guided bomb. The Lizard laser-guided bomb is also available.

Most Su-25s are powered by the 40.2-kN (9,036-lb) Soyuz/Gavrilov R-95Sh, and the engine is retained in the Su-25SM upgrade, rather than being replaced by the more powerful (44.1-kN/9,913-lb) R-195 fitted to a few late-production Su-25s and the Su-25T/TM.

By Soviet standards of the 1970s, the Su-25's cockpit was well laid out, and was praised by pilots. The panel is topped by an ASP-17BTz-8 reflector sight, complete with a canopy strut-mounted gun camera.

the time of writing (October 2002), there were no signs that any upgrade contracts were being seriously negotiated or awarded to the Israeli-Georgian team.

Upgrade market

Russian and some foreign military aviation observers estimated in 2001 that the overall Su-25 upgrade market may be as huge as 400 airframes; however, that is now viewed as a notably inflated figure. As might be supposed, the main reason for the lack of strong interest in both the Su-25KM and Su-25SM upgrades is that the full-standard Scorpion package is too expensive and, sometimes, over-sophisticated for most, if not all, of the existing Su-25K operators. These operators are subject to tight budgets and are much more concerned about maintaining fleet airworthiness than undertaking costly upgrades. They are likely to be interested only in introducing low-cost upgrades to improve navigation accuracy and maintainability (through extension of the airframe and engine time between overhauls), and in some cases in integrating a limited range of guided weapons. The Su-25K's direct operating costs in Eastern Europe are quoted at around US$5,000 per flying hour, which is roughly 60 per cent of those of the Su-22M4 and 80 per cent of those of the MiG-23MF/BN.

SEAD-capable two-seat 'Frogfoots'

The only upgrade package known to have been ordered by an export customer is the Suppression of Enemy Air Defence (SEAD) package. It has been introduced to some,

perhaps all, of the eight Fuerza Aérea del Peru (FAP – Peruvian air force) Su-25UB two-seaters, as well as to a limited number of Belarussian air force Su-25s. These aircraft received the L150 Pastel radar warning receiver/emitter locator system housed in an underfuselage KRK-UP pod (the same size and shape as that used to house the Su-25T/TM's Khod and Kinzhal systems), alongside the launch control equipment for the Kh-58U/E ARM. The Kh-58U/E is the most widely used Russian anti-radar missile; it weighs 1,411 lb (640 kg) including the 328-lb (149-kg) warhead, and can be launched at distances of 4.3 to 54 nm (8 and 100 km).

The rear cockpit's instrument panel of the SEAD-capable Su-25UB received a new IM-3M-14 monochrome CRT display, onto which the target information derived from the Pastel can be displayed. Data (in the form of range, bearing and probable type information) for up to six enemy radars in frequencies between 1.2 and 18 GHz can be displayed, and radar emissions can be detected at a range nearly 20 per cent greater than the radar's own detection range.

The upgrade of the eight or so Peruvian and an unknown number of Belarussian Su-25UBs is known to have been carried out by the Baranovichi-based 558 ARZ facility in Belarus, in close co-operation with the Sukhoi Shturmoviks company. It is likely that some of the Russian two-seat 'Frogfoots' will also receive such a capability in the near future (complementing, and perhaps partially replacing, the MiG-25BMs and Su-24Ms in the dangerous SEAD role), instead of being converted to the much more expensive Su-25UBM standard.

One innovative role of the FAP Su-25 fleet is anti-drug intercept operations. In 1999 and 2000, FAP 'Frogfoots' were rushed into action to reinforce the nation's anti-drug campaign. Due

to its remarkable low-speed manoeuvrability, the straight-winged and heavily armoured jet, featuring an impressive thrust-to-weight ratio, can successfully be used to intercept single- and twin-engined general aviation aircraft smuggling raw cocaine and cocaine paste from the Upper Hullanga Valley in Peru to neighbouring Colombia. In some cases, the Su-25 pilots who intercepted and identified suspected drug-carrying aircraft were ordered to shoot them down, for which they used R-60 air-to-air missiles and the powerful GSh-302 built-in 30-mm gun.

European small-scale upgrade effort

In 1999, two Czech air force Su-25Ks underwent a so-called 'small-scale' upgrade (more precisely, it should be called adaptation) of their communication, navigation and identification equipment to ensure limited NATO and ICAO compatibility and to improve navigation capability in operational missions. This involved the use of affordable and proven Western commercial hardware, the work being carried out by the local CLS company, based at Prague-Kbely. Trimble 2101 I/O Approach Plus commercial GPS receivers, as well as VOR/ILS and DME receivers (TACAN optional), were added to the aircraft's existing nav/attack system, and the new navigation components were integrated via an interface unit developed by CLS. The navigation suite upgrade is designed in such a way that information from the new navigation components can be displayed on the existing 'clockwork' HIS navigation indicator. Anti-collision lights and an additional radio were also installed. However, plans for the type's withdrawal from use in late 2000 prevented the fleet-wide upgrade of the Czech air force Su-25Ks.

It is likely that similar packages may soon be adopted by other under-funded European air arms, such as those of Bulgaria and Macedonia, whose Su-25s are set to soldier on for some years. East European operators are reportedly set to implement some measures for NATO/ICAO interoperability and life-cycle costs reductions.

The newly established air force of Macedonia turned to the Ukraine for its initial equipment, receiving four Su-25s to provide a cost-effective yet hard-hitting attack force. They were thrown straight into action in the fight with Albanian rebel forces.

The second seat makes the Su-25UB a sought-after commodity, especially as it is more adaptable than the single-seater to missions requiring sophisticated systems. However, it is also in high demand for training. Russian Su-25s are expected to serve until around 2020.

The Macedonian air arm is the latest European Su-25 operator, receiving four ex-Ukrainian Su-25s (including one two-seater) in June 2001. They feature the full set of self-protection aids, including scabbed-on flare dispensers. The newly-acquired aircraft were used in anger immediately following their delivery, attacking ethnic Albanian rebels who occupied the villages of Arachinovo and Radusha. There, the Macedonian Su-25s were reported to have used powerful FAB-500M62 1,100-lb (500-kg) high-explosive bombs (mostly due to the great psychological effect of the powerful blast, rather than their destructive power), as well as 57-mm and 80-mm rockets. They also carried out a number of visual reconnaissance and 'force projection' missions over the occupied villages.

In addition to their main role of ground attack, Macedonian Su-25s, armed with R-60 air-to-air missiles, are employed as fighter-interceptors – because they are the only jet aircraft in the air arm's inventory. There were a number of reports in early 2002 that Macedonia was considering the acquisition of at least two more ex-Ukrainian Su-25s, and imminent low-cost upgrades of their nav/attack systems cannot be ruled out.

Future exports prospects

Between 1992 and 1998, some 12 to 15 newly-manufactured Su-25s at TAM were taken on strength by the newly-born Georgian air force. Up to six of these aircraft were later reported to have been lost during the conflict in Abkhasia in 1993 and 1994. Another Georgian air force example was shot down during operations against rebel groups in the republic. Currently, the Georgian air arm has an inventory of six to eight single-seaters and one or two two-seaters. Interestingly, in the mid-1990s, TAM began production of a five-unit Su-25UB batch (TAM designated these two-seaters Su-25U) originally ordered by the Georgian air arm, but lack of funding meant only two aircraft were delivered. The only known TAM Su-25 export deal following the Soviet Union collapse involved 10 single-seaters acquired by Congo in 1999 and 2000.

Between 1979 and 1999, TAM built more than 820 Su-25s (825, according to unconfirmed information) in four main variants: the basic single-seater and two-seater, target-towing Su-25BM single-seater, and anti-tank Su-25T. These were distributed as follows: 582 basic Su-25s for the Soviet/CIS and Georgian air arms, 50 Su-25BM target tugs, between 180 and 185 Su-25Ks, and up to 22 Su-25Ts (some partially completed and still stored at TAM), as well as one to three Su-25U two-seaters built at TAM using uncompleted Su-25T airframes fitted with two-seat nose fuselages supplied by UUAP. Ulan Ude Aviation Plant produced 150 to 200 two-seat Su-25UBs and Su-25UBKs, and three or four Su-25TMs.

By the mid-1990s, 'Frogfoot' export sales amounted to 201 aircraft (185 single-seaters and 16 two-seaters), sold to five states: Bulgaria – 40, Czechoslovakia – 38, Iraq – 73, Angola – 12, and North Korea – 36. Another 46 (35 being second-hand aircraft) were sold between 1998 and 2001 to five other non-CIS states: Peru (18), Ethiopia (four), Eritrea (up to 10), Congo (10) and Macedonia (four).

Regarding the Su-25's future sales prospects, it can be noted that newly-built Su-25Ks and more modern Scorpions, the latter priced at about US$12 million per unit, are viewed as being highly competitive with modern attack aircraft such as the BAE Systems Hawk 200 and Alenia/EMBRAER AMX. The Su-25 is still being offered as a low-cost replacement of the older ground attack types (AT-37, F-6, F-7, MiG-21, MiG-23 and Su-20/22) used by a number of Third World states. UUAP and TAM are active in this combat aircraft market, and the former has offered a number of Southeast Asian states its fully combat-capable Su-25UBK in basic or upgraded form, at a price of about US$10 million.

It would be reasonable to predict that a renewal of the type's export success will take place in the early 2000s, most likely in traditional Soviet/Russian customers such as Syria, Libya, Algeria and Vietnam, whose air arms have to replace their inventories of aged Su-20/22s, MiG-21M/MFs and MiG-23BNs. Existing African and Asian customers – Ethiopia, Eritrea, Iran and Iraq – have air arms that may also opt to acquire more Su-25s. Thus, in the forthcoming decade, 'Frogfoot' exports of both newly-built and second-hand machines may number as high as 60 aircraft.

A similar number of Su-25s belonging to the air arms of some ex-Soviet republics may be upgraded, to greater or lesser extent, and the market for life extension and maintenance rationalisations may be as large as 100 aircraft. Su-25TM low-rate production at UUAP and TAM is likely to continue for some years to come, though it is probable that the new type will have only limited service in Russia and, at best, will enjoy only limited export success.

Alexander Mladenov

Above: In Europe, Bulgaria and Macedonia are the last remaining 'Frogfoot' operators. Bulgaria has a fleet of 39 Su-25K/UBKs, and around half are expected to serve on until 2010/2014. Six Su-25Ks and two Su-25UBKs will undergo a minor upgrade for NATO/ICAO interoperability in early 2003.

Left: Slovakia operated 10 Su-25Ks and a single Su-25UBK (left) until early 2002, but they are now grounded and are to be formally withdrawn in late 2002.

Sri Lanka Air Force

Photographed by Peter Steinemann

Chinese MiGs

Sri Lanka's first fighter equipment consisted of five
Soviet-built MiG-17Fs (CF902 to CF906), which were
delivered in 1971 but saw little service before being
put into storage, there being little requirement for
combat equipment. That changed in 1983 with the
outbreak of fighting between the government and
the LTTE (Liberation Tigers of Tamil Eelam), although
it was 1991 before jet fighters reappeared in the air
force inventory. After two FT-5s and a single FT-7
were delivered, the combat force of four Chengdu
F-7BS (serialled CF704, CF705, CF707 and CF708)
arrived in December for use by No. 5 Squadron at
Katunayake. The F-7BS is based on the J-7II and
retains the low position of the nose pitot. Chaff and
flare dispensers were added in the mid-1990s, and
one of the aircraft was given a very pale grey
scheme. Three F-7BS fighters remain on No. 5
Squadron's books but are currently grounded. They,
and the FT-7, were due to be overhauled in Pakistan
during 2002 and returned to service with the unit.

From 1983 until a Norwegian-brokered ceasefire in December 2001 the Sri Lanka government was engaged in a fierce battle with Tamil separatists in the north of the country. The air force was shaken from its role as little more than a tourist airline to become a sharp-edge combat force. In the last two years the SLAF has embarked on another major procurement round to streamline its training and enhance its combat effectiveness.

No. 5 Jet Squadron

Based at Katunayake air base near the capital, Colombo, No. 5 Jet Squadron was the SLAF's first effective fighter unit, and remains as a principal element of the fighter attack force. Since the grounding of the Chengdu F-7s and Guizhou FT-7, the unit's equipment consists of the MiG-27 fighter-bomber and a single MiG-23UB (*above*) for conversion training. Seven MiG-27s were delivered in 2000, beginning operations in August, although two have subsequently been lost in accidents and one destroyed in the raid on Katunayake. The 'Flogger' was used to great effect in strikes against LTTE forces, but was withdrawn from the fray in July 2001, leaving the Kfir to continue bombing missions until the ceasefire in December. MiG-27 sorties have mostly been flown by Ukrainian pilots.

The MiG-23UB was delivered in December 2000, although it was not used for pilot conversion until mid-2002 – previous 'Flogger' training had been undertaken in the Ukraine. The aircraft was damaged during the LTTE attack on Katunayake air base in September, but was repaired and returned to the air in the spring of 2002, ready to begin the task of training Sri Lankan pilots.

SLAF: Past and Present

To complement its fighter equipment the Sri Lanka Air Force also received a handful of conversion trainers. The first was a MiG-15UTI (CF901) in 1971 which accompanied the MiG-17Fs. This aircraft is now on display in the SLAF museum at Ratmalana. In July 1991 two Shenyang FT-5s (*above*, serialled CTF701 and CTF702) were delivered from China, followed in October by a single Guizhou FT-7 (*left*, CTF703), in preparation for the delivery of the F-7BS fighters. The FT-5s were retired in 1997.

No. 10 Jet Squadron

This unit was established to operate Kfirs from Katunayake. The first batch of aircraft, comprising two Kfir C2s and a single TC2 trainer (*right*) arrived in January 1996. Three more C2s were delivered in April. Two further C2s followed, distinguished by having darker camouflage, and two C7s arrived in early 2001. The latter had not entered squadron service by the time they were destroyed in the LTTE attack on the air base on 24 July 2001. Two more C7s were subsequently delivered to make good the loss.

No. 9 Helicopter Squadron

The first six Mi-24s were delivered to No. 9 Sqn at Minneriya-Hingurakgoda in November 1995, probably from Ukrainian stocks. They were swiftly pressed into action against LTTE forces. Further deliveries raised Mi-24 deliveries to 13, although three were lost in action and three returned to the supplier. Only one (*below*) is currently operational. Seven twin-cannon Mi-35Ps have been received since 1998 to restore the unit's effectiveness.

No. 7 Helicopter Squadron

Based at Minneriya, No. 7 Squadron flies the Bell 206 (*above*) and Bell 212 (*right*). Both types frequently operate with armament as makeshift gunships, two of the JetRangers being able to carry a 12.7-mm HMP (Heavy Machine-gun Pod) on one side and a 2.75-in rocket launcher on the other. The Bell 212s can also carry these weapons and, as here, mount flexible guns (either 7.62 mm or 12.7 mm calibre) in the cabin door.

No. 4 VIP Helicopter Squadron

Based at Katunayake, No. 4 Squadron flies four black-painted Bell 412SPs (*right*) on VIP/staff transport tasks. They have had warning receivers fitted. SLAF rotary-wing types that are no longer in service include the Bell OH-13, Dauphin and Kamov Ka-26.

No. 6 Helicopter Squadron

Workhorse of the war with the LTTE has been the Mi-17 (*below*), flying with No. 6 Squadron at Vavuniya, the closest base to the war zone. The first three were delivered from Kazan in July 1993, and the force built up to at least 11. The fleet wears different colour schemes according to whether it came from the Kazan or Ulan Ude factories, or whether it has been repainted locally (*above*).

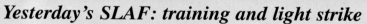

Yesterday's SLAF: training and light strike

The Royal Ceylon Air Force (Sri Lanka Air Force after 1972) was formed on 1 October 1951, its first equipment being the Chipmunk Mk 21. The last Chipmunks (*above left*) were not retired until 1990. They were supplemented by six Cessna 150s, and then replaced by 12 SIAI-Marchetti SF-260Ws in 1990/91. The Warriors (*above*) were ex-Burmese aircraft acquired through a Belgian dealer. The Cessnas augmented the SF-260s or were used for liaison.

First light strike capability was provided by the BAC Jet Provost T.Mk 51 (*below*), of which 12 were delivered in two batches. Following the outbreak of fighting with LTTE forces, six SF-260TPs were acquired for COIN duties from November 1985, two being swiftly lost and then replaced by ex-manufacturer demonstrators in 1987. Four ex-Argentine IA-58A Pucarás were delivered in 1992 with the intention of establishing No. 7 Squadron. In the event, two were written off in the mid-1990s, and one was used for spares, leaving just one airworthy aircraft (*left*, with SF-260TP) for COIN duties, assigned to No. 1 Flying Training Wing.

No. 1 Flying Training Wing

SLAF training has undergone a considerable transformation in the last two years, centred around the delivery of 10 Nanchang PT-6 basic trainers from China (*left*) to No. 1 FTW at Anuradhapura from January 2001. The arrival of the PT-6s allowed the retirement from the training role of the SF-260W and Cessna 150, although two SF-260s are retained for ground handling training. Three Cessnas remain in service (*below*), but are little used. Also assigned to No. 1 FTW are the surviving SF-260TPs (*above*). They do not have any training function, and are still used for COIN/light strike duties.

Until 1988 most flying training was conducted from the SLAF Academy at China Bay, near Trincomalee, but this airfield was vacated as the airspace was felt to be too close to the war zone. Basic officer training, however, remained at China Bay.

No. 14 Jet Squadron

This unit was established at Katunayake in July 2001 to operate six Hongdu K-8 advanced jet trainers which had been delivered from China in April. However, the day before its official establishment it lost three of its new aircraft in the 24 July LTTE attack on Katunayake. The remaining K-8s have since been very busy training graduates from the PT-6 course, providing a 40-hour advanced course for all students. Prospective fighter pilots then undergo a further 60 hours on the K-8.

Sri Lanka Air Force

No. 2 Heavy Transport Squadron

From its base at Ratmalana No. 2 Squadron provides the bulk of the SLAF's transport capability. The most recent additions, and its most capable aircraft, are two ex-RAF Hercules C.Mk 1s (*below and inset*), which were delivered in 2001. Following their arrival the sole remaining Shaanxi Y-8 (*bottom left*) was used only sparingly until it was lost in a crash at Ratmalana in August 2002. Four Y-8Ds were delivered from China, one being shot down by the LTTE on 18 November 1995. During the war they were occasionally used as makeshift bombers. Lighter members of the No. 2 Squadron roster are the Antonov An-32 (*right*) and Avro 748, which are used for intra-island transport duties. Civilian contractors operating An-24RVs and An-32s were also called in to support the transport effort during the fighting.

The SLAF's association with the HS748 (*above*, known as the 'Avro') began in 1979 with the transfer of a single aircraft from the national airline. Indeed, until 1983 much of the SLAF's activity was to provide support to the growing tourism industry. At least seven have served with the SLAF, but only two remain on No. 2 Squadron's books, and only one of the pair is airworthy.

No. 8 Light Transport Squadron

The first type to be supplied by China was the Harbin Y-12II (*below*), powered by Pratt & Whitney Canada PT6A-27 engines. A total of nine was bought, the first delivered in 1987. Today four remain in service, two having been lost and three grounded. They are used for light transport and multi-engine training, although during the conflict were occasionally used as makeshift bombers. They were delivered in a dark green/brown finish, but now wear light grey.

Also serving with No. 8 Squadron at Ratmalana is the single Beech King Air 200 (*right*). Having earlier been used for coastal patrol, it has now been converted for the signals intelligence-gathering role, and bristles with antennas on the underside.

Yesterday's SLAF: transport and patrol

Formed with British assistance, the Royal Ceylon Air Force's initial equipment was mostly of UK origin, and included six de Havilland Doves (*below*) from 1955 and five de Havilland Herons (*below left*) from 1957. Some of the latter were subsequently converted to Riley standard. Two Douglas C-47Bs (*left*) were transferred from Air Ceylon in 1976. A single Convair 440 was also on charge, while the SLAF has also operated a Beech 18 and a Cessna 421 on photo-survey duties.

On the outbreak of war with the LTTE in 1983, many of the SLAF's transport types were pressed into use as makeshift coastal patrol aircraft to staunch the flow of supplies across the Palk Strait from India. These included the Doves and C-47s, which were augmented by a King Air 200 acquired in 1983 and several Cessna 337s, which were used by No. 3 Maritime Squadron at China Bay.

Boeing F-15 Eagle

It is easy to forget that the Eagle is now over 30 years old. Its age notwithstanding, this sleek performer is still widely regarded as the most potent operational fighter in the world and has the kill ratio to prove it. Nevertheless, the proliferation of the Sukhoi Su-27 family and the threat of follow-on designs provided the impetus for an Eagle replacement in the fighter role, and the United States Air Force plans to gradually retire the F-15C/D from active service after the F/A-22 Raptor becomes operational in 2005. However, the Strike Eagle, which went to war within a year of reaching initial operating capability, remains at the forefront of modern tactical airpower, and is slated to continue in USAF service beyond 2030. This review looks at the Eagle in service today, and explores its growth and combat actions over the last decade.

In both 'Mud Hen' (F-15E) and 'light grey' (F-15C) forms the Eagle continues to rule the roost, and will do so until the F/A-22 enters service. An ongoing series of improvements have kept both fighter and fighter-bomber versions at the cutting edge of air combat technology.

Although modern air warfare is mostly conducted at medium/high level, the F-15E community still aggressively practises the low-level mission. Contrary to popular opinion, the F-15E provides a comfortable ride in the low-level regime.

In 1991 the USAF deployed 96 F-15C/Ds to Saudi Arabia as part of Operations Desert Shield and Desert Storm. It also deployed 48 newly acquired F-15E Strike Eagles from the 4th TFW, Seymour Johnson AFB. The air war lasted only a matter of weeks, and the Iraqi war machine was soon crushed. The F-15, built in the same St Louis factory as the 5,000-plus F-4 Phantoms, was the linchpin of the entire operation.

F-15Cs provided 24-hour Combat Air Patrol protection against a foe which may have been technically inferior, but was still capable of packing a hefty punch. Mirage F1EQs and a range of MiG fighters gave F-15C pilots cause for concern – in particular the MiG-29 and its passive Infra-Red Search and Track system. By the end of the war, 33 of the 38 coalition kills were credited to F-15Cs of the 33rd TFW, 1st TFW and 36th TFW, and one to a 4th TFW F-15E (which had directed a laser-guided bomb onto a hovering Iraqi helicopter). This tally was achieved without a single F-15 loss, although three F-15Es never made it home – one was lost in training, and the other two to hostile ground fire. An

additional two Iraqi kills were credited to a Royal Saudi Air Force F-15C of 13 Squadron, bringing a total of 36 kills for zero losses.

A range of factors contributed to this although, technologically speaking, the ability to electronically identify targets at night and beyond visual range gave the F-15C the upper hand in nearly all of its engagements. This capability came from Non-Cooperative Target Recognition, a radar technique possessed by both the E-3 AWACS which guided the F-15s, and the APG-63/70 radar found on the Eagle itself. Additionally, it is very likely that a still classified electronic identification technology was employed to further solidify hostile, neutral or friendly target categorisation.

F-15Es also performed well – extremely well when one considers that the aircraft was so new. Early post-war claims that it had successfully destroyed a number of mobile 'Scuds' were soon found to be mistaken, but the Strike Eagle had still made it much more dangerous for Iraqi mobile 'Scuds' to operate. The Strike Eagle had also heavily persecuted Iraqi armour, air defences and logistics points on a 24-hour basis, and for much of the war, Seymour Johnson's 335th TFS flew daytime missions, while the 336th TFS flew at night.

Gulf War lessons

Desert Storm uncovered faults with the F-15E that would ordinarily have been discovered in the less hazardous conditions of peacetime testing and squadron-level operations. These problems were overcome by a mixture of ingenuity, sheer guts and determination by the 'Rocketeers' and 'Chiefs', for whom the war provided an opportunity to perform their own, very real, operational test and evaluation.

Critically, nearly all of the Strike Eagles in theatre lacked AAQ-14 Target Pods – half of the renowned LANTIRN system which allowed attack and navigation functions to be carried out in bad weather, day or night – and this rele-

plethora of Iraqi SAM systems, the F-15E relied upon the F-4G and similar Suppression of Enemy Air Defence (SEAD) assets for many of the opening sorties. To complicate matters, the ALR-56 Radar Warning Receiver (RWR) was also plagued with problems and was slow to display radar threats, especially when they existed in abundance.

Prior to deployment to the Persian Gulf, the F-15E was limited in the number and type of weapons configurations authorised for carriage. The staple loadout was either Mk 82 and Mk 84 Low Drag General Purpose (LDGP) bombs or Mk 20 Rockeye Cluster Bomb Units (CBUs). The USAF subsequently issued a waiver that permitted crews to experiment with different stores loadouts and mixed-store configurations. While this sounds innocuous enough to the casual observer, it was a source of additional concern as turbulent air flow below the Conformal Fuel Tanks (CFTs) had sucked falling bombs back into the airframe during earlier tests. Needless to say, none of the crews was overly enamoured with the idea of being brought down by their own bombs. One final niggle persisted throughout the campaign: incorrect or ineffective fusing of the bombs. This process required lanyards connected to the airframe to pull arming pins from the bomb's fuse as it left the aircraft. Sometimes the lanyard snapped, leaving the pin inside the fuse, and the bomb remained inert as it fell to earth.

Dedicated fighter

In response to the growing number of F-15E squadrons, the USAF officially deleted the secondary air-to-ground role of the Eagle in 1992. Few people were even aware that the

gated the F-15E to bombing visually by day, or by radar at night. For those who did use one of the scarce pods, it quickly became obvious that tracking, slewing and slew rate deficiencies existed, and most WSOs therefore opted to manually lase and track the target until their bombs impacted, often with excellent results.

Furthermore, the ALQ-135 Internal Counter-Measures Set (ICMS), which was part of the Tactical Electronic Warfare System (TEWS), had been removed from the 4th TFW's aircraft before Desert Storm because of performance problems in Band 3. It was reinstated for those aircraft serving in Desert Storm, but the problems remained. Against the

F-15Cs from the 1st Fighter Wing wheel over Chesapeake Bay – marked for the commanders of the wing and the 71st FS. This unit is slated to be the first front-line wing to give up its Eagles for F/A-22 Raptors. The first F/A-22s are scheduled to arrive in late 2004, and the wing is due to have completed its re-equipment by 2009, with 78 aircraft on charge (24 per squadron plus six reserves).

Far left: The F-15 remains central to the airpower capabilities of its three overseas operators, two of which – Israel and Japan (F-15Js from Dai 304 Hikotai illustrated) – are in the process of major upgrades for their fighter fleets.

Watched from another F-15E, the commander's aircraft from the 391st FS releases an IR decoy flare. Both carry dummy AMRAAMs on the outer shoulder launch rail.

Since a realignment of the Nellis-based trials and operational evaluation units, the 57th Wing has lost its trials mission, but works closely with the co-located 53rd Wing Eagle evaluation squadron. New tactics derived by the 422nd TES are rapidly absorbed by the 57th Wing's Fighter Weapons School, which acts as a 'post-graduate' college for experienced crews. Graduates of FWS return to their units to pass on tactical knowledge, and to assist in sophisticated mission planning.

Eglin's 33rd FW – notably the 58th FS – grabbed the headlines during Desert Storm, accounting for the majority of the air-to-air kills (mostly with AIM-7s). This pair is from the 60th Fighter Squadron, complete with dummy Sparrow rounds.

F-15A/B/C/D had this capability, chiefly because the USAF had no intention of employing the Eagle as a bomber, and it was therefore practised only by a select number of F-15C/D squadrons. To clear up the common misconception that this capability was limited to carrying a few dumb bombs, it is worth explaining that the F-15A/B/C/D was in fact also able to carry and employ the AGM-65 Maverick and the stand-off GBU-15 glide bomb/AXQ-14 Data Link Pod combination.

AAQ-14 Target Pod

While the 335th FS 'Chiefs' remained in Saudi Arabia to police the southern Iraqi 'No-fly' zone at the end of Desert Storm, the 336th FS 'Rocketeers' returned home to North Carolina, where Seymour Johnson AFB was receiving its first batch of AAQ-14 pods (the vast majority of target pods manufactured prior to this had either gone for testing at the 57th Wing or been sent to the jets deployed in Operation Desert Storm). Testing of new software eventually began in an effort to improve the pod's performance. Consequently, the tracking and slewing characteristics of the FLIR tracker were enhanced, and the AAQ-14 finally provided a dependable automatic tracking feature. Proliferation of the Target Pod continued, and systems began to appear at both Luke AFB (then home of the F-15E training wing – the 58th FW) and Seymour Johnson AFB.

By the mid-1990s, despite its relative infancy, the AAQ-14 was starting to show signs of wear. Problems developed with the rotating nose section used to house the IR and laser optics. Cooling lines, which carried vital coolant to these components, were chafing and leaking, prompting the redesign and subsequent modification of a seal inside the pod. The new V-Seal design took some time to materialise, and attempts to provide an intermediate remedy backfired when ground crews unwittingly mixed a new coolant type with the old one. Each pod consequently had to be drained and cleaned before it could be re-attached to the jet – a time-consuming and frustrating affair. V-Seal pods are externally indistinguishable from the early C-Seal types, although ground crews have taken to scrawling 'V-Seal' on the updated pods in order to help them distinguish between the two without having to resort to opening the pod up.

Fighter Data Link

Prior to the Gulf War, the USAF had invested in the development of the Joint Tactical Information Distribution System (JTIDS), a radio-based Link 16 network that allowed secure sharing of information between ground- and air-based Command, Control & Intelligence, Surveillance and Reconnaissance (C²ISR) platforms. A few Air Defense Command Eagle squadrons had received the system following the 1984 F-15 MSIP II (Multi-Stage Improvement Program) modification, but JTIDS did not see widespread distribution and was cancelled in 1989.

Desert Storm prompted the reactivation of the programme, and the Multifunction Information Distribution System – Low Volume Terminal (MIDS-LVT) was subsequently developed and funded as a joint venture between the US, France, Italy, Germany and Spain. MIDS was targeted at the F-15C/D as a priority customer, and worked between 960 and 1,215 MHz, to provide a high-capacity, spread-spectrum, frequency-agile, secure digital communication system that allowed for interoperability between US and NATO forces (the NATO equivalent to Link 16 is TADIL-J).

The F-15E was also set to receive a derivative of MIDS, as attempts to work with E-8 JSTARS (Joint Surveillance Target Attack Radar System) reconnaissance aircraft during Desert Storm had ultimately proved to be a frustrating, if

not futile, affair. The F-15E was a significantly lower priority platform than the Eagle, and was initially set to receive a MIDS unit known as Fighter Data Link (a smaller and lighter LVT), once front-line F-15C/D units had been fully equipped. In fact, the F-15E eventually received FDL in October 2001, having jumped to the top of the list of priority customers following the outbreak of Operation Enduring Freedom. All F-15E wings are now FDL-equipped, while the F-15C/D fleet is, at the time of writing, about 50 percent of the way through installation.

FDL is extremely useful in helping to dissipate the ever-present 'fog of war'. It merges sensor data from a range of individual platforms and then displays a top-down view of the battle zone onto a 'SIT' display in the cockpit of any FDL user logged on to the network. Traditional limitations (such as own-ship sensor range and coverage) can therefore be overcome and, as situational awareness flourishes, enterprising new tactics can be developed.

The F-15C community is highly impressed by FDL, but refuses to be drawn in by the temptation to view Link 16-derived data as infallible. The approach being taken is to view the information as simply another piece in the jigsaw puzzle – a useful tool with which to corroborate or contradict other sensor data – but certainly not enough to be used on its own to satisfy shot criteria. For the F-15E, which derives the added benefit of being able to cue either the radar or Target Pod to ground tracks appearing on the FDL display, the same rule applies, and FDL is used with limitations in the system fully acknowledged.

Weaknesses inherent to data-sharing include the possibility that target locations broadcast onto the network contain positional errors, or may have been misidentified. The end result is that a target track appearing on the SIT display may actually be several miles away from the actual location of the target. One only has to imagine the confusion that may result from this if multiple tracks are erroneous, and one begins to appreciate that FDL, if used incorrectly, could cause more problems than it actually solves.

Expanding the Strike Eagle fleet

In 1992 the USAF began reducing the number of operational F-15A/B/C/D aircraft as part of a force reduction and restructuring process. When completed in 1997, the F-15 fighter force had shrunk from 342 to 252 aircraft. Conversely, the number of operational Strike Eagle squadrons continued to grow, and by roughly the same time, 209 had been delivered by Boeing. F-15E wings at RAF Lakenheath, England (48th FW – 492nd FS/494th FS);

Since being adopted as a replacement for the F-111, the F-15E has emerged as one of the USAF's most important combat aircraft, able to tackle a wide range of missions. A key factor is a healthy range, thanks to conformal fuel tanks, up to three drop tanks, and inflight refuelling. The type's long reach eases basing problems, notably in the Central Command theatre.

Although the AMRAAM has superseded the AIM-7 Sparrow as the F-15's prime medium-range AAM, the older weapon remains in widespread use. Here two F-15Es from the 3rd Wing's 90th FS perform a simultaneous shoot. Sparrow requires the fighter's radar to illuminate the target throughout the fly-out, whereas AIM-120 requires radar support only through the first stages. In the terminal phase the missile's own radar provides guidance, allowing the launch aircraft to turn away from the threat.

**F-15E, 90th Fighter Squadron
3rd Wing, Elmendorf AFB, Alaska**

Both aircraft carry unguided weapons in the form of retarded Mk 82s (above) and CBUs (below). The latter are now rarely used because of the high probability of collateral damage.

**F-15E
48th Fighter Wing
RAF Lakenheath**

Like the Alaska-based aircraft, Lakenheath 'Es' are powered by uprated PW-229 engines. This one is marked for the 3rd Air Force commander.

Two F-15Cs from the 18th Wing cruise near their base at Kadena, on the Japanese island of Okinawa. Both carry the KITS pod for rangeless ACM instrumentation on the outer missile rail. The aircraft in the foreground is marked for the Fifth Air Force commander, while the other is a regular 67th FS 'red-tail'.

Replacing the current generation of Sidewinder missiles as the F-15's short-range AAM is the Raytheon AIM-9X. Using the same body as its predecessor (and therefore the same launch rails), it has a vastly improved imaging seeker, thrust-vectoring and vestigial tail fins. A test round is seen carried by a 422nd TES F-15C.

Elmendorf AFB, Alaska (3rd Wing – 90th FS); and Mountain Home AFB, Idaho (366 Wing – 391st FS), had all converted to the F-15E by 1993. The F-15E Fighter Training Unit at Luke AFB closed in 1995 and all Strike Eagle training has since been conducted by the 333rd FS and 334th FS Replacement Training Units at Seymour Johnson.

Lakenheath- and Elmendorf-based aircraft were fitted with the brand new F100-PW-229 Improved Performance Engine (F-15E 90-0233 onwards), which was rated at 29,500 lb (131.27 kN) thrust and equipped with an Improved Digital Electronic Engine Control (IDEEC) system that quickened engine response times while increasing the engine's life expectancy. Time would show that the PW-229 was actually a maintenance-intensive item, and many ground crews preferred the PW-220, which powered previous F-15Es and many F-15C/Ds. Interestingly, the F110-GE-129 IPE (which offers the same thrust as the P&W IPE) was selected for the F-16C/D but not for the F-15E, despite proving in the IPE competition (held at Edwards AFB by the 412th TS – F-15 Combined Test Force) that it was more reliable than both Pratt & Whitney motors.

'Mud Hens' at Elmendorf and Lakenheath also featured the MSOGS (Molecular Sieve Oxygen Generating System), with the 366th Wing and 4th FW F-15Es later receiving depot-level retrofits during routine inspections. MSOGS provides a self-sufficient source of breathing air to the crew, and does away with the maintenance-intensive bottled oxygen system by drawing engine bleed air through filters to produce a breathable gas. It never made its way into the F-15C/D.

Because of the additional powered generated by the PW-229 over the PW-220, an Autonomous Thrust Departure Prevention System (ATDPS) was installed in 48th FW and 3rd Wing Strike Eagles to help prevent a departure or structural failure which might result from asymmetric thrust at high speeds following a single engine failure. ATDPS accomplishes this by instantly reducing the throttle setting of the remaining engine, although the pilot is able to override this if required.

Just prior to the start of Operation Northern Watch, Boeing began production of a batch of E210 and E222 F-15Es (E210/E222 refers to the 210th/222nd F-15Es to be made for the USAF, and is an alternative designation to using Block numbers – in this case Block 58/Block 59). These aircraft were manufactured as attrition replacements, and were delivered to the 48th FW, RAF Lakenheath, and 3rd Wing, Elmendorf AFB, in 1996 and 1997.

Unofficially termed 'White Paper F-15Es', they caused havoc among the three squadrons, each of which now exceeded its allocation of Primary Authorized Aircraft. While boasting a complement of 24 aircraft may seem advantageous in principle, the reality was that the squadrons were funded only to maintain and operate between 18 and 22 airframes. E210 and E222 aircraft featured small improvements over previous F-15Es, including revised CFT seals, composite speed brakes and air inlet panels, and a host of smaller changes such as anti-skid modifications.

OFP – improving the Eagle

Both F-15 variants have been refined via software changes known as Operational Flight Program, which is run by the F-15 Special Program Office at Wright Patterson AFB. OFP allows central computer software changes via flightline upgrades, and introduces additional features or enhancements to the aircraft's avionics suite.

In the mid-1990s, OFP Suite 3 was installed in the F-15C/D, thus improving air-to-air radar performance in electronic countermeasures environments; individual target break-out among closely formated contacts; improved air-to-air missile Dynamic Launch Zone envelope predictions; radar/TEWS interface improvements; and ALR-56 launch alert improvements. Suite 4 was installed in 2001/2002 and this conferred AIM-120C compatibility and a pre-launch Time Of Flight indication for the AIM-9L/M Sidewinder missile. F-15C/Ds of the 493rd FS, 48th FW, RAF Lakenheath became the first F-15s to shoot the AIM-120C during their annual deployment to Eglin AFB, Florida, for Combat Archer – a live-fire weapons exercise conducted to evaluate the performance of air-to-air munitions.

Right: Two F-15Cs from the 422nd TES cruise over the Nellis ranges. This test squadron has subsequently transferred to the 53rd Wing, with a change in tailcode to 'OT'. Many of the improvements made to the F-15 over the years have been tested by this unit in an operational environment, using the electronic combat range and other facilities at Nellis to simulate a 'real world' combat environment. Similarly, new tactics for fighter, weapon and sensor employment are developed by the squadron.

On top of the scheduled OFP modifications, the Eagle has received over 20 'extra-curricular' software updates since the conclusion of the Gulf War. These are often implemented at the behest of aircrew, and include such alterations as modifying the type of data projected onto the HUD and revising display formats used on the radar screen.

Strike Eagle OFP updates, which have been more prolific by virtue of the airframe's immaturity, carry the 'E' identifier to help distinguish them from F-15C/D OFPs. They have typically added hardware as well as software. For example, the 1998 Suite 3E brought about support for the GBU-28 'bunker-buster' bomb; introduced a GPS navigation system known as EGI (Embedded GPS/INS); and replaced the old Remote Map Reader and Tactical Situation

Northern England, Wales and Scotland provide the 48th FW's F-15Es with good low-level training grounds, and often challenging weather conditions. An EW training range is also available.

F-15E cockpit displays

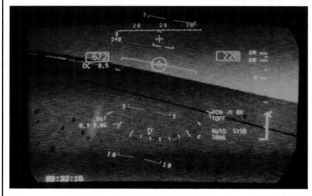

The rear cockpit of the F-15E is dominated by four large MPDs (right), with 'HOTAS'-style hand controllers and a side-mounted 'Up Front' Controller. There is also a control column and rudimentary flight instruments. The WSO can configure the cockpit in many ways, according to preference, and has a wealth of displays at his/her command. Above is the pilot's HUD, with symbology overlaid on the image from the Nav FLIR. This can be repeated in the rear cockpit.

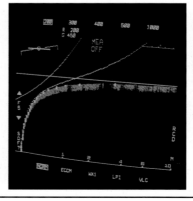

Above is a moving map display, aligned with the direction of flight. At right is the infra-red image from the AAQ-14 targeting FLIR, complete with cross-hairs.

The E-scope image (left) is a visual interpretation of the terrain-following radar's returns, while above is a frozen SAR image, known as a 'patch map'.

Seymour Johnson AFB is the spiritual home for the F-15E community: it was the first operational base, it accommodates four squadrons of Strike Eagles, and it handles conversion to the type. The 4th Fighter Wing is assigned to Ninth Air Force, and its two operational squadrons (335th and 336th) routinely deploy to 9th AF's main area of operations: Central Command. Until early 2002 the 4th FW was not part of the USAF's AEF organisation as the wing had a permanent 'on-call' expeditionary tasking already. However, it (along with the 366th Wing) has been integrated with the AEF programme. Two of the 4th FW's squadrons are assigned the training mission, including the 334th FS 'Eagles', whose aircraft wear a blue fin stripe. Mastering the 'Mud Hen' requires above average crew due to the complexity of its systems, its wide range of capabilities, and the wide range of missions it can be employed for.

Display with an all-digital moving-map called Digital Mapping System. It also added improvements to the ALR-56 to allow better launch and threat type discrimination, in addition to an increase in threat azimuth update rates. These changes augmented similar improvements to the radar modes and air-to-air capability improvements of the aforementioned F-15C/D Suite 3 OFP.

DMS was a real improvement over the old TSD, as it was digitally stored and allowed a range of flexible features to be utilised in flight in order to expeditiously modify the aircraft's flight plan. It is essentially an advanced digital moving map with stroke and raster imagery overlaid to represent the F-15E's flight plan, onboard sensor footprints, threat rings and target locations.

There has been common ground between the Eagle and Strike Eagle suites as most F-15C/Ds are fitted with an air-to-air tailored version of the APG-70 found on the vast majority of F-15Es. The F-15E, which received FDL via Suite 4E in October 2001, is set to receive Suite 4E+ in September 2002. 4E+ falls outside of the normal cyclical OFP programme and will further enhance the air-to-air capabilities of the radar, in addition to upgrading the Programmable Armament Control Set (PACS – used to interface the F-15E's weapons with the central computer). The PACS upgrade will bring the F-15E compatibility with the GBU-31A and GBU-31B Joint Direct Attack Munition; AGM-154A and AGM-154B Joint Stand-Off Weapon; and CBU-103, CBU-97 and CBU-87 Wind Corrected Munitions Dispensers.

While the PACS upgrade will allow communication with these smart weapons via the aircraft's onboard sensors, the hardware change necessary to facilitate this will come in the form of MIL-STD 1760 databus wiring to five weapons stations out of a possible 15. Upgrading all weapons stations fleet-wide is anticipated to cost over $50 million, and is unlikely to be funded in the immediate future.

DAFCS improvements

Some updates have fallen outside of these scheduled OFP updates. One particular example was a change to the

Digital Automatic Flight Control System on the Strike Eagle. DAFCS sends inputs to the F-15E's Control Augmentation System (which influences the manner in which the aircraft handles by adding small control inputs to the rudders and horizontal stabilisers). In 1991 an F-15E had been lost at Luke following loss of control at a high angle of attack; an almost identical mishap occurred in 1994 when a Fighter Weapons School (57th FW) F-15E crashed with the loss of both crew. Both incidents were principally attributed to pilot error, but the Strike Eagle's digital CAS may have been lacking in this portion of the flight envelope.

CAS uses control logic which was taken from algorithms in the analogue F-15D CAS, and subsequently translated into digital format for the F-15E's DAFCS. This in itself was not problematic, but at high AoA, and particularly with asymmetric weapons or fuel loads, the F-15E is more susceptible to departing controlled flight than the D model (especially so if the pilot ignores the cockpit beepers and warnings that indicate a precarious situation). Following the second loss, Boeing rushed to define new control laws that would dampen out the jet's handling characteristics at high angles of attack, and hopefully reduce the likelihood of a departure.

Because CAS simply overlaid small rudder and stabilator inputs onto the primary hydromechanical control movements, Boeing knew that there was no way to actually prevent an over-enthusiastic pilot from spinning the machine. Changes were made and not a single aircraft has been lost for this reason since. Boeing used OFP Suite 3E to install a spin recovery display for the pilot's left Multi Purpose Display – this indicates optimum control positions for a recovery in the event that a departure occurs.

Electronic warfare

Throughout the early 1990s, the range and combinations of stores available to the F-15E was gradually increased thanks mainly to the efforts of the 79th TEG at Tyndall AFB and the CTF at Edwards AFB. Correct fusing techniques were also established. Yet the Strike Eagle continued to experience difficulties with its TEWS, and the 79th Test & Evaluation Group was assigned to evaluate the entire suite amid public criticism of the USAF by the US General Accounting Office. The GAO was so dismayed that the USAF had ordered the ALQ-135(V) without properly evaluating it first, that it recommended the US Defense Secretary limit the Air Force's future ability to autonomously run EW acquisition programmes.

The F-15C/D did not come in for the same treatment as its ALQ-135 combined a comprehensive frequency coverage into one system and had been tested long ago. The Strike Eagle's ECM set-up differed because it broke the RF spectrum down into two distinct bands – Band 3 and Band 1.5. By working this way it optimised the way in which it dealt with evolving and existing air-to-ground threat emitters – the F-15C's ALQ-135 was largely air-to-air focused and was therefore less concerned with combating a

Many F-15Cs carry the 'Eagle Eyes' rifle scope, mounted next to the HUD. This was introduced in the early days over Bosnia to aid pilots identify slow-flying aircraft. Here it is carried by an 18th Wing aircraft.

complete array of air-to-ground emitters. The two different jamming systems of the ALQ-135(V) were theoretically able to dispense dedicated RF countermeasures against several different threat types simultaneously.

Band 3, which was supposed to have reached Initial Operating Capability in 1989, was finally deemed to meet acceptable standards in 1992, but Band 1.5, which had been nothing short of disastrous, was completely inadequate and plagued with problems. Of several issues that arose from this testing, the most worrying was that Band 1.5 ICMS circuits would reset themselves in the middle of an engagement (i.e. cease jamming) without so much as a single indication to the crew.

In 1999 testing was finally drawing to a close, and by 2002 fixes have reportedly been implemented. Operational crews have yet to cast their verdict because less than half of the 227 F-15Es delivered at the time of writing had received the upgrade. The F-15E community privately lobbied the Air Force to cancel Band 1.5 and spend the money on an upgraded AAQ-14, but to no avail.

OPC, ONW and OSW – policing Iraq

Following the successful conclusion of the Gulf War, the Coalition forces instigated Operation Provide Comfort in response to the ongoing persecution by the Iraqi dictator

against the Kurdish people in the north of Iraq. It saw F-15Es from the 'Chiefs' (335th FS) enforce a no-fly zone until July 1991, at which time Operation Provide Comfort II began. During Operation Provide Comfort the 335th TFS had flown escort missions in support of RAF and USAF transports providing Kurdish refugees with humanitarian aid.

For the duration of Operation Provide Comfort II, operations flowed smoothly, with one notable exception. On 14 April 1994, two US Army Black Hawk helicopters were mistakenly identified as Iraqi Mi-24 'Hind' gunships by two patrolling 53rd FS F-15Cs. In the engagement that followed, 26 people on board both UH-60 helicopters were killed when each was downed by air-to-air missiles from the Eagles. An AWACS controller was later found to have been in dereliction of his duties, and both F-15 pilots had transmitted ambiguous radio communications while attempting to visually identify the two helicopters. The 492nd FS, 494th FS and 391st FS subsequently sent F-15Es to Incirlik AB, Turkey, until 1996, at which point OPC II ceased and Operation Northern Watch commenced.

F-15s deployed to ONW are based at Incirlik AB, Turkey; a convenient station from which to launch sorties to patrol the 'No-fly' zone above the 36th Parallel. US assets at Incirlik share the tarmac with aircraft from a number of other coalition air forces and operate under the 39th Air and Space Expeditionary Wing (ASEW) – a centralised command that provides each squadron with a familiar wing structure for the duration of its three-month assignment.

Operation Southern Watch polices the southern 'No-fly' zone that stretches south of the 32nd Parallel. It was established on 27 August 1992, several years before ONW, and provides a buffer zone for Saudi Arabia and Kuwait. OSW Coalition forces are now stationed at Prince Sultan AB (PSAB), but prior to 1996 had been based at a number of

In the aftermath of the Gulf War the Europe-based Eagle force underwent downsizing and reshaping. Once the premier fighter unit in the continent, the 36th FW at Bitburg was disbanded, two of its squadrons joining the 52nd FW at nearby Spangdahlem. This is one of the wing's aircraft, an F-15C assigned to the 53rd Fighter Squadron. It carries two Iraqi flags for kills scored by Captain Benjamin D. Powell on 27 January 1991 against a MiG-23 and a Mirage F1. At the time the 53rd TFS was flying from Al Kharj as part of the 4th TFW (Provisional). Subsequently, the reputation of the 53rd was tarnished by the tragic 'blue-on-blue' incident in 1994, in which two US Army UH-60s were shot down in the mistaken belief that they were Iraqi Mi-24 'Hinds'.

Langley's 27th FS (illustrated) was the first F-15C unit to put the improved APG-63(V)1 radar into front-line service, in 2001. This retains a mechanically-scanned radar, but is more powerful and reliable than earlier units.

The Air National Guard roster includes six operational F-15 squadrons, all flying the F-15A/B. Three are primarily allocated the air defence role as part of the First Air Force organisation. The operational West Coast unit is the 123rd FS 'Red Hawks' at Portland (above).

Above right: Another two ANG F-15 units are gained by the Eighth Air Force and have a general battlefield air defence role. One of these, the 110th FS, is based at St Louis in the heart of the US. All six ANG units are integrated into the expeditionary air force structure, sharing duties in the AEF 9 rotation of late 2001/early 2002 which provided aircraft for Iceland, Northern Watch and in the Caribbean.

Below: Once tasked solely with the air defence of the Hawaiian islands, the PACAF-gained 199th FS now also deploys on global taskings.

Right: No F-15A/Bs remain in active-duty units and all instruction for these squadrons is now provided in F-15C/Ds. The ANG's seventh Eagle squadron, and Oregon's second, is the 114th FS 'Eager Beavers' at Klamath Falls, which provides training for the ANG's F-15A/B fleet. It previously performed the same task for the F-16ADF.

Saudi Arabian military airfields and operated under the Joint Task Force – South West Asia. F-15s figured heavily in both of these operations and, while the F-15C provides Combat Air Patrol cover high above the Iraqi desert, the F-15E gives the Coalition a powerful and versatile armed response to any acts of aggression or Iraqi excursions into the 'No-fly' zone.

F-15C squadrons from the US and Europe deploy for three months at a time to police the skies above Iraq although, invariably, it is the F-15E that sees the lion's share of the action. Strike Eagles, being fewer in number, tend to rotate to these tours more frequently and, although helped by the introduction of the Air Expeditionary Forces initiative (which plans deployments well in advance), tend to be overstretched. This is especially true of the 48th FW, which sends F-15C/Ds from the 493rd FS on a regular basis, but nearly always has 'Bolars' or 'Panthers' F-15Es deployed.

Until 2002, the F-15Es and F-15C/Ds of the 3rd Wing in Alaska were exempt from AEF taskings as they came under Pacific Air Forces (PACAF) command. Instead, they deployed to South Korea during times when the US Navy's carrier battle group departed the Pacific for replenishments and R&R. They also attended a range of multi-national and US-run exercises.

The 366th Wing, Mountain Home AFB, is an Expeditionary Wing that operates F-15Es and F-15C/Ds specifically tasked to deploy together for the opening stages of an unplanned military response. To this end, F-15C/D Eagles from the 390th Fighter Squadron have deployed to OSW alongside 18 F-15E Strike Eagles from the 391st FS as a form of practice for this eventuality. The

Eagles and Strike Eagles of the 'Liberty Wing' at RAF Lakenheath do not specifically train to deploy together, and are the only other wing to operate both F-15E and F-15C/D.

F-15 operations over Iraq reached their busiest in 1999 and 2000, when Iraqi AAA and SAMs fired upon coalition aircraft almost daily, prompting frequent retaliatory strikes. While the Eagles flew guard, the F-15Es carried a mixed load of ordnance to provide the flexibility necessary to respond to an unexpected threat. Most missions are intended to maintain a deterrent to Iraqi misdemeanours, and pre-planned target sets are always carried in the cockpit in case the order to strike is given. This gives theatre commanders several options in the event that the 'No-fly' zone is violated, or a coalition aircraft fired upon.

Most responses have seen F-15Es or F-16CJs attack Iraqi AAA and SAM equipment, but occasionally Strike Eagles have dropped guided munitions onto ammunition dumps and command and control facilities. Local rules of engagement allow the aircraft under threat to defend itself, but Coalition commanders have sometimes deferred an immediate response in preference to a well planned retaliation a few hours later. 500-lb GBU-12 LGBs are used in most instances where a measured response is necessary, and 2,000-lb GBU-10s have been employed against hardened and armoured targets.

While the threat of the Iraqi Air Force actually opposing the 'No-fly' zone enforcers remains minimal, F-15Cs escort all strike packages as far into Iraqi airspace as is necessary. Weapons load-outs for the Eagle depend largely upon the number of each type of air-to-air missile in stock at the forward location, but will typically consist of six AIM-120 AMRAAMs and two AIM-9 Sidewinder missiles. The use of AWACS, IFF and EID (electronic ID) means that Eagle pilots are not required by the rules of engagement to seek a visual ID on fixed-wing targets, although helicopter flights are handled differently in order to prevent another friendly-fire incident.

Since Operation Desert Storm, F-15C combat training has seen a shift from day- to night-time operations, and NVGs (night-vision goggles) have been in daily use for several years now. The Eagle has not received any Global Positioning System upgrade to its navigation suite, and still relies on the superb accuracy of its Ring Laser Gyro INS and vigilant clock-to-ground-to-compass work on the part of the pilot to ensure navigational accuracy.

The shift in emphasis from day to night is a natural consequence of the United States' propensity to fight its battles at night. It is at this time that the enemy is invariably the least effective, as few adversaries have either the training or the equipment to be fully effective at night. As with the F-15E, the F-15C/D received relatively cheap cockpit modifications to bring ambient lighting and instrumentation lighting into line with the levels necessary for effective use. The use of the low light, FDL, BVR missiles and EID tech-

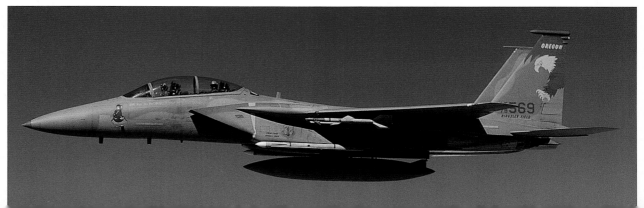

niques certainly make it easier to dominate one's foe, but as one F-15C squadron commander pointed out, it also makes it easier to inadvertently kill other friendlies. The full prowess of the F-15C in night operations was demonstrated with dramatic effect in the Balkans during the 1990s.

Stand-off weapons

The AGM-130 powered air-to-ground munition was first used by F-15Es deployed to ONW. This stand-off, precision-guided, 2,000-lb behemoth features an updated GBU-15 with a rocket booster mounted underneath. It was designed for the F-111 and F-15E following the Gulf War, and interfaces with the Strike Eagle via an AXQ-14 Data Link Pod (DLP). The DLP allows video from the missile's IR or EO sensor to be broadcast back to the WSO on any of the Multi Purpose Displays, and also allows steering data for the missile's control fins to be transmitted to a receiver at the back of the bomb. The AGM-130 uses the rocket motor to maintain a set airspeed for the duration of the booster burn time, following which the empty motor is jettisoned and the missile should have enough kinetic energy to maintain enough airspeed to reach the target.

Early AGM-130s and GBU-15s incorporated an INS to allow them to steer in the right direction of the target area, following which the WSO would visually acquire the target

on his MPD and then refine the target lock. This is termed 'mid-course guidance', and is extremely useful as it allows the WSO to concentrate on other tasks as the bomb makes its way to the target and the F-15E egresses from hostile territory. It also allows the weapon to fly through cloud, which may be obscuring the WSO's view of the target area at the time of launch. In this particular scenario, the WSO can acquire the target via the slewable optical or IR seeker once the bomb penetrates the cloud.

In the final phases of flight, both the GBU-15 and AGM-130 can be manually steered into the target, with video being beamed back to the AXQ-14 until the moment of impact, or the WSO can refine the aimpoint and allow the bomb to automatically guide itself to the intended impact point. The AGM-130 adds another layer of sophistication by means of an integrated radar altimeter, which allows the crew to set a cruise height for the bomb to ingress the target area.

'Terminal' guidance was added to both bombs in the late 1990s via a GPS modification that takes pre-launch target coordinates from the jet and then autonomously steers the weapon to the target. The WSO can still refine the aimpoint or manually steer the bomb, but in the event that a fault in the DLP develops or the weather deteriorates, the weapon will still guide itself into the original location.

The AGM-130 is available in either Mk 84 2,000-lb or BLU-109/B 2,000-lb forms, the latter of which can penetrate

With the end of the Cold War the ANG-manned CONUS air defence network, based largely on a force of F-16ADFs, was dramatically cut to just six dedicated fighter squadrons, divided equally between F-15 and F-16. Defending the southeast of the US is the 159th FS, Florida ANG. Headquartered at Jacksonville, the squadron also maintains a permanent alert detachment at Homestead ARS. First Air Force assets have been heavily involved in homeland defence efforts in the wake of the 11 September terrorist attacks, including detachments to airfields near major cities.

The Fighter Eagle school is at Tyndall AFB, Florida, where AETC's 325th Fighter Wing is located with three squadrons (1st FS – red tail, 2nd FS – yellow tail and 95th FS – blue tail, illustrated above and right). Like the front-line 1st Fighter Wing, the 325th is gearing up to receive F/A-22s from 2004 to initiate the long process of converting from the Eagle.

Flying from Cervia, the 493rd Fighter Squadron bore the brunt of air defence operations during Allied Force, claiming four kills. Missions supported NATO attack aircraft, while the Eagles also maintained combat air patrols over surrounding nations such as Bosnia. The 18-aircraft 493rd FS deployment was bolstered by a personnel detachment from Alaska's 3rd Wing.

F-15C 86-0156 was the 493rd FS aircraft flown by Captain Jeff Hwang when he shot down two MiG-29s over Bosnia on 26 March 1999. The other 'MiG-kill' aircraft were 84-0014 and 84-0015. The former also had a kill from March 1991 against an Iraqi Su-22, while assigned to the 53rd TFS in the aftermath of Desert Storm.

hardened targets, while the GPS modified GBU-15 is known as the Enhanced GBU-15 (EGBU-15). Although firm information concerning the success of the first launches during ONW is limited, it is understood that they were unsuccessful. Boeing tested a new turbojet-powered AGM-130 in 1998, and demonstrated a 102-mile (164-km) range resulting from a 15,000-ft (4572-m) launch. This impressive reach increases the F-15E's stand-off range by a value of almost 300 percent, although the USAF has yet to purchase the weapon.

URITS, KITS and rangeless ACMI pods

In 1997, F-15C/Ds of the 18th Wing, Kadena AB, Japan, became the first USAF F-15s to receive a rangeless air combat manoeuvring instrumentation system. The Kadena Instrumented Training System (KITS) was developed by Cubic Systems Incorporated to allow the Japan-based Eagles to record mock combat exercises using a GPS-based Airborne Instrumentation System (AIS) that records aircraft location and performance information throughout the course of the dogfight.

When the data from each participant's AN/ASQ-T34 AIS pod is merged at the end of the sortie, a complete picture of the exercise can be replayed and critiqued from a variety of computer-generated camera angles. Interfacing with the F-15's PACS allows the KITS to monitor weapons engagement criteria and provide each participant with Real Time Kill Notification. RTKN plays short sound files (such as 'hit' or 'miss') into the pilot's headset in the launch and target aircraft, thus allowing immediate analysis of tactics and performance.

Elsewhere, the 48th FW received a Metric Systems Incorporated rangeless ACMI system known as USAFE Rangeless Interim Training System (URITS) in May 1999. URITS works in a similar fashion to KITS, but employs the AN/ASQ-T38B AIS pod – the first ACMI pod to utilise military-grade GPS signals for greater positional accuracy. It, too, features RTKN, although the USAFE has yet to approve its use. The key benefit of both systems is that the

participating aircraft are able to practise anywhere where a GPS signal can be received, and do not have to rely upon expensive and busy 'tethered' ACMI ranges (which require AIS pods to transmit data to ground stations). The F-16C/Ds of the 52nd FW, Spangdhalem AB, Germany, have also received URITS, while the F-15C/Ds and F-15Es of the 3rd Wing, Elmendorf AFB, received KITS in 1999.

Eagle over the Balkans

From 1993 onwards, F-15s from USAFE wings deployed to Aviano AB, Italy, to provide support for a range of operations conducted under the auspices of the UN and NATO. By July 1993, F-15Cs from Bitburg's 53rd FS had flown over 660 combat air patrol sorties to protect NATO troops on the ground in the regions of Banja Luka and Sarajevo. Later that year, eight F-15Es from the 48th FW's 492nd FS deployed to Aviano as part of Operation Deny Flight. Assisted by a small contingent of 494th FS Strike Eagles, they maintained their presence in Italy for more than a year.

When instability in the region further deteriorated in November 1994, NATO sanctioned a limited strike against Serbian targets in Croatia, in particular Udbina Airfield. Eight 492nd Strike Eagles carried GBU-12 500-lb LGBs as part of a mixed strike package consisting of 30 aircraft. Their target was a cluster of SA-6 SAM sites which posed a significant threat to any future air operation. The sortie was cancelled mid-flight for reasons that remain unclear. F-15Es launched once again later in December, this time to destroy a pair of SA-2 SAM sites which had recently fired upon two Royal Navy Sea Harriers.

Operation Deliberate Force commenced in August 1995, following the mortar shelling of a market square in Sarajevo, and five punitive strikes destroyed Serbian armour and supplies around Sarajevo on 30 August. GBU-10 and GBU-12 laser-guided bombs were dropped by the 48th FW on 5 September, as Operation Deliberate Force strikes became more widespread. On 9 September the GBU-15 was dropped for the first time in anger by the

The USAF's premier wings in the Pacific region (3rd Wing, left) and Europe (48th FW, right) have led the fleet in terms of receiving the latest F-15E variants. Both wings introduced aircraft with Improved Performance Engines, and have received the latest-build aircraft. 48th FW F-15Es saw considerable action over the Balkans during the 1990s, culminating in Allied Force.

F-15E – nine were used to strike air defence targets around Banja Luka.

In 1998 Operation Allied Force came into effect. Following repeated NATO warnings to President Milosevic to remove his armed forces from Kosovo, 15 F-15Cs from the 493rd FS, 48th FW were deployed to Aviano AB, Italy. A five-phase plan was put into effect: NATO flights would initially act to deter Yugoslav president Slobodan Milosevic, but would become more aggressive if demands for Yugoslav troop withdrawals and the cessation of ethnic cleansing were not met. Despite some gains made at the Rambouillet talks in France, an additional 12 F-15Es from the 492nd FS were sent to Italy in February 1999. As F-117 Nighthawks arrived at Aviano AB, Lakenheath's F-15Cs were displaced to Cervia AB and were reinforced to a total strength of 18 aircraft. Six 494th FS F-15Es had arrived in December 1998, allowing the 492nd FS's jets to leave for Turkey for their Northern Watch AEF commitment.

Operation Noble Anvil (the US portion of Operation Allied Force), commenced on the night of 24 March 1999. Following several cruise missile (CALCM) strikes by B-52s, 26 F-15Es struck air defence (SAM, AAA, GCI EW) assets as they followed behind a protective wall of 493rd FS F-15Cs flying Offensive Counter Air sorties to flush out the airspace ahead of hostile aircraft. As the conflict progressed, 'Mud Hen' crews turned their hands to dropping cluster bombs against Yugoslav troops in the open, and the GBU-28 bunker-penetrating laser-guided bomb was used against an underground hangar at Pristina AB. It was at this stage that the AGM-130 was used effectively for the first time – two weapons destroyed two MiG-29s on the ground.

MiG encounters

Four MiG-29s were claimed by the 493rd FS: two were shot down on the opening night of the war, when Captain Mike Shower fired an AIM-120 at a MiG-29 lifting off from its airfield at Batajnica and Lieutenant Colonel Cesar Rodriguez (who already had two kills to his name from Operation Desert Storm) claimed a third when he also used an AIM-120 to kill a MiG-29 beyond visual range. Interestingly, Shower required three AMRAAMs to down his target – the first two missed while still BVR, but the third, fired at closer range (6 nm), found its target.

The final two kills for the 493rd FS came courtesy of Captain Jeff Hwang on 26 March 1999. Hwang was flying a Defensive Counter Air (DCA) sortie over Bosnia when AWACS called out two MiG-29s flying into Bosnian airspace. The MiGs were promptly picked up by Hwang on his radar. Reacting to RWR indications, the MiG-29s changed heading to bring them on an intercept course for Hwang's flight (DIRK). Hwang sought permission to fire, following IFF and radar indications that the contacts were indeed hostile, but AWACS was slow to respond.

As the MiGs closed to within 30 miles, Hwang had little choice but to clear his wingman to fire. Lieutenant Boomer McMurray – who had been tracking the bandits on his own

'Top cover for America' – the 3rd Wing at Elmendorf has a squadron of F-15Cs and one of Es. During the Cold War the Eagles were routinely scrambled to intercept Soviet long-range bombers and patrollers as they snooped around the northern approaches to the US heartland. Today such uninvited guests are encountered less than once a year, but Alaska has taken on a new importance as a rapid deployment base. Eighteen of the 3rd Wing's F-15Cs are fitted with the APG-63(V)2 radar, with a revolutionary AESA 'e-scan' antenna. Pilots have reported a major improvement in radar performance, especially the ability to break out and identify targets at long range. This AESA-equipped aircraft is from the 19th Fighter Squadron.

At Edwards AFB the 416th Flight Test Squadron operates a mix of F-15s and F-16s as the Global Power Fighter Combined Test Force. The unit has both military and contractor staff assigned. Test work involves both tests dedicated to the F-15, such as the WCMD (Wind-Corrected Munition Dispenser) drop being undertaken right, and non-type specific work. Its high performance makes the F-15 a very useful chase aircraft, especially for the F/A-22, and as part of the Test Pilot School syllabus. Permanently assigned general test aircraft wear a high-conspicuity red/white scheme (above). In 1997 a 412th TW F-15D was named Glamorous Glennis, and flown by Chuck Yeager on the 50th anniversary of his first supersonic flight.

Right: The principal flying component of the Air Armament Center at Eglin AFB is the 40th Test Squadron. Here one of the squadron's F-15Es drops a GPS-enhanced EGBU-15 2,000-lb (907-kg) electro-optical guided bomb.

radar – fired a single missile, which was followed moments later by two more AIM-120s from Hwang's aircraft. Closing to visual range, two of the AMRAAMs found their target, and both kills were later attributed to Captain Hwang.

Operation Noble Anvil was the first recorded air operation where Night Vision Goggles were used in the F-15. The devices had been integrated into the F-15E by a select cadre of F-15E aircrew from the 422nd Test & Evaluation Squadron at Nellis AFB. They had suggested adapting NVGs to the Strike Eagle during the mid-1990s and, draw-

Left: For 25 years the Sidewinder has been the standard short-range AAM, but it is now giving way to new weapons. Israel uses the Python 3 and 4, while Japan has adopted the AAM-3. For USAF Eagles the AIM-9X is soon to enter service. Here a test round is fired by a camera-equipped 40th Test Squadron F-15D.

ing on their experiences at night in such aircraft as the A-10, designed a primitive but effective cockpit lighting solution known as 'Christmas tree lights'. These were small, low-wattage bulbs that dimly lit the cockpit and were strung around the cockpit canopy sill by means of Velcro fasteners. Chemical sticks were velcroed above important instrumentation and finger lights were carried by aircrew to illuminate maps and other paper work. Given that the incompatible cockpit lighting could simply be left off at night, these three components provided an effective, if haphazard solution.

The MPDs, which were too bright for the NVGs, were shielded by green filters known as 'Glendale Green', and these were delicately fixed into place along with similar filters for the annunciator panels and master caution lights prior to flight. Treated bulbs for the cockpit storm (flood) lighting soon made their way into production, and these have now superseded 'Christmas tree' lighting kits. Even so, they are far from cost effective, and cost around $90 per bulb.

F-15C modifications are almost identical in nature. It is possible that the F-15E will eventually receive 'covert lighting' – external lights visible to NVGs but not to the unaided human eye. It is unlikely that the F-15C will receive any covert lighting system and the Eagle community does not see it as a priority, in any case.

'Gucci' gear – AESA, JHMCS and EGI

In 1999 18 F-15C/Ds of the 3rd Wing at Elmendorf AFB received the AN/APG-63(V)2 Active Electronically Scanned radar. AESA is the latest version of the APG-63 and was developed in parallel with the upgraded APG-63(V)1. While the 'V1', which made its way into newer F-15Es and debuted on F-15C/Ds of Langley's 27th Fighter Squadron in April 2001, is more powerful and maintenance-friendly than previous radar sets, the V2 employs entirely new technology.

In place of a hydraulically-steered radar dish is a stationary panel covered with an array of hundreds of small transmitter-receiver modules. These modules have more combined power and can perform different detection, tracking, communication and jamming functions in multiple directions simultaneously. The end result is a radar that has exceptional target tracking, identification and engagement capabilities. BAE Systems supplies the IFF system that works in tandem with AESA, and early feedback from the 12th FS and 19th FS is that the radar is working extremely

right side of the JHMCS visor, and the pilot's head movements are identified by shifting magnetic currents detected by cockpit-mounted sensors. These current changes are translated into the pilot's line of sight, and are then used to cue any number of sensors – including missile seeker heads – to the object of interest at the centre of the pilot's helmet display.

JHMCS integrates with the AIM-9X Sidewinder, and is a direct response to the highly agile short-range IR-guided missiles currently in use by a number of potential adversaries. Once in full production, it will allow the F-15C to merge with and visually designate an opponent even if the target falls outside of the gimbal limits of onboard IR or radar sensor systems. While JHMCS brings the F-15C in line with the Russian-manufactured HMS systems installed on the MiG-29 and Su-27, the basic premise of the F-15C's mission will remain the same: to avoid a merge through superior BVR performance. While the AIM-9X is destined for both Eagle and Strike Eagle, the JHMCS will probably not be fielded on the F-15E, despite the fact that it can also be used to cue air-to-ground munitions and FLIR sensors such as the AAQ-14 Target Pod (TP).

The 1998 Suite 3E modification made several important changes to the Strike Eagle's avionics suite, but none as significant as the installation of EGI. Using this Embedded GPS/INS system, the F-15E is now better able to accurately acquire and attack targets via the radar and TP – the crew can simply enter GPS co-ordinates for static targets into the weapons computers. Cockpit workload has been slightly decreased as a result, and navigational errors and velocity errors that used to require Precision Velocity Updates via the APG-70 within five minutes of striking the target, are no longer required. EGI is accessed via the Up Front Controller in either cockpit, and the crew can alternate their PPKS (Present Position-Keeping Source) between GPS or INS in the event that either system fails, or the GPS signal is interrupted.

Advanced Target Pod

Replacing the ageing AAQ-14 Target Pod is seen by most F-15E aircrew as the most important item on the Strike Eagle's list of future upgrades. While the AAQ-14 is still highly effective, the introduction of an improved target pod known as the Advanced Target Pod onto F-16s has whetted the appetites of eager WSOs who are keen to enjoy the fruits of this new system.

Improvements to LANTIRN have been ongoing since it first entered service and, in the late 1990s, an operational evaluation by the 422nd TES of a LANTIRN BDA pod was conducted. BDA was a standard AAQ-14 shell, but housed a hyper-spectrometry sensor in addition to the normal FLIR optics. The concept behind this Bomb Damage Assessment pod was to be able to accurately assess the damage to a target even if the FLIR picture was clouded by dust and debris from the ensuing explosion. Hyper-spectrometry

Above: F-15s joined F-16s in protecting major US cities following the Al Qaeda attacks on New York and Washington. First Air Force's ANG air defence squadrons were to the fore, including the 101st FS, Massachusetts ANG. Here the 'Eagle Keepers' commander's F-15C patrols the skies over New York.

Left: The future for the Eagle is dark grey – if current USAF fighter procurement proceeds as planned, the 'light grey' fighter Eagles will be supplanted by the F/A-22 from 2004. Displaced F-15C/Ds will cascade to ANG squadrons, allowing the retirement of the F-15A/B. No such tangible replacement is in sight for the F-15E, which is expected to serve until 2030, with a host of improvements planned for it, including the addition of the Small Diameter Bomb to its armament options. This weapon can be carried in large numbers and can be delivered with high accuracy. The F-15E is also likely to receive technology spin-offs from other programmes, including AESA radar antennas. Future export campaigns will also be based solely on the 'E'. Replacement for the F-15E, when it comes, is currently subject to some debate. Lockheed Martin has proposed its 'FB-22' to supersede both the F-15E and F-117. Based on the F/A-22, the 'FB-22' has a delta wing and lengthened fuselage for greatly increased range and warload, while retaining the stealth and 'supercruise' properties of the F/A-22.

well. There are currently no plans to extend the number of AESA operators beyond these Elmendorf F-15Cs.

During the mid-1990s the F-15C had its central computer replaced with an IBM Very High Speed Integrated Circuit computer. VHSIC provides the number-crunching power to drive OFP improvements and new hardware, such as the JHMCS. Joint Helmet Mounting Cueing System is a helmet-mounted sight that weighs in the order of 4.3 lb (2 kg) which can designate targets up to 90° off-boresight. It shares many similarities with the DASH helmet used by the Israeli Air Force.

Low-Rate Initial Production (LRIP) of the helmet was approved in May 2000, and the system has been tested over the course of 700 flight hours by the F-15 Combined Test Force at Edwards AFB. Data regarding the aircraft's flight parameters and weapons/radar status is projected into the

Ground crew prepare a 391st FS aircraft for a mission over Afghanistan. The principal warload comprises a GBU-24 Paveway III laser-guided weapon (with BLU-109 penetrating warhead) on the port CFT, and an EGBU-15 on the starboard wing pylon. The latter weapon employs electro-optical guidance in the terminal phase, either by target recognition or direct input from the WSO (via the AXQ-14 datalink pod). In addition, it has a GPS guidance system added as a back-up, ensuring that the weapon strikes within a few feet of its intended aimpoint even if the EO sensor cannot steer it more precisely, or if the link goes down.

Above right: The AGM-130 was also used in Afghanistan (seen here being launched by a 4th FW aircraft). It is essentially a rocket-boosted GBU-15, and was used by F-15Es operationally for the first time over Iraq during Operation Northern Watch duties.

allowed the pod to see through fragmentation debris in order to accurately capture the destruction of the target. The crew received a standard FLIR image in the cockpit, and the BDA was digitally recorded in the pod itself for later download and post-sortie damage assessment. LANTIRN BDA was subsequently re-named LANTIRN Bomb Impact Assessment (BIA), and was expected to be retrofitted to a small number of AAQ-14s (48) during the course of 2002.

ATP features a brand new, third-generation FLIR system and has integrated an electro-optical tracker into the rotating 'NESA' section of the AAQ-14. The level of magnification in the pod is so great that the laser spot can be seen on the MPD in some situations. FLIR picture quality has been improved, and an Automatic Target Recognition (ATR) feature is provided via an additional Line Replaceable Unit housed at the rear of the pod.

ATR was a feature that was destined for the AAQ-14 from the moment that the concept was outlined on paper, but early LANTIRN funding and prototype problems led to its deletion from the programme. The author understands that the air-to-air tracking logic of the ATP is now better able to deal with the demands of 'complex' targets, such as a helicopter main rotor. Tests showed that it has been solidified to the extent that a solid lock can be maintained even when the target passes over a high-contrast, confusing scene such as a busy motorway. The ATP connects to the F-15E's existing MIL-STD 1553B bus via existing connectors, and therefore requires no airframe modifications to equip existing LANTIRN users. There are no firm dates for installation of the ATP at the time of writing.

OEF – Eagle over Afghanistan

In the wake of the September 2001 terrorist attacks in New York and Washington, the USAF rapidly launched Operations Enduring Freedom and Noble Eagle. Noble Eagle saw F-15Cs and F-15Es based in the United States flying combat air patrol sorties over key US cities in order to fend off further airborne attacks. In co-ordination with ground-based radars and AWACS platforms, these aircraft routinely flew nine-hour sorties while carrying a mixed load of IR- and radar-guided air-to-air missiles.

Over Afghanistan the 391st FS F-15Es routinely covered the night hours, using the superior capabilities of the LANTIRN and APG-70 systems to deny Taleban and Al Qaeda fighters the cover of darkness. Due to the distances involved to the war zone and an average sortie duration of 6.1 hours, inflight refuelling was an essential facet of each mission.

Rules of engagement in the event that an aircraft was suspected of falling victim to hijack, or deviating from Air Traffic Control-assigned vectors, were comprehensive. In the first instance, the suspect aircraft would be intercepted and visual and radio communication was to be made with the errant aircraft's crew. In the event that this contact failed to elicit the appropriate response, the F-15 would fly in front of the target and release flares to grab the attention of those on the flight deck. If a change of course was not made following this, the F-15 would fire warning shots from its wingroot-mounted M61A1 Vulcan 20-mm Gatling gun. The final recourse, undertaken as a last resort, was to launch a single AIM-9 Sidewinder at one of the aircraft's engines. Assuming that the target aircraft was multi-engined, successive launches would be made at the remaining powerplants.

Operation Enduring Freedom addressed the need to pro-actively target the infrastructure and leadership of the Al Qaeda network and supporting Taliban in Afghanistan. Among the contingent of hundreds of US assets deployed to the region were 12 F-15Es from the 391st FS, 366th AEW, Mountain Home AFB. It is believed that they were based at Ahmed Al Jaber Air Base, Kuwait, where they commenced operations under the command of the 332nd Air Expeditionary Group. The 'Bold Tigers' replaced elements of the 48th FW which were two weeks away from completing a deployment to Operation Southern Watch at PSAB. A Combined Air Operations Center at McDill AFB, Florida, built the daily list of targets, most of which in the first two weeks were of a fixed category such as buildings, ammunition caches, and Taliban command and control facilities.

In late October 2001 the first wave of 391st FS F-15Es attacked targets around strategically important Afghan cities, and met little resistance while doing so; an almost continuous cycle of F-15E attacks ran for three months thereafter. So poor was Afghanistan's infrastructure – following 10 years of fighting the Russians and years of neglect and bitter in-fighting since – that the list of fixed targets was exhausted within two weeks.

Up until this point, AGM-130 and EGBU-15 weapons had been used to destroy caves leading to underground facilities and to precisely strike targets where specific weapons effects were required. It was the first time that the EGBU-15 had been used in anger, and was one of just a number of 'firsts' achieved by the 'Bold Tigers' during their three-month operation over Afghanistan. The GBU-24A/B, fitted with the BLU-109 penetrator warhead, was used to neutralise reinforced targets and underground Taliban facilities, and the GBU-24A, which featured a standard Mk 84 bomb body, was also used against less fortified fixed targets. There were occasions when deep penetration, several storeys underground, was required, and five GBU-28s were used accordingly.

Most sorties were conducted in pairs of Strike Eagles, but because the F-15E has no VHF radio equipment, the 'Bold Tigers' would often link up with F-16s, which used their VHF radios to pass messages to and from Command and Control aircraft, and to communicate with Forward Air Controllers over the target or on the ground. This proved useful when the few remaining fixed targets were depleted and the CAOC started to assign the F-15E the mission of Close Air Support (CAS) and Time Sensitive Tasking (TST).

These roles required crews to fly over geographically defined 'kill-boxes' for a predetermined 'vulnerability period'. If hostile action or enemy targets was detected within the boundaries of that kill-box, the F-15Es would find the perpetrator and kill it. Using FDL to receive targets from the E-8 JSTARS reconnaissance aircraft and RC-135

Rivet Joint electronic surveillance aircraft, the F-15E was able to quickly locate and identify its targets. Mixed loads of unguided 500-lb Mk 82 LDGP bombs and PGM munitions (GBU-12, GBU-10, EGBU-15 etc.) were carried to allow the destruction of targets over a range of rapidly evolving environments. Noteworthy was the fact that Strike Eagle crews did not carry cluster bombs, a decision made at squadron level due to the indiscriminate nature of the weapon and concerns over the possibility of collateral damage. In one sortie alone, a pair of F-15Es released a GBU-28, two GBU-24s and six GBU-12s

Many 'Bold Tigers' operations were conducted at night, a time when the Taliban mistakenly believed that they were most safe. Vehicles fleeing for the mountains were seen clearly on the monochrome FLIR screen in the cockpit, and the 391st crews perfected the art of lasing GBU-12s into vehicles travelling as fast as 60 mph (96 km/h). There were numerous occasions when friendly Special Forces troops on the ground came up against pockets of enemy resistance and needed swift assistance from the air. The Gatling gun was used on several occasions to dispatch Taliban troops who were too close to friendly forces to even drop a 500-lb laser guided bomb, and there were also occasions where use of the gun was critical to ensuring the survival of a key logistics route or building (such as a road or bridge, respectively).

Gold Pan datalink

Following Desert Storm, the USAF had funded the development of a jam-resistant datalink pod to upload imagery from ground- and air-based transmitters directly to the F-15E's cockpit. This system, termed Gold Pan, utilised the ZSW-1 Improved Data Link Pod (IDLP) and allowed targets of opportunity and updated intelligence data to be sent to the WSO during flight. Gold Pan, which uses the AXQ-14 electrical interface and is 1760 databus-compliant, was first used during Operation Allied Force when target imagery was uploaded to airborne 48th FW F-15Es for attacks against unplanned targets.

While the 391st FS was ready to use it once again in Afghanistan to attack time-critical targets detected by such platforms as the RQ-1 Predator unmanned aerial vehicle, the crews were limited to receiving tasking over radio or FDL because the ground-based stations were not available. There remains underlying dissatisfaction within some elements of the Strike Eagle community with the way in which the USAF has been slow to fully support and implement Gold Pan.

While responding to a succession of TST taskings raised by AWACS and Predator, the 391st FS flew the longest recorded fighter mission ever when the crew of CROCKETT 52 clocked up an incredible 15.5-hour sortie, including nine over the target. The 'Bold Tigers' returned home in January 2002 having achieved a sortie generation rate better than 85 percent, and were replaced by the F-15Es of the 335th FS, 4th Wing from Seymour Johnson AFB. The 'Chiefs' deployment was very similar to that of

the 'Bold Tigers', and they became the first aircraft to deploy the BLU-118/S Thermobaric bomb – a fuel air explosive weapon that penetrated Taliban caves and then filled them with an explosive vapour. The ensuing explosion killed hiding Taliban fighters as the oxygen in the air within the cave was consumed by the burning vapour.

F-15I Ra'am

On 27 January 1994 the Israeli government announced its intention to purchase the F-15I (local name Ra'am = thunder) to fulfil a long-range, all-weather precision strike requirement. The F-15E had competed against the F/A-18 and F-16 for the contract, and would be used to kill high-value targets in all weathers and times of day. McDonnell Douglas (Boeing) was to build an initial batch of 21 F-15Is for the IDF/AF under the Foreign Military Sales (FMS) designation Peace Fox V. An option for four more airframes (Peace Fox VI) was subsequently exercised in November of 1995, raising the total to 25, at a total cost of approximately $2.5 billion. The F-15I would complement the F-15A/B/C/D airframes already owned by the Israelis and, like the older IDF/AF Eagles, would benefit from similar indigenous avionics from Israel's defence sector.

The first F-15I flight took place at Boeing's St Louis plant on 12 September 1997 and Minister of Defence Yitzhak Mordechai officially took receipt of the first F-15I on 6 November 1997. In January 1998 deliveries began to 69 Squadron 'Hammers' at Hatzerim AB in southwest Israel, and continued at a rate of one airframe per month. Martin Marietta provided 30 AAQ-14 and AAQ-13 LANTIRN pods, which had originally been destined for Israel's F-16I fleet, and were said to be identical to those used on USAF F-15Es. The F-15I fleet was powered exclusively by the F100-PW-229 engine, although the Israelis opted to retain

F-15Es from the 391st FS overfly Nevada in an exercise in early 2001. The squadron was particularly well-prepared for operations over Afghanistan, as its planning staff had already identified the country as a potential combat theatre following the October 1998 cruise missile strikes against Osama bin Laden's terrorist training camps. Accordingly, squadron staff had studied the virtually non-existent infrastructure and defences of the country, and concluded that the bulk of operations would be of a time-sensitive or close support nature. In the event, the vast majority of 'pre-fragged' targets had been exhausted after just two weeks of air operations, and the F-15Es were used for weeks thereafter for exactly the kind of TST and CAS operations which the 391st had predicted.

'Bold Tigers' in Enduring Freedom

Within hours of the terrorist attacks on New York and Washington, fingers were pointing at Osama bin Laden's Al Qaeda network, and the Taliban regime in Afghanistan which harboured its leaders and training camps. Over the coming weeks a large (mostly US) force was built up in the surrounding region, ready to begin operations. The 391st FS was one of the units involved, being dispatched to the Gulf two weeks in advance of its scheduled rotational deployment, replacing F-15Es from the 48th FW.

The 391st Fighter Squadron spent 96 days in-theatre during Operation Enduring Freedom. The accompanying aircraft-by-aircraft table highlights the intensity of operations for the squadron during this period, and the unusually long sortie times for what were considered 'fighter' missions.

Aircraft serial	87-0169	87-0173	87-0182	87-0198	87-0201	87-0204	87-0207	87-0208	90-0227	89-0506	88-1667	88-1705	**Total**
Hours flown	289.5	216.6	312.7	252.6	231.4	265.9	185.4	291.4	189.6	304.1	271.6	154.8	**2965.6**
Average sortie duration	6.2	6.8	6.5	7.0	6.8	5.3	5.0	6.6	5.7	6.3	7.3	4.2	**6.1**
No. sorties flown	47	32	48	36	34	50	37	44	33	48	37	37	**483**

Too new to have yet received its 'OT' tailcode, one of the USAF's final batch of 'E227' Eagles (00-3000) launches from Nellis on a trials sortie with the 422nd TES. Beneath its belly it clutches a datalink pod, normally used for guiding the GBU-15/AGM-130 electro-optical guided weapons, but also (in the ZSW-1 version) as a means of receiving targeting information via the Gold Pan system. The aircraft is carrying two of the current generation of GPS-guided weapons: a GBU-31 JDAM and an AGM-154C JSOW.

Long before the final USAF Eagles trickled from the St Louis production line, sizeable numbers of F-15A/Bs began arriving at Davis-Monthan AFB for scrapping or storage. Useful parts are recovered from 'Christmas tree' airframes to keep others flying, while a large number are held in store, heavily protected against the ravages of the Arizona sun and sand.

the 'turkey feather' engine exhaust nozzle covers that had long since been ditched on previous Israeli (and US, Japanese and Saudi) F-15s. F-15Is wear an FS 30219; FS 33531; FS 34424; and FS 36375 paint scheme.

The F-15I does not feature the AN/ALQ-128 Electronic Warfare Warning System or the AN/ALQ-135 ICMS, and therefore has a different set of airframe-mounted antennas. Additional upwards-facing chaff and flare dispensers were installed on the top side of the tail booms and the low voltage formation strips were blanked out. A revised APG-70I radar was incorporated, but this was limited in bandwidth and frequency range, and could not initially generate Synthetic Aperture Radar 'patch maps' to the full 0.67-nm resolution – this capability followed after the US DoD authorised a subsequent upgrade. It is also highly likely that the 'special' radar modes (NCTR and ECM modes) were not exported on the APG-70I. Small switchology changes were made to the Hands On Throttle And Stick

(HOTAS) configuration in the F-15I, and the WSO was able to access the radar's Auto Acquisition dogfighting modes – a feature not found in the F-15E.

In place of the deleted US avionics systems was an Elisra SPS-2100 IEWS (Integrated Electronic Warfare System) that included a Missile Approach Warning system. KY-58 Have Quick anti-jam radios were removed in favour of an Israeli-supplied system, and an improved Sony 8-mm video tape recording system was installed to capture HUD and MPD video, in addition to the same DMS facility used by Suite 3E-upgraded USAF F-15Es.

An indigenous CC (Central Computer) also made its way into the Ra'am, and an Israeli-manufactured GPS navigation system, similar to the Embedded GPS/INS in the F-15E, was installed, although presumably does not use the military-grade GPS signals available to EGI. Combined, the GPS and real-time playback VTRS give the F-15I the capability to perform in-flight bomb damage assessment and to pass accurate target details to other airborne assets as a substitute JTIDS or MIDS.

All of the aircraft's sensors can be directed by an Elbit Systems DASH (Display And Sight Helmet) helmet-mounted sight, which provides both crew members with a quick, accurate and simple targeting mechanism. DASH gives a near-duplicate of the information contained within the HUD, as well as radar parameters. F-15Is can also use the Rafael Python 4, a fourth-generation, high off-boresight, IR-guided missile which was developed in the early 1990s and is intended for use with DASH. It can be cued by simply aligning the target with a reticle inside the helmet visor, and then handing it off to the missile seeker. With the full-scale introduction to service of the AIM-9X to US F-15C/Ds still some way off, this capability makes both IAF F-15 marques the most potent dogfighters in the Eagle family.

Procurement of the Ra'am provided the impetus for a fleetwide modification of the F-15A/B/C/D Baz (buzzard) force, a programme which began in the mid-1990s and resulted in a 'prototype' rolling out in November 1998. Known as Baz Meshopar (Improved Baz), the programme incorporates many F-15I avionics items, such as cockpit displays, HOTAS, data transfer equipment and GPS. Modified aircraft can carry the Python 4 and are DASH-compatible. Upgraded two-seaters are fully capable of launching precision-guided munitions.

F-15S – Saudi Strike Eagle

In the wake of Operation Desert Storm Saudi Arabia looked to swell its air force, initially supplementing its fleet of Tornado IDSs. The Royal Air Force had already demonstrated in the conflict that updating the Tornado with a FLIR and laser designator system was a sufficient modification to allow good medium-level PGM strike capability. Even so, without an extensive and costly refit the Tornado would soon be seen as outclassed. Consequently, on 1 October

1992 the US government announced that it would sell Saudi Arabia 72 F-15XPs (along with associated support contracts and equipment) as an extension of the Peace Sun FMS programme which had already seen the delivery of 74 F-15C/Ds (plus a Gulf War top-up batch of 24). The 'F-15XP', which was subsequently named the F-15S, was a revised F-15E Strike Eagle with a somewhat degraded strike capability.

Saudi Arabia's F-15S fleet was supplied with 48 sets of AAQ-19 Sharpshooter and AAQ-20 Pathfinder pods. The AAQ-20 Pathfinder is a simplified AAQ-13 LANTIRN Nav Pod, and is less effective in ECM environments as it lacks an Electronic Counter-Counter Measures (ECCM) mode. The AAQ-20 is roughly the same as the AAQ-14, although is not compatible with the AGM-65 Maverick air-to-surface missile and cannot hand-off a target to the missile seeker head as it lacks a missile boresight correlator. Some of the air-to-air features of the AAQ-20 are also deleted, although the exact changes are unknown as the air-to-air functions in the AAQ-14 remain classified.

The APG-70S supplied to the RSAF features only 60 percent of radar bandwidth found in the US APG-70 and is limited to 16 channels rather than the 32 of the original radar set. A limited resolution of 4.7 nm in HRM mode was provided although, as with the APG-70I, the US later upgraded this to the 0.67-nm maximum. To further reduce the F-15S's capability, Boeing deleted the nuclear armament circuitry (as it did in the F-15I); altered the AWG-27 Programmable Armament Control Set to prevent carriage of certain stores (possibly including CBUs); and deleted hands-off automatic terrain-following from the ASW-51 autopilot. The aircraft was delivered with the same Ring Laser Gyro INS found in US F-15s, and Saudi Arabia has since added a commercial-grade GPS system to improve the F-15S's navigation and attack accuracy.

TEWS ECM capabilities were de-tuned to provide the jet with good protection against other threats in the Persian Gulf, but to limit their capability against US aircraft. The ALQ-128 Electronic Warfare Warning System was initially deleted altogether, although some F-15Ss have been seen with an EWWS-type antenna on the left vertical stabiliser, leading to speculation that the system has been acquired by the RSAF in recent years. An NVG-compatible, flat-panel Liquid Crystal Display Up Front Controller was added to both cockpits, and this is identical to the system provided on the F-15I.

It is obvious that the extent to which the F-15S was degraded was a matter of delicately balancing the valid needs of the RSAF with the concerns of the Israeli government, which had baulked at the announcement of the F-15S sale. The limited number of LANTIRN pods reduced the scale to which offensive F-15S operations could be conducted although, contrary to popular reports, the Saudis have never designated the remainder of their F-15S fleet as interceptors – their F-15C/Ds are perfectly capable in this arena.

Powered by F100-PW-229 IPEs, the S models also initially lacked BRU-46 and BRU-47 hardpoints on their CFTs (omitted in order to further reduce offensive capabil-

All Israeli fighter Eagles, like this ex-USAF F-15B, are being brought up to Baz Meshopar (Improved) standard, with revised avionics and cockpit systems. Improved aircraft can also carry the Python 4 missile (carried here on the outboard wing pylon launch rail). Unlike the Python 3, which has larger wings, the Python 4 can also be carried on the inboard launch rails when CFTs are fitted.

Below left: Reflecting its primary attack role, Israel's F-15I fleet has three-tone top-surface camouflage applied. Israeli Eagle crews, whether flying fighter or attack aircraft, routinely use the DASH helmet-mounted sight. This is particularly useful for targeting the high off-boresight Python 4. Unlike the F-15S, the F-15I was fitted with a full-spec LANTIRN system, although the APG-70 was initially downgraded for export.

Below: 106 Tayeset is the main operating unit of the F-15C/D in IDF/AF service, while 133 Tayeset has the older F-15A/Bs. Another unit, No. 148, has F-15C/Ds and is thought to be staffed by reservists. This F-15D was participating in a 2002 Red Flag exercise at Nellis, a rare foray outside of Israel for an IDF/AF Eagle. Note the empty shackles for the AIM-7 Sparrow missiles.

Right: Eager to find ways of keeping the F-15 production line open, McDonnell Douglas sales staff targeted Saudi Arabia as a likely candidate for Strike Eagle purchases following the end of Operation Desert Storm. The RSAF was, of course, already an F-15C/D operator, and it had witnessed at first hand the performance of the USAF F-15Es during the war with Iraq. However, arms sales to Saudi Arabia had never been straightforward due to the power of the Israeli lobby in the US Congress. In late 1991 MDC proposed the sale of 24 F-15Fs – a single-seat downgraded derivative of the F-15E. Congress blocked any sale of this variant. Another subsequent proposal of a downgraded two-seat variant, known as the F-15H, was also blocked. Finally, on 10 May 1993, and in the knowledge that Israel was strongly tipped to buy a similar aircraft, Congress approved the 72-aircraft sale of of a third downgraded variant (F-15XP), which subsequently became the F-15S we know today.

ity and to placate Israel's mounting irritation), although tangential carriage CFTs (which incorporated the BRUs) were subsequently delivered between 1998 and 2000. Until then, the RSAF had used unwieldy Multiple Ejection Racks on the wing and centre-line pylons as an interim solution.

Khamis Mushayt AB is home to the RSAF Strike Eagles, and they are operated by 55 and 92 Squadrons. Unlike the IDF/AF, which conducts its own training and occasionally sends aircrew to USAF F-15E squadrons, the RSAF has received regular exchange pilots and WSOs from the USAF for training purposes.

F-15K

South Korea awarded its biggest ever (approximately $4 billion) contract to Boeing on 19 April 2002 when it declared that it would purchase 40 F-15K Strike Eagles. The General Electric F110-GE-129 IPE has been selected to power the Strike Eagle, making it the first F-15 ever to use GE engines operationally. Seoul expects deliveries to span over four years, with the first examples being handed over in 2005.

The F-15E offers a better range/payload solution than the F/A-18, which Boeing also entered into the 1997 F-X competition, and with three fuel tanks can carry 12 JDAMs (Joint Direct Attack Munition – GPS-guided GP bombs) over a thousand miles. The APG-63(V)1 – an upgraded APG-70 with Sea Surface Search, Sea Surface Track and Ground Moving Target Indicator modes – has been purchased, although the US has extended an option for the Republic of Korea Air Force (ROKAF) to operate the APG-63(V)2 AESA radar. New MPDs provide the F-15K with full Joint Helmet Mounted Cueing System (JHMCS) compatibility, while a new PACS, additional chaff and flare dispensers, NVG-compatible cockpit lighting and Advanced Display Core Processor (which replaces the CC and MPD Processor), are also added.

A MIL-STD 1760 data bus will equip the CFTs and wing hardpoints, thus accommodating GPS-guided weapons as per the Suite 4E+ upgrade slated for USAF F-15Es in late 2002. The F-15K is the first Strike Eagle to be mated with the AGM-84D Harpoon and AGM-84E Stand-Off Land Attack Missile, both of which will be used to counter an aggressive North Korean Naval contingent. The newest generation AIM-9X has also been purchased, and this integrates with JHMCS to provide high-off boresight engagements of adversary aircraft closing to within visual range. The AIM-120C AMRAAM missile will provide a first-shoot capability against the Su-27s and Su-30s of the Chinese People' Army Air Force (PAAF) at ranges beyond visual acuity, and would also prove dominant in any engagement against Japanese Air Self Defence Force F-15J aircraft. The AGM-88B HARM anti-radiation missile has been purchased to satisfy a gap in the ROKAF's ability to actively persecute hostile threat emitters. While cleared for carriage by the F-15E, it is not immediately apparent how this weapon will receive target data from the launch aircraft.

The F-15K's TEWS suite will be the most advanced of any Foreign Military Sale F-15, and for the first time ever, an export F-15 will feature some form of NCTR. An improved, microwave power module-equipped ALQ-135M ICMS was developed for increased jamming effectiveness and improved maintainability, and the ALR-56M Radar Warning Receiver will also debut on the F-15K as an alternative to the troubled ALR-56C. The F-15K will receive LANTIRN pods under a $163.7 million contract with Lockheed Martin.

Called Tiger Eyes, the new LANTIRN suite features a mid-wave staring array Forward Looking Infra-red (FLIR) navigation pod with terrain-following radar, and a mid-wave staring array FLIR targeting pod with a 40,000-ft laser and Charge Couple Device (CCD) TV, and a long-range IRST system. The same flat-panel, NVG compatible Up Front Controller found in the F-15I and F-15S will be installed, as will an ARC-232 radio for both UHF and VHF compatibility. Finally, Fighter Data Link will add unprece-

The F-15S was downgraded in several key areas, although the US subsequently relented on radar patch-map capability and conformal weapon carriage, which were restored to full F-15E levels. The nav/targeting pod system is the Pathfinder/Sharpshooter system, which utilises the same pod bodies as the LANTIRN, but has downgraded capabilities. This No. 92 Sqn aircraft is heavily laden with an 'airshow' loadout of retarded Mk 82 500-lb bombs, Sidewinders and Sparrows, with additional weapon options in front. These include GBU-10 and GBU-12 Paveway II LGBs, Mk 84 LDGP and a CBU.

As the bedrock of Marine offensive muscle, the F/A-18 figures prominently in each WTI exercise. Above is an F/A-18D of VMFA(AW)-121, while left is an F/A-18A from Beaufort-based VMFA-312. The red wingtip instrumentation pod allows the aircraft to be tracked during air combat over the Yuma range complex.

Special Category Squadron of the Year, and it also received the Meritorious Unit Citation in 1988, 1990 and 1995.

WTI course

All WTI courses are hosted by MAWTS-1 personnel, and each class usually has approximately 175 students. Almost every Marine aviation unit has at least one student present. The six-week course teaches the students how to plan and execute all manner of aviation operations, integrating the six functions of Marine Aviation, which are:

- Offensive Air Support
- Anti-air Warfare
- Assault Support
- Aerial Reconnaissance
- Electronic Warfare
- Control of Missiles and Aircraft

Each course is broken down into two phases: academics (2.5 weeks) and the flight phase (3.5 weeks). Students receive over 150 hours of formal classroom instruction on weapons and tactics from MAWTS-1 instructors

Logistic Officers Course, the Rotary Wing Crew Chief and KC-130 Navigator, Loadmaster, Flight Engineer Weapons and Tactics Instructor Course, and the MACCS Enlisted Weapons and Tactics Courses. The advanced curriculum includes the Tactical Air Commanders Course, MEU/SPMAGTF ACE Commanders Course and the MAWTS-1 Commanders Course. MAWTS-1 personnel conduct a Mobile Training curriculum consisting of the MEU ACE Training Course, the MAGTF Aviation Integration Course and the Marine Division Tactics Course. MAWTS-1 also maintains close, mutually beneficial liaison with the aviation and tactics schools of the US Navy, Army, Air Force, and several allied nations.

Since its establishment, MAWTS-1 has been staffed by individuals of superior aeronautical and tactical expertise, instructional abilities and

professionalism. MAWTS-1 also rapidly responds to the needs of operational campaigns, and also assimilates tactical developments into its curriculum. Each MAWTS-1 instructor averages over 90 days every year providing support to FMF (Fleet Marine Force) units at their home bases or while deployed, including certification, standardisation and supplementary ground and airborne instruction for every aviation squadron in the Corps.

The WTI Course is recognised as the most comprehensive graduate-level aviation course of instruction in the world today. It is through the dedication and untiring efforts of its personnel that MAWTS-1 has earned its reputation as the vanguard of Marine aviation. MAWTS-1 was awarded the Navy Unit Citation for the period 1982-85. In 1986 and 1995, MAWTS-1 was named Marine Corps Aviation Association

MAWTS-1 history

The origin of MAWTS-1 may be traced to the aftermath of World War II, when Marine pilots were first assigned to Navy Composite Squadrons (VCs). These squadrons, operating from shore bases and carriers, were assigned the special (nuclear) weapons delivery mission. In 1952, with the introduction of improved weapons and aircraft, the Marines were reassigned to Marine Special Weapons Delivery Units (SWDUs) on each coast. These units were short-lived because their mission was given to attack squadrons in 1953. Special Weapons Training Units (SWTUs) were then formed to provide necessary training for the attack squadrons. During the 1960s, conventional weapons delivery was added to the curriculum of the SWTUs, but the emphasis on special weapons delivery remained.

In response to their growing mission, the size of the SWTUs was increased, and they were redesignated as Marine Air Weapons Training Units – MAWTULant at Cherry Point, North Carolina, and MAWTUPac at El Toro, California. These units continued to expand through the 1960s, both in terms of assigned aircraft types and in the development of new curricula.

In 1975, a study group was formed at Headquarters Marine Corps to determine requirements for the enhancement and standardisation of aviation training. A series of recommendations, labelled as numbered projects, were made to the head of Marine Aviation and to CMC (Commandant of the Marine Corps) in early 1976. Project 19 recommended establishment of the Weapons and Tactics Training Program (WTTP) for all of Marine Aviation. The cornerstone of the WTTP was the development of a graduate-level Weapons and Tactics Instructor (WTI) Course and the placement of WTI graduates in training billets in every tactical unit within Marine Corps aviation. In late 1976 and early 1977, separate WTI Courses were conducted by MAWTUPac and MAWTULant. Consolidated WTI Courses were subsequently conducted at Marine Corps Air Station Yuma, Arizona, by a combined MAWTU staff in May 1977 and February 1978.

Due to the overwhelming success of the consolidated WTI courses, the Commandant of the Marine Corps commissioned Marine Aviation Weapons and Tactics Squadron One at MCAS Yuma, Arizona, on 1 June 1978.

Although the replacement of the AV-8B by the F-35C is seen as a pressing need for the USMC, the Harrier is still a capable attacker. This AV-8B II+ of Yuma-based VMA-311 carries a Litening targeting pod.

during the academic phase. They study areas such as threat analysis, tactics, weapons systems employment, aviation training management and integrated operational planning. The academic programme routinely features guest speakers with real-life experiences.

Classroom lectures begin with the 'big picture', covering the six functions of Marine Corps aviation. From there the phase is 'necked down' to specific communities and aircraft-type training.

Flying phase

The flight phase reinforces academic objectives with hands-on experience. The last week of the flight phase is the final exercise (FINEX), during which students plan and carry out a fully integrated combined arms operation. Yuma is the central control site for WTI, co-ordinating operations that include more than 2,500 support personnel throughout the Marine Corps, and around 70 US and Allied combat aircraft. The programme stresses integrated training, includes all facets of aviation and other supporting arms while working within the Marine Corps Command and Control System in support of a Marine Air Ground Task Force (MAGTF) scheme of manoeuvre.

Above: VMFT-401 and its Northrop F-5E/Fs, conveniently based at Yuma, provides the core of the aerial opposition during WTI, although 'guest' adversaries are drawn in, often from other services.

Below: An F/A-18D lands after a WTI sortie. Close to the Mexican border on the banks of the Colorado, Yuma offers superb flying weather year-round. The ranges lie to the north and east of the base.

WTI offers the chance for live and dummy weapon drops. This VMFA-312 'Checkerboards' F/A-18A carries an LGTR (Laser-Guided Training Round), used to simulate a Paveway-series LGB.

During WTI 2-02, held in March/April 2002, there were EA-6B Prowlers, AV-8B Harriers, KC-130F/T Hercules, F/A-18A/D Hornets, AH-1W SuperCobras, UH-1N Hueys, CH-46E Sea Knights and CH-53D/E Stallions representing all four Marine Air Wings. The Air Force brought an E-3B Sentry and an E-8C JSTARS for command and control, while F-16s flew out of Luke AFB as aggressors, complementing VMFT-401's F-5E/Fs in this task. A-10As from Davis-Monthan AFB also participated, as did an EC-130H Compass Call and RC-135 Rivet Joint. The Navy participated with an E-2C Hawkeye and VFA-25 F/A-18Cs flying from Lemoore.

Right: Bell UH-1Ns serve the Marine Corps in many roles, including airborne forward air control and light attack. This HMLA-267 Twin Huey, hover-taxiing past a line of 'Whiskey' Cobras, is armed with door gun and seven-round rocket pods.

Above: Long gone are the days when the Marines would operate alone: almost any operation would involve joint-service participation. This is reflected in WTI, with aircraft and personnel from other services being integrated into the exercise – especially those with reconnaissance and command functions. This is an E-8C from the 93rd ACW's 12th ACCS.

A complete command and control system is operational throughout the Yuma Training Range Complex during WTI. Both inert and live ordnance is used, and an array of different weapons is employed. Ground combat, combat support and combat service support officers are also in attendance. Upon graduation the students become Weapons Tactics Instructors, returning to their squadrons to serve as warfare instructors and planners for their commands. The exercise also includes personnel from Command, Control and Communications, radio battalions, AGS (Air-Ground Support), Intelligence and 'Smokey Sam' teams, the latter being used to emulate SAM launches, providing the aircrews with a more realistic battlefield threat array.

Major James Reed, with 2,700 hours in F/A-18 Hornets, is the MAWTS-1 Operations Officer, a billet he has held for two years. He has served with VMFA-312, VMFA-235, and

Live Mk 83 1,000-lb (454-kg) bombs are released over the Yuma ranges. As well as its WTI commitments, MAWTS-1 acts as the tactics and evaluation organisation for Marine aviation

VMFA(AW)-242, in addition to MAWTS-1. "There is a lot more that goes on here than I ever dreamed of. At the beginning, students and instructors will train with their own communities. As time goes on, they then begin working with other communities and integrate into various larger operations. The culmination of the WTI course is FINEX, where everyone is involved in one large operation. We have many ranges here and have a lot of range time reserved.

"We ensure that everyone does things in a uniform manner – that way all the fleet squadrons are consistent. That is what standardised training is all about. It is an excellent course for Marine Corps aviation, and we have

This F/A-18D carries a typical asymmetric load of three Mk 83s and two seven-round pods for 2.75-in (70-mm) rockets. The latter are mainly used for marking targets in the FAC(A) role.

For all-weather precision attack the F/A-18D is the vehicle of choice for the Marines. The three-screen 'missionised' rear cockpit allows precise navigation and targeting in all weathers. The D-model also provides tactical reconnaissance when fitted with the ATARS sensor suite.

students that come from all of the other services. They see the value in the course and it helps them work within contemporary joint operations worldwide. The MAWTS-1 personnel are the 'cream of the crop' instructors. They do such a good job, it makes my job easy – and working with top-notch people makes it very enjoyable here."

The commander's view

MAWTS-1 Commanding Officer is Colonel Marty Post. He flew A-6s for many years, now flies F/A-18s, and has 4,300 hours of time in various aircraft. He has been both the XO and CO of Fleet Marine Force (FMF) squadrons, and has been the MAWTS-1 CO for the past two years. "WTIs have been happening since 1978. This particular WTI we have been doing some new things – as we do in every class. We are taking a look at GPS and JDAM CAS procedures

A night attack AV-8B of VMA-513 refuels from a KC-130, clutching an inert Mk 83 and AIM-9M under its wing and practice bomb carrier under the centreline.

– a result of lessons learned from the employment of the 'J'-weapons in Afghanistan.

"We are also evaluating the next-generation .50-calibre machine-gun used on rotary-winged aircraft: the M3M. In particular, we are validating it on the CH-46E and CH-53D/E. It has a superior rate of fire, up from about 700-800 rounds per minute on our older guns to 1,100 rounds per minute. We have been making refinements with how it mounts in the windows, and we may even adapt the weapon to the ground-side.

"The people from the Warfighting Lab and Europe are here to help out with the development of it, too. With the M3M, the 100-round cans are replaced with a 600-round integrated can with a chute. This is a big time saver. The fleet crew chiefs have an opportunity to use and critique the system, and by the time it enters production – it will be a proven design.

"We continue to refine procedures and are taking a look at the lessons learned from OEF (Operation Enduring Freedom). In fact, some of our MAWTS-1 instructors flew with the units over there. The procedures we have developed here over the years were used daily in OEF. In OEF, the helicopters ran into severe brown-out conditions and had to operate in high altitudes. In the fall we may do a high-altitude long-range

assault mission to simulate similar conditions.

"We are also looking at a new IR marker for the UH-1Ns and AH-1Ws, as well as a new FAC(A) system. We are also exploring a new battery-operated krill light to replace IR chem lights. They can be turned off and last for days. We have many guest speakers, and one who was memorable was a Special Forces MSGT controller who was one of the first to go into Afghanistan. He directed airpower to targets and directed over 850 JDAM drops alone. He talked about his equipment, different techniques, directing different types of aircraft, what worked and what didn't work.

"Yesterday one of our scenarios here was to set up 15-20 various Soviet-style vehicles around the Twentynine Palms ranges. We sent F/A-18s, using an armed reconnaissance method, to find and destroy the vehicles. We

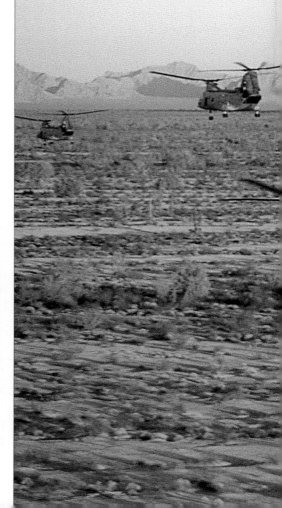

had a JSTARS on station and they would pass along the information to the Hornets. The Hornets then would locate and engage the targets. This was a great exercise and was pertinent to the way we did business in Afghanistan. We also used the 'skids' [AH-1Ws and UH-1Ns] to escort LAVs and LARs – fly slightly ahead and ensure the area is clear. The 'skids' also give the ground troops instant on-call CAS if needed.

"We strive to integrate joint assets in WTIs and use platforms such as AWACS, JSTARS, NRO and U-2 imagery, Patriot missile batteries, and more. We teach the students to understand the capabilities of the joint systems and how we interface with other services. When you have 70 aircraft and 2,500 Marines, the amount of logistical support is intense. The fuel, water, food, FARPS – this is all part of the training here. The ranges all around here are excellent for what we do.

"Our people have excellent communications with the fleet units, and always explore the latest and greatest gadgets and ideas. We adjust the curriculum often to reflect contemporary systems and methodology. All the people working for me here have been hand-picked by their various communities, and I get the best of the best. Having such quality people makes my job easy – they are always looking for a better way of doing something and they are proactive. It is very easy for me to come to work every day!"

Ted Carlson

At the core of Marine aviation is the helicopter assault – vertical envelopment – for which the CH-46E is the prime vehicle. During WTI CH-46 crews practise advanced assault techniques, often at night and usually at low level, while the facilities on the ranges allow crew chiefs to undertake live gun firing at night (above). The gaggle of 'Frogs' setting out on a dusk insertion exercise are from a mixed bag of squadrons, including HMM-364 'Purple Foxes' (foreground, right) and HMM-163 'Ridgerunners' (foreground, below).

Air Mobility Command

Headquartered at Scott AFB, Illinois, Air Mobility Command (AMC) was established on 1 June 1992 and combined the airlift and combat search and rescue assets previously assigned to Military Airlift Command (MAC), with many of the tanker aircraft operated by Strategic Air Command (SAC).

Air Mobility Command, United States Air Force

AMC's structure has undergone several major revisions in the 10 years since the command was created. Its combat rescue forces were transferred to Air Combat Command (ACC) control on 1 February 1993, and just two months later, on 1 April, the operational support airlift fleet underwent a reorganisation that transferred many of its aircraft to individual user commands, including ACC, Air Education Training Command (AETC), Air Force Materiel Command (AFMC) and Air Force Space Command (AFSPC). The most significant change, however, took place on 1 October 1993, when the tactical (intra-theatre) airlift assets were transferred to Air Combat Command (ACC). With the exception of a single wing that was based in Panama, all of these units were based within the CONUS and were equipped with C-130s.

In exchange, AMC gained control of the majority of ACC's remaining tanker aircraft. This structure remained in place until 1 April 1997, when the intra-theatre airlift units were returned to AMC control. The command finally gained control of the last of the tankers in June 2002, when ACC reorganised its expeditionary 366th Wing. Type conversion training for AMC aircrew has been conducted by AETC since that command was created on 1 July 1993, a move which entailed several airlift training units leaving AMC control.

As the Air Force component of the USTRANSCOM, AMC is the single manager for air mobility and is charged with providing inter- and intra-theatre airlift, air refuelling, special air missions, operational support and aeromedical evacuation for US forces. Its units fly nearly 30,000 flying hours each month. It also provides forces to Central Command (USCENT-COM), European Command (USEUCOM), Pacific Command (USPACOM), Southern Command (USSOUTHCOM) and the newly established US Northern Command (USNORTHCOM), in support of peacetime and wartime tasking. It also provides humanitarian support around the world.

Today, Air Mobility Command assets and personnel are forward-deployed to a number of locations in support of these Unified Commanders and the Expeditionary Aerospace Force (EAF). Examples include Operations Northern and Southern Watch, and Enduring Freedom (USCENTCOM, USEUCOM and USNORTHCOM) and Operation Coronet Oak (USSOUTHCOM).

Air Mobility Command

Reporting directly to AMC are the Air Mobility Warfare Center (AMWC) and the Tanker Airlift Control Center (TACC). The AMWC, which was activated near McGuire AFB at Fort Dix, New Jersey, on 1 May 1994, consolidated the functions of seven geographically separated units. The centre's seven components comprise the Operations, Education and Resources directorates, the 33rd Flight Test Squadron (FLTS), 421st Ground Combat Readiness Squadron (GCRS), Combat Aerial Delivery School (CADS) and the AMC Battle Lab. With the exception of the GRCS and the CADS, which are located at Little Rock AFB, Arkansas, all are headquartered at Fort Dix/McGuire AFB, which serves as the primary air mobility hub for the US east coast.

In addition to providing advanced training in areas that pertain to mobility, the AMWC is

Right: AMC's 'flagships' are the two VC-25As used by the Presidential Airlift Squadron, one of which is seen here over Mount Rushmore. When carrying the President, 'Air Force One' is supported by a variety of other aircraft, including the little-seen Gulfstream C-20C which is a 'war-readiness' platform.

Left: The C-17A Globemaster III is well advanced along the path to replacing the C-141, with two active-duty wings completely re-equipped (62nd AW on the west coast and 437th AW in the east). Here Spirit of Berlin from the 437th Airlift Wing disgorges paratroops during an exercise.

Direct Reporting Units

Air Mobility Warfare Center – Fort Dix, New Jersey

33rd FLTS	McGuire AFB, New Jersey
	no aircraft assigned
Det. 1	Charleston AFB, South Carolina
Det. 2	Keesler AFB, Mississippi
	no aircraft assigned
421st GCRS	Little Rock AFB, Arkansas
	no aircraft assigned
CADS	Little Rock AFB, Arkansas
	no aircraft assigned

Tactical Airlift Control Center – Scott AFB, Illinois
no aircraft assigned

Air Mobility Command

A trickle of new-generation CC-130J Hercules IIs is reaching ANG and AFRC units. This aircraft serves with the California Guard's 115th AS at Channel Islands ANGS, the first unit in 15th AF to get the 'J'.

Having been assigned to Air Combat Command until 1997, the tactical C-130 fleet is back with AMC, comprising three active-duty wings and numerous ANG/AFRC squadrons. This C-130E is assigned to the 43rd Airlift Wing at Pope AFB.

responsible for the evaluation, development and exploitation of doctrine, technology, defensive systems and tactics to support the development of air mobility, advances in force mobility and the development of combat delivery systems. It is also tasked with conducting operational test and evaluation organisation for air mobility-unique weapons systems, subsystems and mission equipment.

The Tanker Airlift Control Center (TACC), located at Scott AFB, Illinois, which became operational on 1 April 1992, is AMC's command's hub for planning and directing tanker and transport aircraft operations around the world. Created to centralise command and control responsibilities that were previously

located within numbered air forces and airlift divisions, the centre schedules and tracks strategic tanker and airlift resources worldwide.

The command's two numbered air forces comprise the Fifteenth Air Force (15th AF) and the Twenty-First Air Force (21st AF), which are both made up of gained Air National Guard (ANG) and active component airlift, air mobility and air refuelling groups and wings. In addition, AMC is supported by two numbered air forces that are components of the Air Force Reserve Command (AFRC). These comprise the Fourth Air Force and the Twenty-Second Air Force.

AMC has approximately 51,500 active-duty personnel and employs more than 8,000 civilians. In addition, more than 85,000 airmen from the ANG and AFRC come under its control. More than 750 strategic and tactical airlifters, and 500 tankers, are assigned to AMC, and more than 50 percent of the fixed-wing fleet is

operated by the two reserve components. The command also operates 15 helicopters that support VIP operations around the District of Columbia.

Fifteenth Air Force's main area of operations covers the region west of the Mississippi River as far as the east coast of Africa, from the North Pole to South Pole. It is composed of seven airlift, air mobility and air refuelling wings and groups, and a further 18 are gained from the Air National Guard. The command is responsible for all of AMC's operational support airlift (OSA) aircraft. Mission support for the 15th AF is provided by two Air Mobility Operations Groups (AMOG).

The Twenty-First Air Force is composed of a single air refuelling group, and seven airlift and air mobility wings and groups, including one dedicated to special air missions in support of the US Government. A further 26 wings are gained from the ANG. Two Air Mobility

Air Mobility Command bases – active-duty and reservist units

Air base key
- ⬡ Active-duty
- ★ Active-duty (lodger)
- ▲ Air National Guard
- ▼ Air Force Reserve

Fourth Air Force, AFRC

HQ: March ARB, California

Fourth Air Force units gained by AMC	
349th AMW (349th OG)*	**Travis AFB, California**
70th ARS	KC-10A
79th ARS	KC-10A
301st AS	C-5A/B/C
312th AS	C-5A/B/C
433rd AW (433rd OG)	**Kelly Field/Lackland AFB, Texas**
68th AS	C-5A
434th ARW (434th OG)	**Grissom ARB, Indiana**
72nd ARS	KC-135R
74th ARS	KC-135R
445th AW (445th OG)	**Wright Patterson AFB, Ohio**
89th AS 'Rhinos'	C-141C
356th AS	C-141C
446th AW (446th OG)*	**McChord AFB, Washington**
97th AS	C-17A
313th AS	C-17A
728th AS 'Flying Knights'	C-17A
452nd AMW (452nd OG)	**March ARB, California**
336th ARS 'Rats'	KC-135R
729th AS 'Pegasus'	C-141C
730th AS 'First Associate Reserve'	C-141C
507th ARW (507th OG)	**Tinker AFB, Oklahoma**
465th ARS 'Okies'	KC-135R
931st ARG/18th ARS 'Kanza'**	KC-135R (at McConnell AFB, Kansas)
916th ARW (916th OG)	**Seymour Johnson AFB, North Carolina**
77th ARS 'Totin Tigers'	KC-135R
927th ARW (927th OG)	**Selfridge ANGB, Michigan**
63rd ARS 'Flying Jennies'	KC-135E
932nd AW (932nd OG)	**Scott AFB, Illinois**
73rd AS	C-9A
940th ARW (940th OG)	**Beale AFB, California**
314th ARS 'Warhawks'	KC-135E

Associate unit shares aircraft with co-located active component wing

Two squadrons of KC-10As on each coast provide AMC with a flexible and capacious tanker/transport. The type can support USN/USMC aircraft (VMFA-314, right), some KC-10s having triple-point drogue tanking capability (above).

Below: KC-10 tankers played a vital part in Enduring Freedom, which involved combat aircraft having to undertake long transits to the war zone. This pair is seen operating from Diego Garcia, in the British Indian Ocean Territory.

Powered by TF33 engines, KC-135Es serve with 11 ANG and two AFRC squadrons, the latter including the 314th Air Refueling Squadron, 940th ARW, shown here.

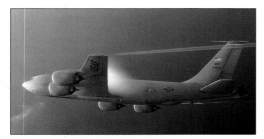

F108-powered KC-135Rs form the bulk of the tanker fleet, 14 active-duty squadrons being augmented by 10 from the ANG and six from AFRC. Left is an aircraft of the 6th AMW at MacDill, while below is an aircraft from the 92nd ARW at Fairchild.

Twenty-Second Air Force, AFRC

HQ: Dobbins ARB, Georgia

Twenty-Second Air Force units gained by AMC	
94th AW (94th OG)	**Dobbins ARB, Georgia**
700th AS	C-130H
302nd AW (302nd OG)	**Peterson AFB, Colorado**
731st AS	C-130H
315th AW (315th OG)*	**Charleston AFB, South Carolina**
300th AS	C-17A
317th AS 'First in Reserve'	C-17A
701st AS 'Turtles'	C-17A
403rd Wing (403rd OG)	**Keesler AFB, Mississippi**
53rd WRS 'Hurricane Hunters'	WC-130H/J
815th AS 'Jennies'	C-130J
439th AW (439th OG)	**Westover ARB, Massachusetts**
337th AS	C-5A
440th AW (440th OG)	**General Mitchell IAP/ARS, Milwaukee, Wisconsin**
95th AS 'Flying Badgers'	C-130H
459th AW (459th OG)	**Andrews AFB, Maryland**
756th AS	C-141C
512th AW (512th OG)*	**Dover AFB, Delaware**
326th AS 'Flying Bunnies'	C-5A/B
709th AS	C-5A/B
514th AMW (514th OG)*	**McGuire AFB, New Jersey**
76th ARS 'Freedom's Spirit'	KC-10A
78th ARS	KC-10A
732nd AS 'Rams'	C-141B
908th AW (908th OG)	**Maxwell AFB, Alabama**
357th AS 'Deliverance'	C-130H
910th AW (910th OG)	**Youngstown/Warren RAP/ARS, Ohio**
757th AS 'Blue Tigers'	C-130H
773rd AS 'Quiet Professionals'	C-130H
911th AW (911th OG)	**Greater Pittsburgh IAP/ARS, Pennsylvania**
758th AS	C-130H
913th AW (913th OG)	**NAS Willow Grove JRB, Pennsylvania**
327th AS	C-130E
914th AW (914th OG)	**Niagara Falls IAP/ARS, New York**
328th AS	C-130H
934th AW (934th OG)	**Minneapolis-St Paul IAP/ARS, Minnesota**
96th AS 'Flying Vikings'	C-130E

Associate unit shares aircraft with co-located active component wing

A New York ANG (109th AS) speciality is polar support, including airlift for the large US Antarctic research effort (Operation Deep Freeze). Ski-equipped LC-130Hs are in use for this mission.

Based at Andrews AFB, the 201st Airlift Squadron, DC ANG, provides staff transport on behalf of the ANG. For several years it has operated three Boeing C-22Bs, but these are to be retired in 2002/03, following the receipt of Boeing C-40Bs (BBJs).

Two IAI Astra SPX executive jets were purchased for use by the 201st AS in the VIP transport and medevac roles. They were outfitted to reconfigurable C-38A standard by Tracor and were delivered in April 1998.

Below: Once the workhorse of the airlift effort, the C-141C StarLifter is reaching the end of its useful life. Only eight squadrons remain, of which seven are ANG/AFRC units. The last active-duty unit is the 6th Airlift Squadron 'Bully Beef Express' at McGuire AFB.

Air Mobility Command – tanker and airlifter tail markings

Active-duty units

Unit	Marking
6th AMW	'MacDill' in white on blue band with lightning bolts
19th ARG	'Black Knights' in black on yellow band
22nd ARW	'Keeper of the Plains' in black on yellow band
43rd AW	'Pope' in white on green/blue band
60th AMW	'Travis' in white on red or blue band, or in black on yellow band
62nd AW	'McChord' in white on green band with silhouette of Cascade mountains
92nd ARW	'Fairchild' in white on black, blue, green or yellow band
305th AMW	'McGuire' in gold with P-38 on blue band
317th AG	'Dyess' in white on stylised Texas flag or a red or blue band
319th ARW	'Grand Forks' in yellow on blue band, black on yellow band, white on red band, or blue on white band
436th AW	'Dover' in black on yellow/blue band
437th AW	'Charleston' in black on a yellow/blue band with crescent moon and palmetto tree
463rd AW	'The Rock' in white on red band, or black on green band

Air Force Reserve Command units

Unit	Marking
94th AW	'Dobbins ARB' in black on a blue band
302nd AW	Colorado flag and Rocky Mountain skyline
433rd AW	Stylised Texas flag
434th ARW	'Grissom' in white on red or blue band
439th ARW	'Westover' and 'The Patriot Wing' in white on red band
440th AW	'Flying Badgers' in white on a red band
445th AW	'Wright Patterson' in red on gray band
452nd AMW	'March' in black on yellow band with prewar national insignia
459th AW	'Andrews' in black on yellow/black checkerboard band
507th ARW	'Tinker' in yellow on blue band
908th AW	'Maxwell' in white on blue band
910th AW	'Youngstown' in black on red, or in white on blue band
911th AW	'Pittsburgh' in black on yellow tail band
913th AW	'Willow Grove' in black on grey band
914th AW	'Niagara Falls' in gold on blue band
916th ARW	'First in Flight' in yellow on green band
927th ARW	'Selfridge' in white on a purple band
934th AW	'Flying Vikings' in white on purple band
940th ARW	'940 ARW' in black on red band with Eagle's head and red band

Air National Guard units

Unit	Marking
101st ARW	'Maine' in green on white band
105th AW	'Stewart' in white on blue band
107th ARW	'Niagara Falls' in white on light blue band
108th ARW	'New Jersey' in yellow on blue band
117th ARW	'Alabama' in red on white band
118th AW	'Nashville' in white on red band
121st ARW	'Ohio' in white on blue band
123rd AW	'Kentucky' in blue on white band
126th ARW	'Illinois' in black on white band
128th ARW	'Wisconsin' in red on white band
130th AW	'Charlie West' in yellow on blue or purple band
134th ARW	'Tennessee' in red on white band
135th AG	'Baltimore' in black or yellow
136th AW	'Texas' in white on blue band
137th AW	'Oklahoma City' in white on light blue band
139th AW	'St. Joseph' in black
141st ARW	'Washington' in white on green band
143rd AW	'Rhode Island' in white on maroon band
145th AW	'Charlotte' in black on light blue band
146th AW	'Channel Islands' in white on green band
151st ARW	'Utah' in white on blue band
152nd AW	'High Rollers' in red on white band
153rd AW	'Wyoming' in black on a yellow band
155th ARW	'Nebraska' in red on white band
156th AW	'Puerto Rico'
157th ARW	'New Hampshire' in white on blue band
161st ARW	'Arizona' in black and stylised state flag
163rd ARW	'California' in white on blue band with white stars
164th AW	'Memphis' in white on a red band
165th AW	'Savannah' in white on red band
166th AW	'The First State' on light blue band
167th AW	'Martinsburg' in white on red band
168th AW	'Alaska' in yellow on blue band
171st ARW	'Pennsylvania' in black on yellow band
172nd AW	'Mississippi' in yellow on a blue band
179th AW	'Mansfield'
182nd AW	'Peoria' in orange on blue band
184th ARW	'Kansas' in blue on white stylised Jayhawk
186th ARW	'Mississippi' in yellow on black band
189th AW	Arkansas Razorback on a red band
190th ARW	'Kansas' in yellow on blue band
191st AG	'Michigan' in yellow on black band

Ramenskoye
Russia's test centre

Ramenskoye airfield, better known by the name of the nearby city (Zhukovskiy) which grew up around it, has played host to virtually all of Russia's post-war aviation achievements, from the first jet flight, through vertical take-off, to spacecraft. This review details the facility, and looks at important aspects of its work over the years.

This view of Myasishchev's area of the LII dates from the early 1970s. Visible top right is the dumped M-50 'Bounder', which was later transferred to the Monino museum, while the broken up remains of the sole M-52 are visible top centre (see photo on p. 88). Other types visible include An-12, An-24, Il-18, Il-62, M-4, Mi-4, Mi-6, Mi-8, Mi-10, MiG-21, Tu-104, Tu-110 (four-engined Tu-104), Tu-134 and Yak-28.

Until the mid-1930s all of the more important Soviet aviation institutions were located in Moscow. The principal design work and aircraft ground tests were carried out at the TsAGI (Tsentralnyi Institut Aero- i Gidrodinamiki, Central Institute of Aero- and Hydrodynamics) located in Radio Street. Flight tests were undertaken at the Central Airport at Khodynskoye field, popularly known as Khodynka. By the beginning of the 1930s it was recognised that Khodynka's location was becoming a constraint on trials work. On the one hand, the city was getting larger and larger, and on the other the aviation centres needed larger laboratories and longer runways.

On 13 August 1933, the Labour and Defence Council accepted the proposal for the so-called 'new TsAGI' near the city of Ramenskoye, 45 km (28 miles) to the south-east of the centre of Moscow. The selected location near Otdykh railway station provided a good rail connection with Moscow, and a large area for the construction of laboratories and an airfield. As the Moscow River flows nearby, it would also be possible to build a pool for seaplane testing.

On 6 November 1935, the building of the first set of TsAGI's new wind tunnels began. At the same time as construction of the institute began, so did building of the adjacent settlement which, in 1939, was given the name Stakhanovo. By this time it had grown to a small town of 10,800 people. On 23 April 1947,

100 years after the birth of the 'father of Russian Aviation' – Nikolay Zhukovskiy – Stakhanovo was granted the status of a city and given the name Zhukovskiy. At present, 120,000 people live in the city, and most of them are involved to a greater or lesser extent in aviation-related activities. Despite the renaming of the town, the

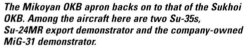

The Mikoyan OKB apron backs on to that of the Sukhoi OKB. Among the aircraft here are two Su-35s, Su-24MR export demonstrator and the company-owned MiG-31 demonstrator.

airfield is officially still called Ramenskoye, although more commonly referred to as Zhukovskiy. New prototypes spotted there by NATO in the 1970s and 1980s were allocated codenames in the 'Ram-' series.

Construction of the airfield began in 1935. When commissioned in June 1939 it had three runways (VPP, *vzlyotno-posadochnaya polosa*, take-off and landing strip): VPP-1 that was 1200 m (3,937 ft) long, VPP-2 that was 900 m (2,953 ft) long and VPP-3 that was 650 m (2,133 ft) long, corresponding to the requirements of the aircraft of the day. Together with the taxiways and aircraft parking lots, the total area covered with concrete amounted to 265000 m² (515 acres). The hydroplane pool was, however, never built.

By the autumn of 1939 the first upgrade had been undertaken, extending VPP-1 to 2000 m (6,562 ft) and installing a 1.2-m (4-ft) high 'sky-jump' mound at each end to assist heavy aircraft into the air. Other areas were extended so that, by the time of the German attack on the

Sukhoi's compound bristles with hardware, including a hump-backed Su-25T. At the bottom are two 'Flagons', an Su-15TM and Su-15UM. In the centre is the P-42, a stripped 'special' T10S (Su-27) used to set a string of time-to-height records between 1986 and 1988.

The full enormity of Ramenskoye's VPP-4 runway becomes apparent in this satellite photograph. VPP-4 is the world's longest concrete runway, although the lakebed runway at Edwards is longer. The US base has a concrete runway of a 'mere' 15,000-ft/4570-m. Partially obscured by the cloud shadow is Myasishchev's ramp, with a VM-T Atlant and an M-17 or M-55 visible.

USSR in June 1941, the LII airfield had 420000 m² (104 acres) under artificial coverage.

From an organisational sense the airfield was initially subordinated to the 8th Department of the TsAGI. On 8 March 1941, according to the decision of the government and the Central Committee of the VKP(b), part of the 'new TsAGI' – together with the airfield – was separated to form the LII (Lyotno-Issledovatelskiy Institut, Flight Research Institute). Test pilot Mikhail Gromov was nominated its commander, and today the LII bears his name. During the war years the airfield served as a base for two heavy bomber divisions with Il-4s and TB-7s, launching bombing raids against targets which included Berlin and Königsberg. The LII was transferred to Kazan and Novosibirsk away from the war zone, with only a subsidiary remaining at Stakhanovo.

Immediately after the war, at the dawn of the jet era, the airfield was extended again, absorbing the adjacent village of Novoye Selo. Runways VPP-2 and VPP-3 were neglected, while VPP-1 was extended by 250-300 m (820-984 ft) at each end, with the 'sky-jumps' deleted. In June 1950 the construction of a massive new runway (VPP-4) with a planned length of 4000 m (13,123 ft) was launched. The runways were also developed in depth to cater for growing aircraft weight. In 1949 the thickness of concrete cover of some segments of the VPP-1 achieved 30-32 cm (11.8-12.6 in), while the thickness of the new VPP-4 runway amounted to 40 cm (15.7 in).

VPP-4 was put into operation in stages, beginning with a 1200-m (3,937-ft) segment in October 1951, which was extended to 3700 m (12,139 ft) at the end of 1952. Finally, in 1959, the runway was extended to 5403 m (17,726 ft). This last extension was related to the initial tests of the Myasishchev M-50 'Bounder' supersonic strategic bomber. In 1980, when work on the Buran space shuttle was under way, a decision was made to reconstruct the main VPP-4 runway, so that the flying analogue of the Buran could be tested there. This operation

A regular sight at Zhukovskiy is the Il-80 'Maxdome' airborne command post, of which four fulfil a similar role to the US Navy's E-6. The Il-80s fly rarely, usually operating from nearby Chkalovskaya.

was finished in 1984. The length and the width (84 m/276 ft) remained unchanged but the surface was improved and thickness rose to 85 cm (33.5 in).

In 1984 another runway was built, a short strip for testing running take-offs by Yak-38 and Yak-41 VTOL fighters, covered with cast iron and steel plates. In 1986-1990 VPP-1 was rebuilt, changing its lateral section from concave to convex. After years of reconstruction and modernisation, Zhukovskiy's runways now boast an astonishing structure. For instance, 10 layers of concrete have been laid on VPP-1 and in its central parts it is 180 cm (5 ft 11 in) thick.

Today two runways are used at Ramenskoye, VPP-1 and VPP-4. There are no weight limitations and they can easily accommodate any aircraft in the world, including a space shuttle. The runways are aligned on magnetic 122° and 302°, while field elevation is 115 m (377 ft)

AMSL. The geographic coordinates are 55° 33' North and 38° 09' East. Altogether, artificial coverage of the Ramenskoye airfield (runways, aircraft parking lots and taxiways) has reached a staggering 2.5 million m² (618 acres). In spite of considerable financial difficulties, the airfield is maintained in good condition.

Operators

Ramenskoye airfield is used not only by the LII institute, but also houses the flight test bases of all Russian design bureaux except for Beriev, Mil and Kamov. It is used for initial tests of all new Russian aircraft. Other work has included evaluation of runway technology, such as snow- and ice-clearance (by blowers using jet engines) and metal FOD removal (by movable

Partnering the Il-80 are two Il-82s, which are Il-76 transports converted for the radio relay mission. The LII has its own fleet of specialised Il-76s in the form of the '976' test-tracking aircraft. Note also the Mi-28.

Above: Tupolev's 'Aircraft 73' was the three-engined forerunner of the Tu-14 (note the third engine intake mounted forward of the fin). In 73LL form it is seen here with an air-dropped test article.

Above and below: Partially built by Siebel before being captured at Halle by the advancing Russians, Germany's rocket-powered DFS 346 programme was restarted at Ramenskoye in October 1946. Flown by a German pilot, Wolfgang Ziese, it was first dropped (unpowered) from a captured Boeing B-29 in 1947.

electromagnet), as well as surface grooving processes.

Presently, Ramenskoye is also a civilian cargo airport and site for several Russian operators, mostly established by design bureaux and other aviation institutes, including Gromov LII, Elf Air, Tsentrospas MChS, Yak Service, Alros-Avia, Tupolev-Aerotrans, Ilavia, Sukhoi, Krylo, Antares-Air, Aervita, Remeks, MChS Rossii, Aviast, ASTAIR, Artuvera and Aerokoncept. A customs and immigration post has also been set up in the airport area. As an airport, Ramenskoye works at daytime only. Every two years in August Ramenskoye plays host to the MAKS airshow.

ATC and measurement systems

At the beginning of 1950s an OSP-48 radio instrument landing system was installed, which was later replaced with a more modern SP-50 system and complemented with a Globus radar system. In 1959-1960 they were subsequently replaced with, respectively, the RSBN-4N and Globus-2 systems. In 1987-1988 the Platsdarm-1N microwave instrument landing system, compatible with ICAO requirements, was installed. In 1957 the KDP-2 (*komandno-dispetcherskiy punkt*) control tower was put into operation. In 1980 a new building was constructed for KDP-2, familiar to all photographers working at the MAKS air shows. In 1972-1975 the Kasimov auxiliary control post was put into operation, and the Start automated air traffic management system was installed.

The main task of the LII institute is to carry out flight tests of experimental and prototype aircraft. This implies the need to have an extensive set of measuring instruments. Such a

Below: This remarkable scene on Myasishchev's ramp was recorded on 11 August 1971. The central subject is a standard 3MD 'Bison-C', but nothing is known of the 'Phantom' and 'Mirage III' which lie, heavily bagged, immediately behind it. They could be reconstructed aircraft captured in the Middle East, or replicas. At the time the Phantom and Mirage represented the peak of Western fighter capability, especially in the Middle East where both types were fielded by Israel. Of equal significance, though, is the aircraft in the background – the forlorn hulk of the sole Myasishchev M-52 – which remained unflown and is seen here, rudderless, awaiting the scrapman's torch.

Forward-swept wings were evaluated for some time after the war, spurred on by the capture of the Junkers Ju 287 second prototype. Most tests were made using the Tsybin series of aircraft (above, this is the LL-3 with 40° wing sweep), which were towed aloft by Tu-2s but had rockets for powered flight. At right is an aerodynamic drop-test vehicle for the current Sukhoi S-37 Berkut, about to be released from a helicopter.

system was created at the Ramenskoye airfield in 1946 when a set of German Askania theodolites, US anti-aircraft artillery radars and the first indigenous telemetric stations were installed. In the mid-1950s new instruments were introduced, including optical theodolites, Amur radars and RTS-8 radio telemetric stations. Since 1959 these instruments have been used to register aviation records at Zhukovskiy. In the 1960s the measured data began to be processed by computer.

From the 1970s, due to the growing flight range of aircraft and missiles under test and the requirement to collect telemetric information during the whole flight, LII ground measurement posts began to be installed outside

Zhukovskiy. Also, the first airborne control and data recording stations were installed in the cabin of the Il-18SIP (*samolyotnyi izmeritelnyi punkt*, airborne measurement post) aircraft. At the end of the 1980s five '976' aircraft were converted from military A-50 'Mainstay' airborne early warning and control aircraft (possibly only one is in current use). The radar of the '976' can track aerial vehicles at a distance of 600 km (373 miles) and space vehicles at 1000 km (621 miles). Six experimental vehicles can be tracked simultaneously. The '976' also carries a satellite-based communications and data-link system, which transfers data to ground posts.

Piotr Butowski

The Soviet Union's first VTOL craft was the Turbolet testbed, similar to the UK's 'Flying Bedstead'. The Turbolet was powered by a 24.53-kN (5,511-lb) RD-9BL from the MiG-19, and first flew in mid-1956 with Yuriy Garnayev at the controls.

Sukhoi T-4: Mach 3 bomber

One of the most remarkable aircraft to be seen at Ramenskoye was the T-4, an all-titanium bomber prototype. It flew only 10 times, and is seen in these ciné stills over the airfield, which is visible in the photo below. It was the world's first fly-by-wire aircraft to be designed as such from the outset.

Only one T-4 'Sotka' ('Hundred' – from its OKB project designation) was completed, first flying from Zhukovskiy on 22 August 1972. During its few flights it reached Mach 1.3, and showed considerable promise. However, its all-titanium structure and futuristic systems made it fantastically expensive, and it was dropped in March 1974 in favour of the cheaper and altogether more prosaic Tupolev Tu-22M 'Backfire'.

Ejection seat trials

The jet era brought with it many new pieces of equipment which had hitherto been unnecessary. With the airflow forces encountered at high speeds, the crew could not bale out in the traditional method, so the ejection seat became a vital piece of equipment. Most of the Russian ejection seats were tested by the LII institute, and the whole history began on 24 July 1947, when Gavriil Kondrashov ejected himself from a Pe-2 test-bed. This was the first ejection of a pilot from an aircraft in the USSR. On 16 January 1949, A. Bystrov became the first in the USSR to eject from a fighter aircraft (a MiG-9UTI flying at a speed of 764 km/h/475 mph). Ten years later P. Dolgov was the first to eject at over 800 km/h (497 mph), Yuriy Garnayev was the first at over 900 km/h (560 mph), Vasiliy Kochetkov the first at over 1000 km/h (620 mph) and, finally, E. Andreyev was the first Russian to eject at supersonic speed.

Improvements to ejection seats followed the processes of increasing the speed and altitude ranges for safe ejection, system automation, as well as the reduction of g-loads acting on the pilot. As well as performing air tests, the LII built several ground-based rigs for testing escape equipment: a 250-m (820-ft) sled track with carriages accelerated by rockets, a centrifuge and an aerodynamic stand for testing the influence of the air flow on the pilot.

In the 1950s and 1960s, each aircraft design bureau independently designed the ejection seats for its own aircraft. At Zhukovskiy these seats were tested using a variety of flying test beds, including the MiG-15, Tu-14, 3M, Yak-25M and Su-9U. A

The MiG-15UTI was a versatile seat testbed. This escape system uses the cockpit canopy to form a protective shield for the pilot.

Yak-25M (c/n 04-18) was used during 1960-1965 for testing the KM-1 (*Kreslo Mikoyana*, Mikoyan Seat) ejection seat for the MiG-21, MiG-23 and MiG-25 fighters, Sukhoi's KS-3 and KS-4 ejection seats for the Su-7 and Su-17, as well as Yakovlev's K-5 and K-7 ejection seats for the Yak-25/Yak-28. From 1967 tests of the KM-1 and KS-4 were continued with a two-seat Su-9U supersonic fighter (c/n 10-18). This aircraft was also used for testing Yakovlev's new KYa-1 ejection seat for the Yak-36 and Yak-38 VTOL fighters. Tests of the Czechoslovakian VS-1 and VS-2 ejection seats for the L-39 Albatros training aircraft were also conducted by the Su-9U.

Over the years the LII used several large multi-purpose flying test-beds. Following Pe-2 and Tu-2 piston-engined bombers, in 1953 the institute received an Il-28 tactical bomber (c/n 701), which was used as a multi-purpose laboratory for 25 years, retiring in 1978. As well as other tasks, the Il-28 tested various ejection seats, including the Zvezda K-22 for the Tu-22 'Blinder', the Tupolev KT-1 for the Tu-22M 'Backfire', the Yakovlev KYa-1, and the KK – *Kreslo Kosmonavta* (Cosmonaut's Seat) – designed for spacecraft (the first Soviet cosmonauts did not land with their craft, but were ejected at several thousand metres altitude). The Il-28 was fitted with two launching positions for the ejection seats, one in the central part of the fuselage (the seat could be ejected upwards or downwards) and the other in the tail.

After 20 years of work scattered between various manufacturers, the time had come for unification, and ejection seat development was entrusted to the specialist Zvezda design bureau in Tomilino near Moscow, under the direction of Guy Severin. Tests began of the unified K-36 ejection seat, installed first in Su-24 tactical bombers. This seat is now the only type used in Russian combat aircraft. About 12,000 K-36 series seats have been built in 15 major variants, and is regarded by many as the best in the world.

K-36 seats were first tested on the sled track, at speeds up to 1360 km/h (845 mph). During 1975-1991 tests were undertaken using an Su-7U aircraft (c/n 16-03). In the mid-1970s the Buran space shuttle became the top priority Russian aerospace programme. Accordingly, in 1979 the LII institute received a MiG-25RU (c/n 01-01) for testing the

The Soviet Union's first ejection seat testbed was this Petlyakov Pe-2. This type was a natural choice for this work as its twin-finned configuration minimised the risk of the brave seat occupant hitting any part of the aircraft if the seat failed to work as intended.

This sequence captures a seat being jettisoned from the nose position of a Tupolev Tu-14, graphically illustrating the tumbling effect. Seat designers worked hard to find ways of stabilising the seat on ejection.

K-36M11F35 ejection seats designed for the Buran (11F35 is the Buran code).

Operations began in 1987 at Zhukovskiy of the next ejection-seat test bed, an An-12BK (c/n 59-02) transport. The tested seat, installed in a universal rotating stand in the aircraft's tail, can be launched at any angle – including sideways or downwards – thus simulating the use of the seat when banked or inverted. The An-12 was used mainly for testing the new Zvezda K-37 system for rescuing combat helicopter crews, applied to Ka-50 'Hokum' helicopter. S. Pereslavtsev was the first man to use the K-37 seat from the An-12 test bed. The K-37 seat, which is extracted rather than ejected, can be used at speeds between 0 and 350 km/h (217 mph) and at altitudes from 0 to 6000 m (19,685 ft), covering the full range of helicopter performance.

A significant part of the LII institute tests involved systems for identifying emergency situations in which the pilot has insufficient time for decision-making, and commanding automatic ejections. Such systems are vitally important for shipborne fighters and VTOL aircraft. Most of the tests were carried out with a Yak-38 aircraft with the SKEM-K-36VM automatic ejection system: when flight parameters are too dangerous for continuation of flight, the computing unit generates a command for ejection. During 15 years of service, 36 Yak-38s crashed. In 31 cases the pilots were ejected, including 18 ejections generated by the automatic system. All ejections were successful.

Since about 1993, Zvezda and LII have been working on the K-36D-3.5 ejection seat, with variable trajectory of flight and a greater range of safe ejections from high-manoeuvrability fighter aircraft. Unlike the present seats, the K-36D-3.5 is equipped with more powerful rockets and more advanced control systems. Tests of the K-36D-3.5, and its K-93 (K-36LT-3.5) lightweight version for light training aircraft, are carried out by the LII using both current seat test beds, the MiG-25 and the An-12.

Another canopy-enclosure style seat is tested from a MiG-15UTI, this time from the front cockpit. The seat has been fitted with a drogue parachute to stabilise it and inhibit tumbling. The seat is launched along an inclined rail.

Several Yak-25s were used for seat tests. Above is a zero-altitude test – among the most dangerous – while at right canopy separation is evaluated. With a large single fin and a large single-piece canopy, this could have had disastrous results if it did not function correctly.

Among the hardest-working seat testbeds at Zhukovskiy was this dedicated Il-28, modified with seat positions in the central fuselage (above) and tail (left). The central position, occupying the weapons bay, could be used for downward ejections as well as upward firings.

Sukhoi products used for seat-testing included an Su-9U and an Su-7U. The latter (above) was the primary trials vehicle for the Zvezda K-36 seat which became the standard unit for Soviet/Russian combat aircraft. The Su-9U (left) is seen with an early version of this seat, displaying the telescopic stabilising arms which are one of the key features of the system. Markings on both aircraft aided accurate measurement of seat performance.

Right: Long-term Ramenskoye resident is this modified An-12BK, 'Red 43'. Having seen service as an avionics testbed with equipment in a modified tailcone, the An-12LL now sports a cylindrical, rotating fairing from which seats can be tested. The barrel fairing can accommodate seats facing both forward and aft, and can eject them from any angle, allowing a massive launch envelope. In an extreme case, the An-12LL can simulate ejections from an inverted aircraft flying backwards with a sink rate of up to 25 m (82 ft) per second. As well as its barrel fairing, the aircraft is fitted here with a trials mid-air and air-ground parachute recovery system.

Left: This MiG-25RU was acquired by the LII in 1979 for ejection seat trials, initially in support of the Buran space shuttle programme. The aircraft is still in use for high-speed, high-altitude tests.

Engine testbeds

Implementation of jet propulsion was the greatest challenge for aviation in the first years after World War II. The first flying engine testbed in the USSR was a Tu-2 tactical bomber carrying a German Jumo 004 engine under its forward fuselage. Some time later, Russians began series production of this engine to power the first jet fighter: the Yak-15. In 1947-1954 the LII institute used four Tu-2s for testing first-generation engines, including the German Jumo 004 and BMW 003, as well as the British Rolls-Royce Nene and Derwent.

More powerful jet engines for bombers and transport aircraft, as well as for supersonic fighters, were tested using six Tu-4 'Bull' bombers. Three were used in 1951-1962 for testing the AM-3 engine for the Tu-16 and M-4 bombers, AM-5F and RD-9B for the MiG-19 and Yak-25, R11-300 for the MiG-21 and Yak-28, RU19-300 for the Yak-30, AL-7 and AL-7F for the Su-7 and Su-9, AL-7P for the Be-10, VK-3, VK-7 and VK-11 for experimental MiG fighters, VD-5 and VD-7 for the Myasishchev 3M, as well as the D-20 for the Tu-124. The tested engine was suspended centrally under the fuselage and it could be dropped in flight in case of emergency. The remaining three Tu-4s were used in 1951-1960 for testing turboprop engines; one or two tested engines were installed in the aircraft instead of its standard engines. The first tested turboprops were

The Soviet Union's first jet engines were tested under the versatile Tu-2 bomber. This aircraft, known as the Tu-2N, flew with a Rolls-Royce Nene I installed in 1947. Other engines tested included the BMW 003 and Jumo 004.

the TV-2, TV-2M, 2TV-2M and TV-12 (NK-12) for the Tu-95 'Bear'. Later, much time was devoted to testing NK-4 and AI-20 engines for the An-12 and Il-18 transports.

An Il-28 'Beagle' was used in 1953–1957 for testing S-155, RU-013 and U-19 rocket engines, which were designed for accelerating high-altitude interceptors. The rocket engine was installed in the tail part of the fuselage instead of the rear-gunner post. The RD-550 ramjet was installed in the fuselage of another flying test bed – a Tu-12 (1949-1951), whereas the RD-900 ram-jet engine for the La-17 drone was installed under the fuselage of a Tu-14 (1953-1956).

The next significant engine testbed was the Tu-16 'Badger' bomber – as many as nine aircraft of this type were used. They bore the main burden of new engines tested at the LII from 1956 until 1990. Almost all new engines developed during this period were tested under the fuselage of a Tu-16, such as AL-21F for the Su-17 and Su-24, AL-31F for the Su-27, AL-41F for the MiG 1-42, R27-300 for the MiG-23, R15-300 for the MiG-25, engines for Yak-38 (R27V-300) and Yak-41 (R79V-300) VTOL aircraft, RD-33 for the MiG-29 and D-30F6 for the MiG-31. In one unique trial a whole Yak-38, minus wings, was suspended under the test bed to test the complicated propulsion system of a R27V-300 cruise engine with swivelling nozzles and two RD36-35 lift engines.

Within the same period (1957-1990) more powerful engines were tested using one Tu-95 and two Tu-142s. They were used for testing 15,000-shp (11190-kW) versions of the NK-12M turboprop, as well as NK-25 jet engines for the Tu-22M3 and NK-32 for the Tu-160 (both 245.26 kN/55,115 lb thrust) and others. Tests of R95-300, TRDD-50 and Lyulka 36 small turbojet engines for strategic and tactical cruise missiles were carried out in 1987-1996 using the Tu-134A testbed c/n 1510.

Today the Tu-16, Tu-95 and Tu-142 are no longer used, replaced by the Il-76 'Candid' transport. Its four-engined high-wing monoplane configuration is ideal for engine evaluation, especially the more powerful ones. The tested engine is installed in the place of the standard inboard engine under the port wing. This solution does not adversely effect the aircraft's aerodynamics as is the case with installation of the test powerplant under or above the fuselage. The Il-76 has expanded measuring and recording equipment installed on board the aircraft.

Il-76LL engine test beds

Regn	c/n	Engine on test	First flight
CCCP-86712		NK-86 turbofan	1975
CCCP-86891	1607A	D-18T turbofan	1982
CCCP-76492	3908	PS-90A (D-90A) turbofan	1986 (26th Dec)
CCCP-76529	0807	D-236T and D-27 propfans	1989
CCCP-06188	1609	TV7-117 turboprop	1989

Apart from the general-purpose flying testbeds, which could be used for testing various types of engines, many engine tests have been carried out with the aircraft for which they were intended. These tests were generally undertaken over a long period of time with the purpose of increasing reliability and service life. Among large aircraft, such tests were carried out with the M-4, Il-28, Il-18, An-24, Yak-40 and others. Tests of dedicated engines were carried out by the LII institute with MiG-21, MiG-23, MiG-25, MiG-29, MiG-31, Su-15, Su-17, Su-24 and Su-27 fighters. Among these were the tests of the new-generation AL-41F using a MiG-25 (LL-84-20); the engine was installed in the place of the normal port R15-300 engine. The LL-84-20 made 22 flights, reaching a speed of 1,080 kt (2000 km/h) and an altitude of 65,620 ft (20000 m). The most recent of such testbeds, an Su-27 with AL-31F turbofans uprated from the standard 122.63 kN (27,557 lb) thrust to 129.89 kN (29,190 lb) flew on 22 January 2002.

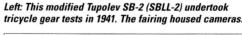

The B-29/Tu-4 'Bull' provided an ideal vehicle for engine tests, with six Tu-4LLs used for various programmes at the LII. Some of them were original Boeing-built B-29s. Jet engines were carried in jettisonable nacelles under the fuselage (above), while turboprops replaced the Tu-4's ASh-73s. Above right is a Tu-4 with two different engines under test (for the An-12, starboard and the Il-18, port), while at right is the NK-12 testbed for the Tu-95's engine.

Left: This modified Tupolev SB-2 (SBLL-2) undertook tricycle gear tests in 1941. The fairing housed cameras.

Above: This Ilyushin DB-3LL was used for testing laminar-flow aerofoil sections on a dorsal mount.

Left: Under test at Ramenskoye in 1940, before the LII's move away from the base, was the Besnovat SK-2, a fighter with tiny wings and an M-105 engine.

Also known as the MiG-8, the Utka (duck) was an unusual canard light aircraft. Built by the MiG OKB, it had little to do with the bureau's principal fighter work and was the product of students at the VVS academy at Ramenskoye. It nevertheless provided some data on the low-speed characteristics of swept wings and canard foreplanes. It was flown at Ramenskoye in late 1945.

Above: The Nene I-powered Tu-12 (Tu-77) was the Soviet Union's first jet bomber, flying in June 1947. This example was designated 77LL, and was used to test the RD-550 ramjet in a dorsally-mounted nacelle.

Above: In the 1950s, when bomber attack was the main threat, there were many schemes to fit rocket engines to interceptors to boost their climb rates and reduce interception times. Rockets were tested in the modified tailcone of this Il-28.

Left: This Tu-14 was used for engine trials, carrying the test item semi-recessed in its belly. The undernose rig was associated with this aircraft's other work as an ejection seat tesbed. The main engine tested by this aircraft was the RD-900 ramjet.

Above: Today the Il-76LL is the main testbed, able to carry engines across a wide size range on its port inboard pylon. This is the Progress D-18T engine, used by the An-124 and An-225.

Left: Tu-16LLs were used widely for powerplant tests in the 1970s, the engine being carried on a retractable pylon in the former weapons bay.

The configuration of the Il-76 allows it to undertake testing of either jets or propeller engines with equal ease. Below is the Klimov TV7-117 turboprop for the Il-114 airliner, while at right is the Progress D-27 used by the An-70 transport. Trials of the D-27 followed on from tests of the less powerful D-236 propfan, initially tested on an Il-76 and later on a Yak-42E-LL.

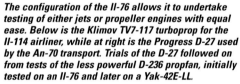

Helicopter testing at Ramenskoye has involved several unusual experiments. In one test (left), the rotor blades were deliberately blown off a remotely controlled Mi-4 by explosive charge to investigate the effects. A mid-air retrieval system was tested by this Mi-8 (right), seen at the moment of snagging a parachute with its pole-mounted cable.

Right: A Mil Mi-4 'Hound' was converted to Mi-4LL status with smoke generators in the blades for research into blade stalls and vortices, and for trials of various blade configurations.

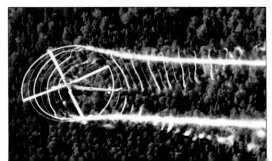

Tanker trials

The LII played host to many refuelling trials over the years, early tests using Tu-4 'Bull' tankers for both drogue tests (above, with MiG-17s) and Tupolev's unusual wingtip-to-wingtip method. The concept was first explored using a Tu-2 and a Yak-15 (right), before being tested by the Tu-4 (left). Its main service application was the Tu-16 'Badger'.

This unusual aircraft is thought to be a Tu-2 which was heavily modified with jet engines and tricycle undercarriage for the high-speed reconnaissance role. It is seen here as the vehicle for a guided bomb test.

Above and below: This Tupolev 73LL was fitted with a gantry under the starboard wing for drop tests of various missiles and sub-scale test vehicles.

Right: Designated ShR-1, this Tu-4LL was used in 1952 to test the bicycle undercarriage for the Myasishchev M-4, an example of which is behind. It was later modified to this configuration, known as ShR-2, with a four-wheel nose gear, with which it flew in 1953.

Below: Many sub-scale drop-test models have been tested at the LII. This unidentified model, also seen on the 73LL (left), could have been a Besnovat design, although it bears some vague similarities to the Lavochkin La-190 fighter.

Icing test rigs

The effects of ice build-up on aerodynamic surfaces and in engines can have disastrous results, and icing properties need to be examined in detail, along with the various remedial methods. Ice tests can be performed using simple spray rigs mounted on the aircraft itself, as demonstrated by the spray rig in front of a Yak-25's engine (right), or by purpose-built trials platforms such as the Il-18LL (below) and An-12BK (below right). These have spray rigs permanently installed ahead of a test fixture which can mount aerofoil sections and other test items.

Zero-length launch MiG-19

In 1956, Artyom Mikoyan's OKB-155 converted two MiG-19 fighters into the SM-30 for 'spot take-offs' using rocket assistance to launch from a vehicle-mounted ramp. Georgiy Shiyanov made the first launch on 13 April 1957. Shiyanov was awarded with the title of Hero of Soviet Union for these dangerous flights.

Above: Known as izdeliye SYe, this MiG-15 was one of three built with revised tail and wing surfaces in an attempt to cure a roll-reversal problem encountered at high speed. It was also fitted with a landing skid for tests of this unorthodox undercarriage.

Above: This 'Fitter' testbed (LLSu-7B) was used in 1965 for supersonic tests of laminar-flow wing sections fitted to the leading edge wing root.

Left: This Su-9 was fitted with sideforce generators and a fly-by-wire yaw control system. It tested FBW systems for the T-4 bomber. Similar surfaces were mounted laterally on an Su-7U, known as the 100LDU, to test pitch control laws.

Left: Designated A-144 and known as the Analog, this MiG-21 was converted to test the ogival wing planform of the Tu-144 SST prototypes. The Analog flew chase on the Tu-144's first flight. Subsequently, the Tu-144 adopted a double-delta wing, similar in planform to that used by the Sukhoi T-4.

Right: Another 'analog' was this Su-9, designated 100L to test a variety of wing planforms for the T-4 Mach 3 bomber between 1967 and 1969. Eight different wings were tested, including this gently curved leading-edge shape (heavily tufted to observe airflow patterns), and the 60°/70° double-delta planform fitted to the bomber.

Bor and Buran - aerospace craft

In 1965 Artyom Mikoyan's design bureau began designing the Spiral system for transporting an aerospace fighter into orbit. The heavy carrier aircraft '205', designed by Andrei Tupolev, was to accelerate to Mach 6 at an altitude of 30000 m (98.425 ft) to launch the small '105' fighter designed by Mikoyan. The fighter, capable of manoeuvring at an altitude of 200 km (124 miles), was to be used for attacks against other spacecraft, for reconnaissance, or for dropping bombs from space. It would come back to earth as a glider. Mikoyan's design bureau prepared a subsonic piloted replica of the '105' intended for landing tests. It was designated 105.11. On 11 October 1976 Aviard Fastovets made the first short flight with the 105.11, and on 27 October 1977 it was released from a Tu-95K for the first time.

The LII institute in Zhukovskiy prepared a series of Bor pilotless vehicles for testing the Spiral system in space. Bor-1 was only a wooden model of the Spiral orbital module, equipped with telemetry instruments. Bor-2 and the improved Bor-3 were one-third scale models intended for testing aerodynamics, control and thermal protection at altitudes up to 100 km (62 miles) and speeds up to Mach 13. The Bor-4 was equipped with full thermal protection and was to be tested in open space. Several vehicles had been made when the Spiral programme was cancelled. However, on 17 February 1976 the Soviet Cabinet and Central Committee of the USSR Communist Party launched another programme – Buran-Energiya, similar in concept to the American Space Shuttle. The Spiral designers were transferred to NPO Molniya; Spiral's manager, Gleb Lozino-Lozinskiy, was appointed Director General and Chief Designer of NPO Molniya.

Completed Bor-4 aircraft were to be used for testing the thermal protection tiles of the Buran. Bor-4S took-off on 5 December 1980 and – like former Bors – made only a suborbital hop over a short distance. The first vehicle to fly in open space, Bor-4 #404, was carried into orbit on 4 June 1982 (as Kosmos-1274) by a strategic missile from Kapustin Yar military space port. It reached an altitude of 225 km (140 miles) and then, after deceleration, it glided down to an altitude of 4000 m (13,125 ft), and then landed by parachute in the Indian Ocean, where it was photographed by aircraft of the Royal Australian Air Force. The next Bor-4 (#403/Kosmos-1445) landed in the Indian Ocean on 16 March 1983, this time watched by the Americans. The two last tests with Bor-4 (#405/Kosmos-1517 on 27 December 1983, and #406/Kosmos-1614 on 19 December 1984) ended in the Black Sea, free from outside observers. However, one of the vehicles sunk and was never recovered.

Bor-4 was still only a half-scale model of the Spiral orbital module, whereas the next vehicle, Bor-5, was a one-eighth scale model of Buran, weighing 1450 kg (3,197 lb) and used for aerodynamic tests of the new space shuttle. During 1984-1988 there were five Bor-5 space missions, of which three were successful. The rocket carried the model to an altitude of about 210 km (130 miles), where it was released. The distance between missile launch pad to the landing place was around 2000 km (1,250 miles), allowing the whole flight to be conducted over USSR territory. The next vehicle, Bor-6, was prepared for tests of communication in the space (propagation of radio waves in plasma), but it was never flown.

Right: For transporting the Buran, and other major space vehicle components (such as the Energiya launch vehicle), Myasishchev adapted two M-4 'Bisons' into VM-T Atlant configuration. The bomber's single fin was replaced by two large endplate fins.

The Bor-2 (above) provided research data for the Spiral programme at speeds of up to Mach 13. The Mach-18 Bor-5 (right) was a scale model of the Buran shuttle.

In the meantime, NPO Molniya began construction of the Buran, and LII prepared several flying laboratories for testing the control system of the space shuttle during its return to earth. Most of the tests were carried out using a Tu-154 aircraft equipped with an experimental digital fly-by-wire system; future pilots of the Buran used it for exercises in descent, approach and final landing.

The atmospheric section of the Buran flight profile was tested using the analogue BTS-002, a direct copy of Buran but with a propulsion system comprising four AL-31 engines from the Su-27. At that time Russia did not yet have an aircraft capable of carrying Buran into the air, so the BTS-002 had to be equipped with engines enabling its take-off from the ground. BTS-002 made 24 flights from Ramenskoye, the first one on 10 November 1985 with Igor Volk and Rimas Stankyavichus at the controls, primarily for training the crews. A MiG-25PU #22 (c/n 05-78) was converted into the MiG-25SOTN – Samolyot Optiko-Televizionnogo Nablyudeniya (Aircraft for Optical-TV Observation) and was used as a chase aircraft and recording platform for the Buran tests. On 15 November 1988, the MiG-25SOTN assisted in landing the Buran during the Russian Shuttle's only flight into space, which was made in automatic mode, without crew.

Prospective Buran pilots trained in a Tu-154 (above and above right) which featured a fly-by-wire control system allowing it to mimic the approach characteristics of the shuttle. Indeed, it was instrumental in the development of the system. Approaches were often monitored by the MiG-25SOTN chase aircraft (above). For more realistic training a full-size 'analogue' of the Buran was built (right), designated BTS-002. This made several flights from Zhukovskiy, launching under its own power thanks to four AL-31 engines clustered around the tail.

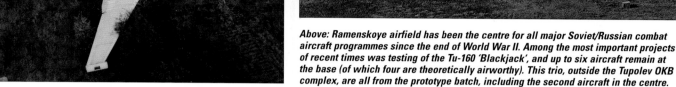

Above: Ramenskoye airfield has been the centre for all major Soviet/Russian combat aircraft programmes since the end of World War II. Among the most important projects of recent times was testing of the Tu-160 'Blackjack', and up to six aircraft remain at the base (of which four are theoretically airworthy). This trio, outside the Tupolev OKB complex, are all from the prototype batch, including the second aircraft in the centre.

Tupolev designs have dominated Zhukovskiy's large combat aircraft inventory, although many have languished on several dumps around the vast field, including in the wooded area to the east of the main complex. Above is the prototype Tu-126 'Moss', while below is the Tu-22LL (with lengthened nose). This 'Backfire' (right) is one of the Tu-22M3 prototypes, with a refuelling probe in the nose, which was used for laminar flow tests.

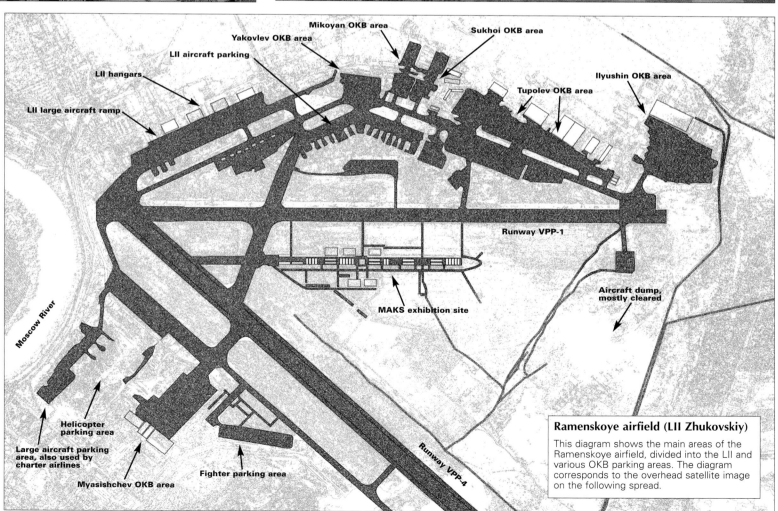

Mikoyan OKB area

Yakovlev OKB area

Sukhoi OKB area

LII aircraft parking

LII hangars

Ilyushin OKB area

Tupolev OKB area

LII large aircraft ramp

Runway VPP-1

MAKS exhibition site

Aircraft dump, mostly cleared

Moscow River

Helicopter parking area

Large aircraft parking area, also used by charter airlines

Myasishchev OKB area

Fighter parking area

Runway VPP-4

Ramenskoye airfield (LII Zhukovskiy)

This diagram shows the main areas of the Ramenskoye airfield, divided into the LII and various OKB parking areas. The diagram corresponds to the overhead satellite image on the following spread.

OV-1 Mohawk
Reconnaissance over the Battlefield

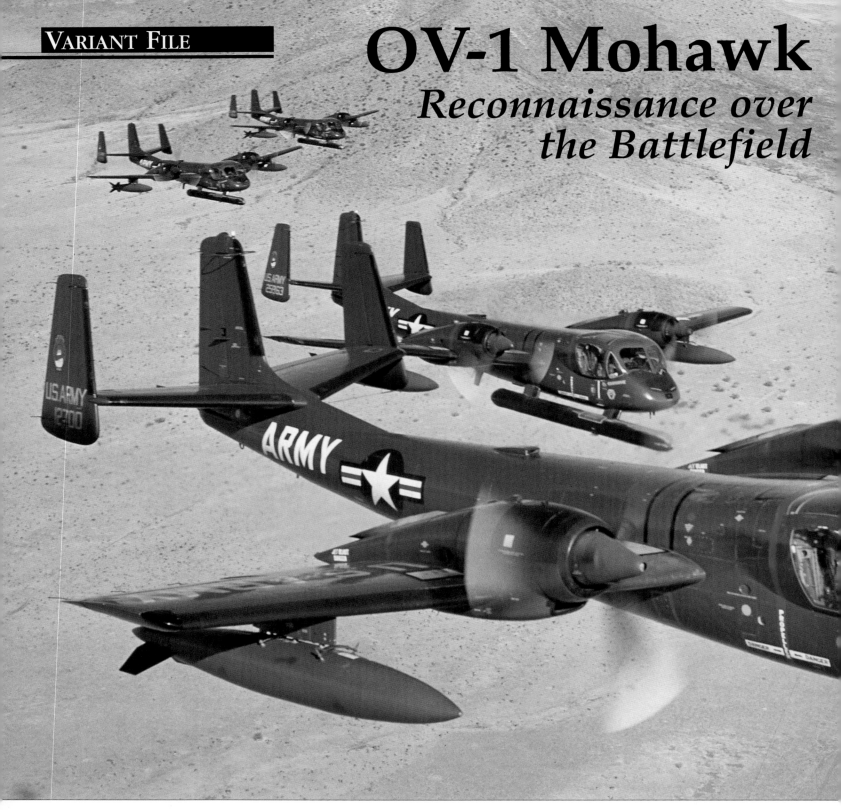

Designed for visual and photo reconnaissance, with a light attack capability, the Mohawk soon evolved into a multi-sensor platform which was to become the eyes and ears of the US Army. From the jungles of Vietnam to the Cold War stand-off in Germany, the Mohawk kept commanders on the ground fully briefed.

Intended as a replacement for the US Army and US Marine Corps Cessna L-19 and OE-1 Bird Dog liaison/observation aircraft, Grumman's Model G-134 was designed to fulfill a joint services requirement. The aircraft was planned as a battlefield surveillance platform for the Army and the close air support role with the Marines. Its design incorporated short take-off and landing (STOL) capabilities that would allow it to operate from small, unprepared airstrips near the front lines and small escort

aircraft-carriers. The Marine Corps actually planned to equip their aircraft with water skis that would have allowed them to land on water and taxi onto island beaches. The aircraft was to be tasked with a variety of missions including visual observation, air control, spotting for both land-based and naval artillery, liaison, emergency resupply and radiological monitoring.

Although a joint Department of Defense directive limited the Army to operating aircraft

that weighed less than 5,000 lb (2268 kg), the service was granted an exemption to this requirement for its new observation aircraft. Grumman was one of six contractors that responded to the joint service request. The field was later narrowed to include designs submitted by North American Aviation and Grumman, and the latter was declared the winner in May 1957. Acting as the procuring agent for both the Army and Marine Corps, the US Navy Bureau of Aeronautics issued a contract to Grumman covering the development and production of nine engineering and service test aircraft for the two services. The breakdown comprised five YAO-1 aircraft for the Army and four YOF-1 aircraft for the Marines.

Grumman built mock-ups of both variants, which featured a single 'T'-type vertical and

Mohawk, myth or legend?

There is an almost urban legend concerning the popular name assigned to the YAO-1/OV-1. In accordance with a policy adopted by the Chief of Transportation, Department of the Army, that continues to this day, the service assigns the names of American Indian tribes to its aircraft. According to the legend Grumman proposed the names Montauk and Mohawk, which were both Indian tribes from New York State. While the company preferred Montauk, since the tribe was from Long Island, the Army selected Mohawk because the tribe, whose locale included the upper central region of New York State, was better known as warriors.

In reality the service had tentatively selected the name Mohawk in late June 1957 but asked Grumman if that name was acceptable. In mid-July 1957 the company's chairman and namesake, Leroy R. Grumman, responded to this question personally when he advised that Mohawk was satisfactory but proposed the name Montauk and suggested it was more fitting as the tribe's habitat and Grumman Aircraft were both located on Long Island. He also felt that Mohawk could be confused with an airline from New York with the same name. Unfortunately, by the time the letter reached Washington the Army had already accepted Mohawk as the popular name. The service did, however, reserve the name Montauk for any future aircraft that Grumman might build for the Army.

The Mohawk's bug-eyed tadpole shape stemmed from the requirement to give the two crew members the best possible all-round view, as dictated by the primary roles of battlefield observation and close air support. The nose was very short to give an excellent view forward and down, while the bulged cockpit allowed the lines-of-sight of the crew to overlap beneath the aircraft even at very low level. Upward visibility was not ignored either, important for when the aircraft was manoeuvring hard over the battlefield. This is the prototype, wearing the Mohawk badge on the nose.

T-tailed mock-ups represent the Army's AO-1 (below) and OF-1 for the Marines (right), the Army aircraft being fitted with skis and armament. The T-tail did not provide sufficient directional stability.

horizontal stabiliser with a larger rudder. This later changed to a more conventional low-set tailplane with three vertical fins. In January 1958, however, the Marines opted out of the programme for financial reasons and the four YOF-1 prototypes were redirected to the Army. A production contract was issued even before the prototype Mohawk made its first flight on 14 April 1959.

Testing

Although the Marines had withdrawn from the programme, the Navy retained responsibility for engineering tests while the Army conducted the service testing. Trials were conducted at a number of locations and the Navy discovered several relatively minor problems during Board of Inspection and Survey (BIS) trials, which began in June 1960. Once these were rectified the first Mohawks were assigned to the US Army Aviation Center (USAAVNC) at Fort Rucker, Alabama, in late 1960. Weapon tests were conducted by the Naval Air Test Center's Weapons Division at Naval Air Station Patuxent River, Maryland, and the aircraft was cleared to carry a variety of conventional, napalm, cluster and smoke bombs, rockets, gun pods and Sidewinder air-to-air missiles.

Cold weather testing was conducted under Operation Great Bear at Fort Greely, Alaska, in February 1962 and verified the Mohawk's ability to operate in temperatures as cold as -50° F (-46° C). Two test aircraft were deployed, comprising serials 60-3743 and 60-3744. As part of its evaluations at Fort Rucker, the US Army

Aviation Test Board confirmed that the Mohawk could be operated from dirt roads, as well as wet or dry fields that had been ploughed to a depth of 12-in (30.5 cm).

Anatomy of the Mohawk

The Mohawk was the first turboprop-powered aircraft designed for the Army, as well as its first to be equipped with ejection seats. The fuselage was 41 ft 4 in (12.60 m) long and featured a two-place cockpit with side-by-side seating for a pilot and co-pilot or observer. A bubble canopy provided both crew members with unobstructed vision 20° down over the nose. Similarly, the upward-hinged, bubbled side entry hatches provided sufficient downward visibility that the pilot's and co-pilot/observer's line of sight converged at a

Above: The penultimate YAO-1 unleashes 2.75-in
rockets from a 19-round LAU-3/A pod during weapon
tests from Patuxent River. The armed capability
stemmed from the original Marine Corps need to
provide very close air support for the 'grunts'.

Field performance was one of the main design criteria,
resulting in the Mohawk being fitted with large trailing-
edge flaps and leading-edge slats, as displayed by this
AO-1AF. The slats were subsequently deactivated and
ultimately deleted from later production aircraft.

Record-setting flights

The Mohawk was always considered a 'hot ship'
and on 16/17 June 1966 Grumman test pilot Jim
Peters proved this by flying OV-1B serial 64-14240
to an altitude of 9,842 ft (3000 m) in just 3 minutes
41 seconds, and then to 19,685 ft (6000 m) in
9 minutes 9 seconds. He subsequently flew the
aircraft to a sustained altitude of 32,000 ft
(9754 m). Each of these events set world records
for turboprop aircraft in the 13,227- to 17,636-lb
(6000- to 8000-kg) weight class. In July US Army
Colonel Edward Nielsen set a world speed record
when he flew the same OV-1B at an average
speed of 254 kt (470 km/h) over a 54-nm (100-km)
course in just 12 minutes 48.8 seconds. Both of
these record-setting flights originated from
Grumman's Calverton, New York, facility.
   The earlier time-to-climb records and sustained
flight records were broken by CW4 Thomas Yoha
and Capt. Richard Steinbock of the 293d Aviation
Company (SA) at Fort Hood, Texas, in June 1971
with OV-1C serial 67-18923. After taking off from
Fort Hood on 8 June the Mohawk climbed to
9,842 ft (3000 m) in just 2 minutes 46 seconds,
and then to 19,685 ft (6000 m) in 5 minutes
46 seconds, and to 29,258 ft (9000 m) in
11 minutes 14 seconds. The next day the aircraft
climbed to a maximum altitude of 39,880 ft
(12155 m) and demonstrated sustained flight at
36,352 ft (11080 m).

point 36 ft (11 m) beneath the aircraft. Although
both positions could be equipped with control
sticks and rudder pedals, the controls were
generally not installed in the right seat.

A single set of propeller and power controls
was provided on the centre console, accessible
to either crew member. Instrumentation was
mounted in the left cockpit but visible to the
observer in the right seat. Major communications, navigation and photographic panels were
also mounted in the centre console between
the crew members, while IFF, HF communication and ventilation controls were installed in a
centrally mounted overhead panel. Engine, fuel
and electrical controls were installed on a panel
above the left windshield.

Crew protection included 239 lb (108 kg) of
armour plating, and the 1-in (25-mm) thick
windshield was both bullet- and flak-resistant.
The cockpit floor was constructed of 0.25-in
(6.35-mm) Dural (an alloy comprised of

aluminium, copper and magnesium) and
removable flak curtains could be installed on
the forward and aft bulkheads. The crew was
provided with Martin-Baker Mk-J5 ejection
seats capable of operating at any altitude at
airspeeds ranging from 100-450 kt (185-833
km/h). The seats were later replaced as part of
a modification programme by the rocket-
assisted Mk-J5D model, which reduced the
minimum ejection airspeed to 60 kt (111 km/h).
Although the transparent overhead panels
could be jettisoned, the normal ejection
sequence fired the crew members through the
frangible Plexiglas.

Tough landing gear

As previously related, the Mohawk was
designed for both shipboard and field operation and its hydraulically operated tricycle landing gear was designed for sink rates of 17 ft
(5.2 m) per second. Both the pneumatic nose
and main struts were equipped with low-
pressure tyres to facilitate field operations, and
a high-pressure pneumatic system was
provided for emergency extension.
Hydraulically operated nose wheel steering

provided the aircraft with a minimum turning
radius of 28 ft (8.5 m). The landing gear was
also equipped with provisions for mounting
skis that retracted along with the gear, permitting the Mohawk to operate from snow and ice
as well as prepared strips. The skis underwent
testing at Grumman's Calverton, New York,
facility and at Allen Army Airfield, Forts Greely
and Wainwright (Ladd Army Air Field) in
Alaska, and a number of sites around Kenora,
Ontario, between 1961 and 1963.

Mounted in the shoulder position, the broad-
chord wing had a span of 42 ft (12.6 m), an
area of 330 sq ft (30.66 m²) and 6.5° of dihedral.
The wing was equipped with mechanically
operated outboard ailerons, as well as hydraulically operated flaps and leading-edge slats. A
second pair of hydraulically operated ailerons,
mounted on the inboard section of the wing,
was interconnected with the flap system and
provided additional lateral control whenever
the flaps were extended. As part of the Marine
Corps legacy the wing design incorporated
provisions for six wing stations. The two

OV-1 missions

With the exception of the armed surveillance mission and the addition of the electronic intelligence role assumed by the RV-1D, the missions flown by the Mohawk at the end of its career were quite similar to those conducted when it was initially deployed in 1961, and included visual/photo, SLAR and infrared reconnaissance.

Prior to the deployment of the OV-1D, which was equipped with three cameras, a typical photo mission called for two OV-1A/Cs to fly in a lead/trail formation. The camera aboard the lead aircraft was set to obtain vertical photography while that aboard the trail aircraft was set to obtain 15 or 30° oblique views. Prior to the 1965 roles and missions agreement between the Army and Air Force, pairs of armed OV-1As and JOV-1Cs carried out most daylight visual/photo missions and provided mutual protection equipped with machine-gun pods and rockets. The deployment of the OV-1D allowed a single aircraft to carry out the photo-reconnaissance mission. Early night missions were aided by upward-ejecting flares mounted in a pair of dispensers carried on the left and right upper wing roots, but after 1966 these were replaced by the LS-59 flasher pod, which continued in use until the last Mohawks were withdrawn from service.

Although the SLAR-equipped OV-1B or OV-1D was not limited by weather or daylight, most missions were flown at night. A typical SLAR mission involved a single aircraft flying a fixed pattern to locate its target(s). The system was capable of recording data on either the left or right side, or both sides of the aircraft's line of flight simultaneously and provided images of both fixed and moving targets on land or water. SLAR missions were flown at altitudes between 7,000 and 14,000 ft (2134 and 4267 m), which determined the overall area that could be mapped.

While the use of oblique cameras and side-looking radar provided a degree of protection and stand-off

From the start the Mohawk was designed to be operated in the field, moving with the ground forces as they advanced or retreated. Accordingly, the aircraft was not only built tough, but was designed to be easy and quick to maintain. These are AO-1CFs.

SLAR- and IR-equipped aircraft initially operated side-by-side in platoons assigned at battalion level. The SLAR aircraft, like this AO-1BF, flew mostly at night and at medium altitude, from where the radar could achieve a good slant range and area coverage.

capability for aircrew conducting those missions, infrared missions required the crew to conduct a penetration mission and fly directly over the target area. Most AN/AAS-14 infrared missions were flown at altitudes between 1,000 and 2,000 ft (305 to 610 m) by a single aircraft under the cover of darkness and crews were required to use terrain-masking as often as possible. At an altitude of 2,000 ft (610 m) the IR system was able to view an area that was approximately 0.5 nm (1 km) wide. The OV-1D's AN/AAS-24 allowed the aircraft to operate as high as 3,000 ft (915 m). The system was capable of detecting heat generated by camp/cooking fires, vehicle engines and groups of enemy forces.

outboard stations on each wing were stressed to carry 500 lb (227 kg), while single inboard stations could carry up to 1,000 lb (454 kg).

The horizontal tail was equipped with mechanical elevators and mechanically driven rudders were installed in each of the three tails. The flight control surfaces were connected to the control stick and rudder pedals by a series of cables, pulleys, pushrods and bellcranks. Two hydraulically operated speed brakes were installed in the aft fuselage. The outboard vertical stabilisers and rudders, as well as the horizontal stabilisers and elevators, were inter-

Grumman's long experience in building carrier aircraft stood it in good stead when it came to designing the Mohawk's undercarriage. To emphasise its strength, this AO-1CF taxis through thick mud during tests. In the event, the OV-1 was destined to operate throughout its long career from the comfort of concrete runways.

changeable between the left and right sides of the aircraft, aiding field maintenance and battle damage repair. The powerplants were also interchangeable between the left and right sides. Maintainability was further enhanced by the location of avionics bays in the fuselage below the wings.

Power for the Mohawk was provided by the Lycoming T53, which also equipped the Bell HU-1 (later UH-1A). It was equipped with a single-stage power turbine that drove a combination axial-centrifugal compressor and an external vaporising combustor. The engines were partially protected against ground fire by being mounted on top of the wings, while the exhausts were located in the top rear portion of the nacelle. This feature provided some protection against heat-seeking surface-to-air missiles by using the wing to mask the hot exhaust

gases. Each engine was equipped with a 400-amp starter generator that powered the 28-volt DC electrical system and a variable-volume hydraulic pump that supplied the 3,000-lb/sq in (211-kg/cm²) system. A single 297-US gal (1124-litre) self-sealing fuel tank was installed in the centre fuselage; however, the aircraft was capable of carrying a pair of 150-US gal (568-litre) external fuel tanks or 300-US gal (1136-litre) ferry tanks. The basic Mohawk airframe was designed for a service life of 8,000 flight hours but was certified for a 7,000-hour service life.

Production

Issued in March 1959, the initial production contract covered 18 AO-1AF and 17 AO-1BF models. The former was a dedicated visual observation and photographic reconnaissance aircraft, while the latter was also equipped to

Variant File

Mohawks for Europe

Grumman attempted to sell the Mohawk to foreign operators as early as 1963, when both West Germany's Heeresfliegertruppe (army air corps) and France's Aviation Légère de l'Armée de Terre (army light aviation corps) evaluated examples of the OV-1B and OV-1C. The German demonstration took place in late 1963 and the two Mohawks carried the German serials QW-801 and QW-802. Although the aircraft flew more than 385 hours during the evaluation, in the end Germany was not able to purchase the aircraft. A much shorter evaluation followed in France using the same two aircraft, which wore the French serials ABU and ABV, but the result was the same.

In 1981 the company was granted an export license for the Mohawk and once again tried to market the aircraft internationally. At that time it planned to refurbish surplus Mohawks for observation and counter-insurgency roles in the Philippines, Thailand and Australia. As in previous attempts, no sales resulted.

Until 1974, when Israel received two OV-1Ds, the nearest the Mohawk came to serving with an overseas air arm was in 1963, when a pair of Mohawks, comprising an OV-1B and an OV-1C, went on an extended European tour for evaluation by the West German (below) and French (above) armies.

conduct electronic surveillance using a side-looking airborne radar (SLAR) system. In 1960, 25 additional AO-1AFs were ordered, along with 17 AO-1CFs, which differed from the earlier two models in being equipped with an infrared reconnaissance system as well as the camera equipment. A total of 54 aircraft was ordered in 1961 but these were all built to the AO-1CF configuration.

Formal flight training began at Fort Rucker in April 1961 and during that summer the first Mohawks were deployed with the 7th Army at Sandhoffen Airfield in Mannheim, West Germany. In response to tensions with the Soviet Union, the first OV-1Bs were deployed to Europe in November 1961 when 11 aircraft arrived in West Germany aboard the former escort carrier USNS *Card* (T-AKV 40).

During 1962 a number of milestones were achieved for the Mohawk programme; not the least of which included the assignment of new designations to all three variants. In accordance with the Department of Defense's new tri-service designation system, the Mohawks were assigned the Mission Design Series (MDS) OV-1, indicating its use as an observation platform with STOL capabilities. Each of the models was redesignated as follows: YAO-1 to YOV-1A; AO-1AF to OV-1A; AO-1BF to OV-1B; and AO-1CF to OV-1C.

Mohawks on the attack

During 1962 the Army began developing an armed surveillance concept for the AO-1AF. The first Mohawks arrived in Southeast Asia as part of the 23d Special Warfare Aviation Detachment (SWAD) in September 1962, following transfer of the unit from Fort Rucker to Nha Trang in the Republic of Vietnam (RVN). The unit's six JOV-1As quickly entered combat and a short time later the USAF began to voice its objections to the Army equipping its

Mohawks with so-called offensive weapons. Officially, the Mohawks were not equipped to conduct offensive missions and were tasked to mark targets and suppress enemy fire in order to conduct photo reconnaissance missions. Additional aircraft were deployed to Vietnam in 1963 and that year the Army placed orders for just 21 OV-1As.

Fiscal Year 1964's budget authorised the purchase of 36 aircraft, and all were built to the OV-1B configuration. A Grumman designed air-refuelling capability – comprising a fixed refuelling probe – was installed on OV-1B serial 62-5868. During 1964 and 1965 the aircraft conducted air refuelling demonstrations with a Marine Corps KC-130F tanker and an army C-7B Caribou that had been equipped with a pod-mounted hose and drogue system in its cargo bay. The Mohawk configuration was, however, deemed to be unsuitable and further testing was cancelled.

Mohawk funding was suspended in 1965 while the Army conducted an internal battle over roles and missions assigned to the type, and battled with the Air Force over its use in offensive roles. The USAF preferred that the Army cease all fixed-wing operations, claiming the armed Mohawks violated the 1948 Key West Agreement that assigned the fixed-wing close air support mission to the Air Force. The ongoing battle escalated inside the Pentagon, however, the armed Mohawk programme came to an official end later in 1965, when the two services came to an agreement over their respective roles and missions.

As part of the final arrangement the Army agreed to stop arming its Mohawks and to transfer the majority of its de Havilland Canada C-7A Caribou transports to the USAF. As a result the Department of Defense allowed the Army to retain its OV-1s, but it would not be permit-

59-2621 was the first production SLAR-carrying AO-1BF. The aircraft retained its photo reconnaissance capability, as evidenced by the overwing illumination flare launchers. The first SLAR aircraft retained the short wings of the AO-1AF, but in later aircraft the wings were extended for better range and load-carrying.

ted to modify additional aircraft to carry armament, beyond the 59 OV-1A/Cs that had been
so equipped. Although the programme had
come to an 'official' end, the units operating in
SEA continued to equip their OV-1s with gun
pods and rockets for 'self defence' until the last
Mohawks were withdrawn.

During this period Grumman relocated the
remaining production from Bethpage to its
facility in Stuart, Florida, where it was already
conducting final assembly and flight testing.
The company also prepared a proposal for a
counter-insurgency aircraft, for the Marines and
the Army, based upon the OV-1. Design 134R
was known as the light armed reconnaissance
airplane (LARA) and, along with all of the
Mohawk's air-to-ground capabilities, the aircraft
featured two additional stations that would
have been added to the fuselage for carrying
two XM-19 7.62-mm gun pods. The design also
contained a redesigned cockpit that provided
tandem versus side-by-side seating along with a
wing that was mounted slightly lower than that
of the OV-1.

Army takes command

In 1966 the Army took over the procurement
function from the Navy, when funding for new
Mohawks resumed, but just 16 OV-1Cs were
purchased. Consolidation of the production
and flight test functions at Grumman's facility in
Stuart, Florida, was completed simultaneously.
This facility had already been home to a significant portion of Mohawk production, and the
company had conducted flight testing there as
early as 1961. Grumman had also received
authorisation to modify two OV-1Cs with
updated systems that equipped the latest
production machines, and a contract was
subsequently issued to modify 31 early OV-1Cs
with these advanced systems.

By 1967 increased funding associated with
the war in Vietnam enabled the Army to order
36 OV-1Cs, while the contractor modified two
OV-1Bs and two OV-1Ds with new equipment
in support of operations in Southeast Asia as
part of the SouthEast Asia MOhawk
REquirement (SEAMORE) Project. Besides the
36 OV-1Cs, the fiscal year 1968 orders included
10 aircraft built to a new configuration known
as the OV-1D. This latest variant was the result
of a 1967 Army decision to reduce the number
of Mohawk variants. The D model combined all

Although little publicised, the Mohawk saw extensive service in Vietnam, first arriving in-country as early as September 1962. Above an armed OV-1A from the 73rd Aviation Company (Surveillance) lets fly with rockets against a target, while below another 73rd SAC aircraft lands past C-7s at Vung Tau. The aircraft has flare packs installed and nose-mounted radio aerials. Note also the working airbrakes, a feature which was deleted from later aircraft.

Below: OV-1B 62-5868 takes on fuel from C-7B Caribou 62-4187 during trials of a bolt-on refuelling probe. Later in the test series the aircraft refuelled while carrying its SLAR pod, and with one engine stopped and the propeller feathered. The refuelling pod, on loan from the Navy, was carried in the rear door.

105

Building on small-scale programmes undertaken in West Germany with the Silver Lance and RV-1C aircraft in the 1960s and 1970s, the Army procured a fleet of RV-1D Elint aircraft, of which this was the first to be converted.

of the reconnaissance equipment of the three earlier variants in a single aircraft, although the SLAR and IR systems could not be carried simultaneously. The initial four YOV-1Ds were conversions of OV-1Cs 67-18898/18901 that had been accepted by the Army between September and December 1968. The OV-1D also featured a number of new systems that were the result of the South-East Asia MOhawk REquirement (SEAMORE) programme that had been initiated in 1967.

The final batch of new Mohawks was included in the 1969 defence appropriation which authorised the production of 27 additional OV-1Ds. The last production Mohawk was accepted by the Army in December 1970.

Remanufacture

Although new production had ended, the company continued to operate a repair, modification and overhaul facility for older Mohawks at its plant at Martin County Airport in Stuart. Beginning in 1973, the first of 78 OV-1Bs, RV-1Cs and OV-1Cs were returned to Stuart and modified to OV-1D configuration. The final remanufactured OV-1D was delivered in January 1987. In 1974 Grumman began modifying OV-1Bs to carry the AN/ALQ-133 Quick Look II airborne emitter locator identification system. Like all other modification programmes, this project was also carried out at Stuart. Approximately 20,000 man hours were required to convert an OV-1B into an RV-1D. The army accepted the first RV-1D in February

1977 and the 31st and final aircraft left Stuart in April 1984.

In an attempt to compensate for the increased weight of the later Mohawk variants, during 1985 OV-1D serial 68-15932 was equipped with a pair of uprated AVCO Lycoming T53-L-704 turboprop engines that provided 1,800 shp (1342 kW). The modified aircraft made its initial flight at Stuart in January 1986 and was subsequently ferried to Edwards AFB, California, where a two-month flight test programme was carried out.

During September 1986 the Army awarded the company a contract covering the development of a block improvement programme for the OV/RV-1D. The programme, which later became known as the multi-stage improvement programme (MSIP), was designed to provide the aircraft with a new digital multi-function display system that replaced many of the older instruments. Also included was a new autopilot, updated radios and navigation equipment that included a global positioning system (GPS). A single OV-1D, serial 67-18922, was modified to the MSIP configuration and it flew for the first time on 19 May 1988 at Stuart. It was officially rolled out on 24 June 1988 and in November 1988 was flown to the company's facility in Calverton, New York, where its avionics were tested in the facility's anechoic chamber. A production MSIP would have combined the avionics modifications and a separate re-engining programme based upon the successful T53-L-704 flight test programme. Grumman had

Calvin's Crew: It's a Phaze Thing – an OV-1D from the 2d MIB's A Company – wears eight camel markings (and the alternative name Barf Bucket) on its nosewheel door in testament to its involvement in Desert Storm. This unit flew on surveillance duties both before and during the conflict. Shortly after its return to Germany it became the first Mohawk victim of the 'peace dividend'.

expected to incorporate the MSIP on 33 to 99 Mohawks, beginning in 1989, however, as a result of the Army's decision to run down the Mohawk fleet the programme was cancelled.

Although the MSIP programme foundered, modifications of OV/RV-1Ds continued at Stuart under the Programme Aircraft Restoration (PAR), at a rate of two aircraft per month. As part of the PAR, which had begun in 1986, the aircraft underwent service life extension that increased the airframe life to 12,000 hours. The PAR programme ended in September 1991 when the last updated Mohawk departed the Stuart facility.

Desert Storm

In 1990 the Mohawk began its last combat deployment when A/15th Military Intelligence Battalion (Aerial Exploitation) [MIB (AE)] and the RV-1D platoon of A/224th MIB(AE) deployed to Saudi Arabia. The RV-1D platoon was attached to A/15th, which reported to the 525th Military Intelligence Brigade (MIBDE), in support of XVIII Airborne Corps, for the duration of Operations Desert Shield and Desert Storm. Early in 1991 the 207th MI BDE and A/2d MIB(AE) deployed from Stuttgart,

Grumman RV-1D Quick Look II
2d MIB(AE), Stuttgart

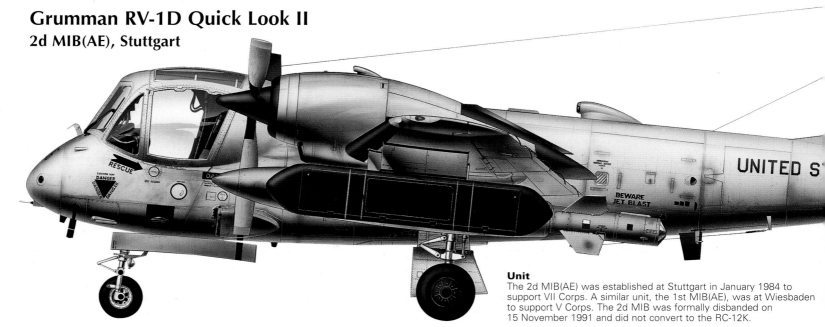

Unit
The 2d MIB(AE) was established at Stuttgart in January 1984 to support VII Corps. A similar unit, the 1st MIB(AE), was at Wiesbaden to support V Corps. The 2d MIB was formally disbanded on 15 November 1991 and did not convert to the RC-12K.

In addition to routine operations in Germany and Korea, in the 1980s the Mohawk also operated in Central America on detachment from CONUS-based units. As well as monitoring rebel factions, it supported the US war on drugs trafficking and production.

Germany, to Saudi Arabia. Prior to the commencement of hostilities crews from these units combined to fly SLAR, IR and ELINT missions along the Saudi borders with Kuwait and Iraq. The SLAR systems were used to pinpoint the location of Iraqi vehicles, while the RV-1Ds ferreted out the locations of Iraqi radar sites in preparation for the strikes that began on 16 January 1991. A/15th, and A/224th's RV-1D platoon, remained in the theatre until April 1991, while A/2d MIB returned to its base in Germany in May.

Retirement

The first of the post-Cold War Mohawk units to deactivate was A Company/2d Military Intelligence Battalion at Echterdingen Army Air Field in Stuttgart, Germany, which stood down shortly after returning from Arabia. Unlike later

Mohawk retirements, this battalion was unique in that the entire organisation was inactivated as a result of the drawdown of forces in Europe and the deactivation of VII Armored Corps. The unit's Mohawks were initially transferred to A/1st MIB at Wiesbaden Army Air Field in mid-1991 before being returned to the CONUS.

The first of two Army National Guard units gave up their OV-1Ds in September 1991 when Georgia's A Company/151st Military Intelligence Battalion at Dobbins Air Reserve Base in Marietta, Georgia, retired its last aircraft and transitioned to helicopters.

Mohawk operations at Wiesbaden began to wind down during 1992 and the last RV-1Ds assigned to the A Company/2d Military Intelligence Battalion departed Wiesbaden in June 1992. That September the last OV-1Ds assigned to A Company of the Oregon Army National Guard's 641st Military Intelligence Battalion were similarly retired at McNary Field in Salem. September 1992 also brought an end to flight test operations at Cairns AAF, Fort Rucker, Alabama, where two OV-1Ds were operated by the Aviation Technical Test Center (ATTC) as part of the Army's 'Lead the Fleet' programme. One additional OV-1D assigned to the SEMA Platoon of US Army Electronic Proving Ground (EPG) at Fort Huachuca, Arizona, continued to support electronic testing until it, too, was retired in August 1993. More than 30 years of Mohawk operations in Germany ended at Wiesbaden in September 1993, when the last of 11 OV-1Ds in service with the 2d Military Intelligence Battalion was flown across the Atlantic.

Operations at Robert Gray Army Air Field on Fort Hood, Texas, ended on 17 September 1993, almost simultaneously with those in Germany. Unlike the other Mohawk units, however, A Company/15th Military Intelligence Battalion was not deactivated after it retired its last OV-1Ds and subsequently received RQ-5A Hunter unmanned aerial vehicles as replacements. Mohawk observer and pilot training came to an end during 1994 when both the 304th Military Intelligence Battalion (Training) at Libby Army Air Field, Fort Huachuca, Arizona, and the 1-223d Aviation Battalion at Cairns Army Air Field, Fort Rucker, Alabama, retired their last OV-1Ds. Over the 34 years that Mohawk training was conducted at Fort Rucker the 1-223d and its predecessor organisations trained 2,302 army aviators to fly the 'Hawk'.

The final chapter in the US Army's Mohawk book was closed in September 1996 when both A Company/224 Military Intelligence Battalion at Hunter Army Air Field in Savannah, Georgia, and A Company/3d Military Intelligence Battalion at Desederio Army Air Field, Camp Humphreys, Republic of Korea, retired the OV-1D. The former unit held a formal ceremony on 13 September and the latter unit followed on 26 September. Like its sister company at Fort Hood, A Company at Camp Humphreys survived the Mohawk retirement and currently operates the Northrop Grumman RC-7B Airborne Reconnaissance Low-Multifunction (ARL-M) aircraft over the DMZ – in the same roles previously undertaken by the OV/RV-1D.

Tom Kaminski

Argentina has breathed new life into two Grumman veterans in the form of the Army's OV-1Ds and the Navy's S-2UP Turbo-Trackers. Twenty-three Mohawks were delivered, of which 10 remain in service with Escuadrón de Aviación de Apoyo de Inteligencia 601 at Campo de Mayo near Buenos Aires, and with the co-located army aviation school.

SEMA – Special Electronic Mission Aircraft
In Germany the RV-1D served on Elint-gathering duties, partnered by the Beech RC-12D Improved Guardrail V which flew the Comint mission. Both aircraft were replaced in the early 1990s by the RC-12K Guardrail Common Sensor (GRCS) System 4 platform, which combined both Elint and Comint functions in one airframe. In Korea the Elint/Comint function was assumed by the RC-12H GRCS System 3.

OV-1 Mohawk variants

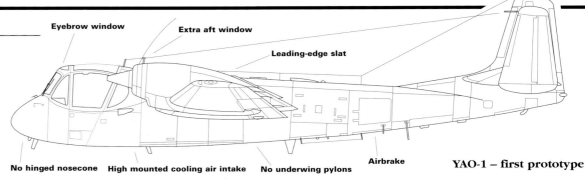

Eyebrow window

Extra aft window

Leading-edge slat

No hinged nosecone   High mounted cooling air intake   No underwing pylons

Airbrake

YAO-1 – first prototype

YAO-1/YOV-1A

Wind tunnel testing of the initial YAO-1/YOF-1 design revealed that, during certain operations, the single 'T' tail would have provided inadequate directional control, and the prototype YAO-1 was accordingly equipped with a conventionally located horizontal stabiliser and three vertical tail surfaces. The prototype YAO-1, serial 57-6463, flew for the first time at the company's Bethpage, New York, facility on 14 April 1959. Company test pilot Ralph Donnell was at the controls for the flight, which ended at Naval Weapons Reserve

Industrial Plant in Calverton, New York. This facility served as the home to Grumman's Calverton production and

57-6463 departs Bethpage on the Mohawk's first flight (left), while above is the second YAO-1, finished in a typical Army olive drab scheme. Both have air data instrumentation booms fitted.

flight test activities and was often referred to as the Peconic plant. Nine prototypes, comprising serials 57-6463/6467 and 57-6538/6541, were delivered between 14 April 1959 and 29 March 1960.

The prototypes were powered by 960-shp (716-kW) Lycoming T53-L-3 turboprop engines that drove fully reversible, three-bladed Hamilton Standard 53C51-23 propellers with a diameter of 10 ft (3.05 m). During initial flight tests the YAO-1 demonstrated its STOL capabilities by taking off and clearing a 50-ft (15.2-m) obstacle in approximately 900 ft (274.3 m). During landing it cleared the same 50-ft obstacle and came to a complete stop in approximately 300 ft (91.4 m). It also demonstrated a roll rate of 180° per second.

In 1962 the YAO-1 was redesignated as YOV-1A. Two aircraft were armed for service in Southeast Asia, becoming JOV-1As in the process, as described in the next entry.

Cruising over Long Island, the first prototype has its central fuselage, inner wings and engine nacelles heavily tufted to record airflow patterns.

AO-1AF/OV-1A/JOV-1A

The two initial lots of production AO-1AF aircraft were powered by the same T53-L-3A engine that equipped the YAO-1. However, the final lot, which were ordered as OV-1As following the adoption of the tri-service designation system in 1962, received the 1,005-shp (749-kW) T53-L-7 turboprop but retained the original propeller.

Several changes were made to the production aircraft as a result of flight testing. These included the incorporation of a new nose gear strut that was equipped with power-assisted steering. Structural changes included the deletion of small windows aft of the cockpit and eyebrow windows above the windshield, relocation of the heat exchanger intake and the incorporation of a hinged nose cone. Deicer boots were also installed along the wing, horizontal and vertical stabilisers and, on the final 21 aircraft, the wing leading-edge slats were deleted. Minor changes

were also made to the avionics, instrumentation and cockpit layout.

The AO-1AF/OV-1A was equipped with a KS-61 photographic system that comprised a KA-30 camera mounted in the mid-section of the fuselage. Two removable flare pods installed on the upper wing roots were each capable of carrying 52 upward-firing A4 and B6 flare cartridges used for night photographic missions. Remotely controlled from the cockpit, the KA-30 provided horizon-to-horizon coverage and was capable of taking pictures from five positions, including vertical, and 15° and 30° oblique positions to either side of the vertical.

59-2603 was the first production AO-1AF, the most notable difference from the prototype being the deletion of the aft side windows. The aircraft is seen carrying M4A resupply containers on Stations 3 and 4.

The upward-firing illumination flare packs above the wingroots were later replaced by underwing flasher units.

Right: 59-2620 was one of the AO-1AFs tested with ski undercarriage. The skis themselves incorporated small wheels.

Production comprised 64 aircraft that were delivered between January 1960 and February 1965.

Type	Serials	Number	Build no.
AO-1AF	59-2603/2620	(18)	1A/18A
AO-1AF	60-3720/3744	(25)	19A/43A
OV-1A	63-13114/13134	(21)	44A/64A

Owing to the original Marine Corps requirement, the basic Mohawk design featured the ability to carry six underwing stores pylons, although only two were initially installed. However, as part of the Mohawk flight test programme the Naval Air Test Center's Weapon Division at Naval Air Station Patuxent River, Maryland, had cleared the Mohawk to carry a variety of weapons. Initially, the service planned to equip the aircraft to carry rockets for marking targets and suppressing enemy fire, but, as the Mohawk's observation role developed the Army considered arming the aircraft. The additional pylons were eventually installed on 59 OV-1A and OV-1C models, along with Mk 20 gunsights. These aircraft were cleared to carry XM-14 0.50-in (12.7-mm) machine-gun pods, and seven- and 19-round 2.75-in (70-mm) rocket pods.

Six prototype and early production aircraft that had been modified were designated JOV-1A and assigned to the 23d Special Warfare Aviation

Detachment, which deployed to the Republic of Vietnam in September 1962. Serials were 57-6538, 57-6540, 59-2603, 59-2605, 59-2608 and 60-3076.

Twenty-seven of the 64 OV-1A aircraft built, including the six JOV-1As, were modified into the so-called 'gunship' configuration for use in close air support (CAS) operations in Vietnam.

OV-1A 59-2610 is loaded with rockets during field trials. Note the nose-mounted antennas for communications with ground forces.

JOV-1A – armed Mohawk

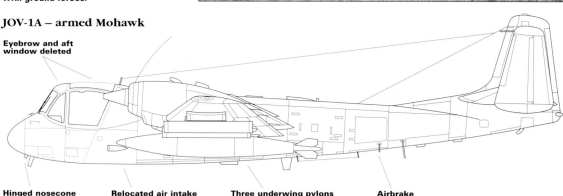

Eyebrow and aft window deleted

Hinged nosecone

Relocated air intake

Three underwing pylons

Airbrake

AO-1BF/OV-1B

The final prototype YAO-1, serial 57-6541, was modified to AO-1BF configuration and flew for the first time on 13 September 1960. Although the aircraft retained the short wing and cameras associated with the OV-1A, it was enhanced by the installation of the Motorola AN/APS-94 Side Looking Airborne Radar (SLAR). The antenna for the SLAR system was contained in an 18-ft (5.49-m) fibreglass pod carried on the lower right side of the forward fuselage. The initial 17 AO-1BFs also retained the short wing, speed brakes and wing leading-edge slats; however, the latter two features were deactivated and deleted on later aircraft.

Ten of the initial aircraft were later modified with longer wings, which were extended by 6 ft (1.83 m) and featured just two store stations, which were intended for carrying external fuel tanks. The wing area was accordingly increased to 360 sq ft (33.45 m²). The initial lot was powered by the same T53-L-3A engine that equipped the earlier versions of the OV-1A. However, the two subsequent batches featured the more powerful T53-L-7 that provided 1,005 shp (749 kW). The third and final

batch was ordered under the designation OV-1B. A number of OV-1Bs were later equipped with the more powerful T53-L-15 via a modification programme.

Operated by the observer from the right seat, the APS-94 provided an imaging reconnaissance capability during the day or night and in any weather conditions. The SLAR directed

its radar signals to either or both sides of the antenna and the returned echoes of fixed and moving objects were recorded onto photographic film strips that were automatically developed and could be viewed in flight in near real-time. The system was also equipped with a moving target indicator (MTI) mode that was useful in highlighting vehicles in motion and was capable of

transmitting this data to a ground station via a VHF data link. The first of 101 AO-1BF/OV-1Bs was delivered in August 1960 and the final aircraft left the production line in May 1966.

Type	Serials	Number	Build no.
AO-1BF	59-2621/2367	(17)	1B/17B
AO-1BF	62-5859/5906	(48)	18B/65B
OV-1B	64-14238/14273	(36)	66B/101B

Built as the ninth and last YAO-1, 57-6541 served as the prototype for the SLAR-carrying AO-1BF. All six pylons were retained initially, but were soon deleted, leaving the AO-1BF/OV-1B able only to carry fuel tanks

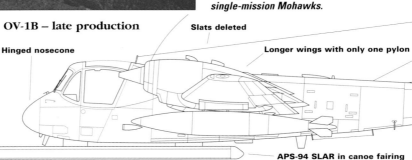

The longer-span wing introduced early in OV-1B production is noticeable on this Georgia National Guard aircraft. The 151st Military Intelligence Battalion was the last user of the first-generation of single-mission Mohawks.

59-2621 was the first production OV-1B, seen here during an early test flight with airbrakes deployed and flare packs fitted. OV-1Bs typically operated at this sort of altitude, whereas As and Cs undertook their photo and IR duties at lower levels. Production Mohawks featured a simplified cockpit glazing arrangement which deleted the small 'eyebrow' and aft windows of the prototypes.

OV-1B – late production

Slats deleted

Hinged nosecone

Longer wings with only one pylon

Airbrake deleted

APS-94 SLAR in canoe fairing

AO-1CF/OV-1C/'Super-C'/JOV-1C

Based on the AO-1AF airframe, the AO-1CF became the most numerous production variant of the Mohawk. Also intended for the battlefield surveillance role, it differed from the earlier version in being equipped with the AN/UAS-4 Red Haze Infrared (IR) detection system and its associated datalink. While the IR system was primarily intended for night-time missions, this variant retained the KS-61 photographic system and removable flare pods. The initial 81 aircraft retained the airframe and T53-L-3A powerplant combination of the AO-1AF, including its speed brakes. The model was redesignated OV-1C in 1962.

During 1965 serial 61-2728 was equipped with a number of new features that were incorporated in Fiscal Year 1966 and subsequent aircraft. The 88 aircraft that followed had the longer wing of the OV-1B and featured the more powerful T53-L-15, which offered 1,150 shp (858 kW). These were also the first Mohawks to sport an air conditioning system. The later versions

This ski-equipped OV-1C was undertaking snow trials at Bemidji, Minnesota in 1963. Further ski trials also took place in Canada – at Kenora in western Ontario.

were also equipped with updated vertical tape instrumentation that replaced the earlier round dial or 'boiler'-type gauges, and were unofficially referred to as 'Super-Cs'. The wing leading-edge slats were later deactivated and deleted effective with the 72nd production OV-1C. These were also the first aircraft equipped to carry the LS-59 flasher pod under the wing in place of the flare pods that had been carried atop the wing at the fuselage wing juncture. Thirty-two OV-1Cs were modified as armed gunships, including a number of JOV-1Cs that were equipped with dual controls and had their IR systems removed.

In 1966 Grumman modified OV-1C serials 61-2683 and 62-5849 with several new systems then being

installed in production aircraft. These included a new identification friend or foe (IFF) set, gyromagnetic compass, marker beacon, updated radios, a Doppler navigation system and an updated AN/AAS-14 system. The aircraft were also equipped with T53-L-7 engines and their wing leading-edge slats were deactivated. Later OV-1Cs

were also equipped with a second panoramic KA-60 camera in the nose. This installation was eventually retrofitted to earlier models as well. Beginning in 1966, 31 early production OV-1Cs were returned to the company's Stuart, Florida, facility where they received these modifications. Armament provisions were also removed from those aircraft that were so-equipped. The first of these modified OV-1Cs was returned to the Army on 29 July 1966.

Grumman delivered 169 AO-1CF/OV-1Cs between February 1961 and December 1969.

Type	Serials	Number	Build no.
AO-1CF	60-3745/376	(17)	1C/17C
AO-1CF	61-2675/2728	(54)	18C/71C
AO-1CF	62-5849/5858	(10)	72C/81C
OV-1C	66-18881/18896	(16)	82C/97C
OV-1C	67-18897/18932	(36)	98C/133C
OV-1C	68-15930/15965	(36)	134C/169C

In June 1968 one OV-1C was modified with direction-finding equipment and was assigned briefly to the 156th Radio Research Company at Can Tho, Vietnam, on Elint duties. The programme, known as Homing Pigeon, was not successful and the Mohawk was withdrawn in September.

Despite the significant addition of its Red Haze infrared sensor, the OV-1C was outwardly little different to the OV-1A. This example is engaged in field exercises with troops and OV-1Bs.

Above: The Georgia Army National Guard unit (151st MIB) had two Mohawk companies for some time – the 158th and 159th MICAS. They operated a sizeable fleet of OV-1Bs and Cs (illustrated)

Below: Resplendent in a newly-applied scheme, N6744 was one of four OV-1Cs assigned to the US Customs Service after being made surplus to Army requirements in 1973. They were modified to house a FLIR turret in an elongated nose, which was used for detecting and tracking drug smugglers.

OV-1A/B/C underwing stores

The JOV/OV-1A and JOV/OV-1C models that had been modified to carry underwing armament were cleared for a variety of forward-firing weapons, bombs and other stores. Stations 1/2, and 5/6, were stressed to carry loads up to 500 lb (227 kg) and were the outboard stations. Wing stations 3 and 4 were located inboard of Stations 1/2 and 5/6, and were stressed to carry loads up to 1,000 lb (454 kg). However, carriage of 300-US gal (1136-litre) ferry tanks exceeded this limit and required special flight restrictions. The OV-1B was only equipped with the inboard wing stations and normally only carried external fuel tanks.

Station 1/6, 2/5
Aero 15C or 15D bomb rack
250-lb (113 kg) bomb
500-lb (227 kg) bomb
LAU-32/A (Aero 6A) or XM157 2.75-in (70-mm) rocket pod (seven-round)
LAU-3/A (Aero 7D) or XM159 2.75-in (70-mm) rocket pod (19-round)
LAU-10/A 5-in (127-mm) Zuni (four-round) rocket launcher
XM13 grenade launcher

XM14 armament system: 0.50-in (12.7-mm) (SUU-12) machine-gun pod (M3 machine-gun with 750 rounds)
XM18 armament system: 7.62-mm (SUU-11) Minigun pod (1,500 rounds)
Sidewinder 1A or 1C missile
LS-59A flasher (Station 5)

Station 3/4
Aero 65A bomb rack
LAU-25 (Aero X5A) with Mk 24 parachute flares (eight)
Mk 81 250-lb (113-kg) bomb
Mk 82 500-lb (227-kg) bomb
750-lb (340-kg) bomb
XM13 40-mm grenade launcher
XM14 armament system: 0.50-in (12.7-mm) (SUU-12) machine-gun pod (M3 machine gun with 750 rounds)
XM18 armament system: 7.62-mm (SUU-11) Minigun pod (1,500 rounds)
Mk 83 1,000-lb (454-kg) bomb
Mk 79 1,000-lb (454-kg) fire bomb (112-US gal/424-litre napalm)
Mk 12 smoke tank
CBU-19 (E-153) chemical cluster bomb
150-US gal (568-litre) Aero 1C external fuel tank
300-gal (1136-litre) Aero 1D ferry fuel tank
M4A resupply container

Silver Lance

Two Mohawks, comprising YOV-1A serial 57-6464 and OV-1C serial 60-3748, operated along the borders between western and eastern Europe during the early 1960s under a project known as Silver Lance. The aircraft were equipped with a passive electronic intelligence (ELINT) system that was referred to as Hot Pipe and Battleaxe, and designed to intercept radar transmissions and locate the emitters. The same missions would later be allocated to the RV-1D. The aircraft were assigned to Detachment 14, 314th Army Security Agency Battalion (ASA Bn) at Herzogenaurach, but operated from Fuerth Airfield in Nürnberg.

The Silver Lance aircraft carried its mission equipment in its avionics bays, and antennas were installed on the fuselage. To reduce the aircraft's weight they were not equipped with speed brakes. Cockpit equipment installed in the observer position included a prototype Doppler navigation system.

RV-1C Quick Look

Late in 1969 two OV-1C aircraft (serials 68-15964 and 68-15965) were modified to a new configuration known as RV-1C Quick Look. These aircraft were equipped with an electronic surveillance and target acquisition system. The AN/APQ-142 electronic intelligence system was designed to intercept and pinpoint the location of radar emitters by using an inertial navigation system that precisely indicated the aircraft's location. The Quick Look mission equipment was primarily carried in two pods carried on the outboard wing stations 1 and 6.

After undergoing testing in the CONUS the aircraft were deployed to Fliegerhorst Army Air Field in Hanau, West Germany, during December 1970, where they underwent Maintainability Feasibility Testing (MFT) with the 131st Military Intelligence Company (MI CO). The RV-1Cs were transferred to the Army Security Agency (ASA) in 1973 and underwent development testing/ operational testing (DT/OT) at Fort Huachuca, Arizona, between February and May 1974.

The Quick Look aircraft subsequently deployed to Kitzingen, West Germany, in September 1974 and were operated by Detachment 1 of the 330th ASA Company. The aircraft relocated to Stuttgart, West Germany, in 1975 and remained there until replaced by the RV-1D Quick Look II during 1979. Both RV-1Cs were eventually modified to OV-1D configuration.

Project 452 SEAMORE

Although the Mohawk had proven to be a capable reconnaissance platform, operations in Southeast Asia required that the aircraft be able to detect the infiltration of personnel and equipment. Since its systems were not specifically designed for these tasks, an effort was made to increase the target detection capability of the infrared (IR) and side-looking airborne radar (SLAR) sensors, improve the target location accuracy, and cockpit sensor displays. Field commanders were looking for a number of capabilities including: the ability to detect slow moving targets, improved resolution, near real-time cockpit display and the ability to provide target coordinates for strike aircraft.

Beginning in March 1967 Grumman responded to the army requirement by modifying four Mohawks as part of the South-East Asia MOhawk REquirement (SEAMORE) project. Developed under design G-134T, the programme involved the modification of two OV-1Bs and two

OV-1Cs. The KS-61 camera system, flare ejector system and AN/UPD-2 systems installed in OV-1B serials 62-5903 and 64-14242 were removed. The AN/APS-94 SLAR was replaced by a modified APS-94C system that included a larger display and an AN/AYA-5 data annotation system, SST-181X ground track beacon and an AN/ALR-25/26 radar warning system were installed.

In addition, the KS-61 camera system, flare ejector system and AN/ART-441A radio transmitting set installed in OV-1C serials 61-2718 and 62-5857 were removed. The AN/AAS-14 infrared surveillance system was replaced by an AN/AAS-22 and an AN/AYA-5 data annotation system, SST-181X ground track beacon, CDC-449 target location computer and an AN/ALR-25/26 radar warning system were installed.

The aircraft were deployed to Southeast Asia in May 1968 and were operated by the 244th Surveillance

Airplane Company at Can Tho in support of the 224th Army Security Agency Battalion until transferred to the 131st Aviation Company at Phu Bai, RVN, in October 1968. One of the OV-1Bs was lost during October 1969 but the surviving three remained in Southeast Asia until 1972. Upon returning to the CONUS the remaining OV-1B (serial 62-5903) was bailed to the National

62-5903 was one of two OV-1Bs upgraded under the SEAMORE programme, featuring RWR and an improved SLAR with better displays.

Aeronautics and Space Administration's Lewis Research Center and supported an ice reconnaissance programme over the Great Lakes during the winter of 1972/73.

OV-1D

First procured in 1968, the OV-1D was developed as a result of a 1967 Army decision to develop a single model capable of being rapidly converted to fly any of the three reconnaissance

The OV-1D introduced a number of new systems, including a beacon tracker installed in an underbelly fairing. This aircraft carries a LS-59 flasher pod and ALQ-87 ECM pods during tests, for which an air data 'barber's pole' boom was added to the nose.

Four OV-1Cs were converted to YOV-1D status to test the quick-change multi-sensor platform configuration. They were also fitted with forward-facing nose-mounted camera.

missions, and thereby reduce the overall number of Mohawks variants it operated. It was the final production model of the Mohawk and, although 41 aircraft were delivered, the initial four YOV-1Ds were conversions of OV-1Cs. They were accepted by the army between September and December 1968. The 37 newly manufactured OV-1Ds were delivered between September 1969 and December 1970.

In addition to the vertical and forward panoramic KA-60 cameras then comprising the KS-61 camera system, the new variant was equipped with a KS-113A surveillance system that featured a KA-76A vertical camera. It was also capable of alternately carrying an updated AN/UAS-4 infrared detection system and its associated AN/AAS-24 infrared sensor, or the improved AN/APS-94D SLAR system. Swapping from one system to the other is reported to have taken approximately one hour to accomplish. Although the YOV-1Ds were powered by the 1,150-shp (858-kW) T53-L-15 engines, the production aircraft were equipped with the T53-L-701, which produced 1,400 shp (1044 kW). The production aircraft were also equipped with upgraded 53C51-27 propellers. The fuselage centre section and wing carry-through structure were strengthened to support the OV-1D's higher operating weights. The aircraft was also capable of being equipped with a louvred, scarfed shroud suppressor (LSSS) that provided passive infrared signature suppression for the engines.

Type	Serials	Number	Build no.
YOV-1D*	67-18898/18901	(4)	1D/4D
OV-1D	68-16990/16999	(10)	5D/14D
OV-1D	69-17000/17026	(27)	15D/41D

* Converted from OV-1C airframes

OV-1D mission systems

The OV-1D's photographic equipment included the KS-61 panoramic camera surveillance system and KS-113A airborne photographic surveillance system. The aircraft was also capable of alternately carrying the AN/UAS-4 Infrared or AN/APS-94 SLAR systems. The latter system was part of the AN/UPD-2A radar surveillance system. Controls for both systems were alternately installed in the observer's station, while associated equipment was carried in equipment bays within the fuselage. The two interchangeable systems could be swapped out by maintenance personnel in approximately one hour.

AN/UAS-4 Infrared Detection Set (IDS)

The Infrared Detection Set weighed 602 lb (273 kg) and was composed of the AN/AAS-24 high-resolution, electronically roll-stabilised infrared sensor, a cockpit display/control console and a remote film recorder. Infrared images of terrain beneath the aircraft were simultaneously displayed on the 10.5 x 8.5-in (27 x 21.5-cm) cockpit display and were recorded in 'real-time'. The recorder imprinted the image on 5-in (127-mm) wide film, along with identifying data provided by the aircraft's central data annotation system. Cockpit controls allowed the observer to slew and expand the view, make adjustments to the image and mark points of interest on the film. The sensor was carried in a blister on the centreline of the aft fuselage.

This view of One Bad Bitch, an OV-1D from A Company, 2d MIB at Stuttgart, clearly shows the nose-mounted KA-60C panoramic camera, which swept from side-to-side ahead of the aircraft, angled down at 20°. Also visible is one of the handles on the APS-94 SLAR pod which facilitated the rapid change of mission equipment between this fit and the IR system.

AN/UPD-2A Radar Surveillance System

Weighing 776 lb (352 kg), the OV-1D's AN/APS-94F Side Looking Airborne Radar (SLAR), and AN/AKT-18B radar data transmitter system were the airborne component of the AN/UPD-2 Radar Surveillance System. The SLAR was designed to locate fixed and moving targets in all weather conditions on either or both sides of the aircraft. The system was composed of an external antenna, processors and a recorder-processor-viewer in the cockpit. The AN/AKT-18B digital datalink system transmitted SLAR surveillance data to the AN/TKQ-2B radar data receiving set on the ground, in 'real-time' via UHF through line-of-sight up to 100 nm (185 km), or via an airborne relay. The recorder-processor-viewer provided the observer with a 9.5-in (241-mm) wide filmed recording of the CRT imagery. The display showed moving targets and fixed terrain imagery on opposites sides of the film and its speed was controlled either automatically, via the aircraft's inertial navigation system (INS), or manually. Like the IDS, the system also recorded identifying data from the central data annotation system. The system was capable of mapping swaths 15.5, 31 and 62 miles wide (25, 50 and 100 km). SLAR-equipped aircraft were readily identified by the large antenna carried under the starboard fuselage in a long rectangular box, with handles for quick removal. Additional equipment was carried in three avionics bays beneath the wing.

KS-61 and KS-113A camera equipment

Three independent cameras were capable of being operated by the pilot or observer, and included two 180° panoramic KA-60C cameras and a single KA-76A serial frame camera. The KS-61 panoramic camera surveillance system comprised a KA-60C that was mounted in the nose and aimed below the normal line of flight at a depression of 20°. A second KA-60C was installed in a blister on the aft fuselage centreline. Its optical axis was perpendicular to the normal line of flight.

The KS-113 Airborne Photographic Surveillance System comprised a KA-76 camera that was installed on a remote-controlled rotary mount in the aft fuselage just forward of the second KA-60C. It was adjustable via a control in the cockpit and could be rotated laterally to one of five positions (vertical down, 15° and 30° depression left or right).

Both of the fuselage-mounted cameras were protected by automatically closing doors. While the KA-60s were generally used at altitudes below 5,000 ft (1524 m), the KA-76 was equipped with four interchangeable lens cones and could be used at any altitude. The installation of an LS-59A electronic flasher pod under the wing enabled the KA-76 to be used for low-level night photography. The AN/AYA-10 airborne data annotation system (ADAS) supplied each camera with a velocity/height signal for automatic control of camera pulsing, film speed, image motion compensation and overlap along the flight line. It also provided each camera with real time navigation data and other information that was recorded on each frame of film.

When operating at an altitude of 2,000 ft (610 m) the vertical KA-60C and KA-76A were capable of photographing flight lines that were respectively 120 miles (193 km) and 220 miles (354 km) long with 60 percent overlap. At altitudes above 5,000 ft (1524 m), which was the limit of the radar altimeter, or if the automatic velocity/height signal failed, a synthetic signal could be selected by the observer, who could also operate the cameras manually during automatic modes or when the cameras were in standby mode.

68-16988 was one of the 37 OV-1Ds built from new, and was completed without the airbrakes (which were disarmed on OV-1Ds converted from earlier aircraft). The store on the intermediate wing station is the LS-59 flasher pod, powered by a nose-mounted ram air turbine.

Grumman delivered the last of these new OV-1Ds on 16 December 1970. Later, the company-operated modification and overhaul facility at Martin County Airport in Stuart, Florida, brought the older Mohawks to the new standard. Beginning in 1973 Grumman modified two RV-1Cs and 59 OV-1Cs to OV-1D configuration at Stuart. The modified aircraft comprised RV-1C serials 68-15964/15965, and OV-1Cs 67-18900, 67-18903/18904, 67-18906/18912, 67-18916/18921, 67-18922/18927,

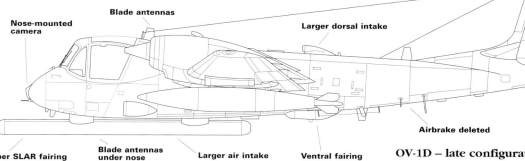

Blade antennas

Nose-mounted camera

Larger dorsal intake

Airbrake deleted

Deeper SLAR fairing

Blade antennas under nose

Larger air intake

Ventral fairing

OV-1D – late configuration (SLAR)

OV-1D and RV-1D countermeasures

The OV-1D and RV-1D were equipped with the AN/APR-39(V)2 passive radar signal detection set, which provided visual and aural indications of air defence radar threat signals. In addition, the AN/APR-44 provided warning of continuous wave radar illumination.

Active infrared countermeasures included the AN/ALQ-147 Hot Brick pod, which provided a false infrared signature to confuse and defeat heat-seeking missiles. When the pod was activated, fuel carried in a 15-US gal (57-litre) reservoir within the pod was ignited to provide a modulated infrared frequency at the exhaust. On the OV-1D a single dedicated AN/ALQ-147A could be carried on Station 1 or 6 or a removable pod could be carried on Station 3 or 4 in place of an external fuel tank. On the RV-1D the pod could only be carried on Station 4.

Additional equipment included the internally installed AN/ALQ-136, AN/ALQ-156 and AN/ALQ-162 electronic jamming systems, which replaced the pod-mounted AN/ALQ-67 and AN/ALQ-80 jamming pods that had equipped the early OV-1Ds.

The Sanders AN/ALQ-147 Hot Brick IR jammer was an important counter to the heat-seeking missiles which posed a major threat to the Mohawk at low altitude over the battlefield. The ALQ-147A(V)1 was a self-contained unit, seen below on a 2d MIB OV-1D (below), while the ALQ-147A(V)2 was mounted on the rear of a fuel tank, as carried by a 2d MIB RV-1D (right, with ALQ-133 pod alongside).

When using SLAR, the OV-1D was less vulnerable as it could operate from stand-off ranges. This became of increasing importance as battlefield SAMs proliferated in Europe.

68-18929/15935, 68-15937/15943, 68-15945/15948, 68-15950/68-15955 and 68-15956/15963. The first of these was returned to the Army in September 1974 and the last in September 1981.

A further 17 OV-1Ds were delivered through the conversion of OV-1Bs comprising serials 62-5865, 62-5867, 62-5872/5876, 62-5878, 62-5885/5890, 62-5898/5899, and 62-5902. The first of these was delivered to the Army in August 1982 and the final conversion

returned to service in January 1987. The process to modify an OV-1C to OV-1D configuration required an average of 16,000 man hours to accomplish and the modification of an OV-1B required approximately 24,000 man hours.

The later OV-1Ds were equipped with an upgraded AN/APS-94F SLAR and were also capable of being equipped with the M130 chaff/flare dispensers, AN/ALQ-147 Hot Brick infrared jamming pod, AN/APR-39 radar warning receiver, AN/APR-44 continuous wave radar warning receiver, AN/ALQ-136 radar jammer, AN/ALQ-162 continuous wave radar jammer and the AN/ALQ-156 missile detection system (RV-1D).

OV/RV-1D specifications

Dimensions	
Length (overall)	44 ft 11 in (13.69 m) (OV-1D)
	41 ft 4 in (12.60 m) (RV-1D)
Wingspan	48 ft (14.63 m)
Wing area	360 sq ft (33.45 m²)
Height	12 ft 8 in (3.86 m)
Tail span	15 ft 10 in (4.83 m)
Wheel base	11 ft 8.125 in (3.56 m)
Static tread	9 ft 2 in (2.79 m)

Weights	
Empty	11,747 lb (5328 kg) (OV-1D)
	11,501 lb (5217 kg) (RV-1D)
Maximum	18,000 lb (8165 kg) (OV-1D/SLAR)
	17,826 lb (8085 kg) (OV-1D/IR)
	17,667 lb (8014 kg) (RV-1D)

Fuel capacity
Internal 297 US gal/1,930 lb (1124 litres/875 kg)
External Two 150-US gal (568-litre) external fuel tanks or 300-US gal (1136-litre) ferry tanks

Powerplant	Two Lycoming T-53-L-701
	turboprop engines each rated at 1,400 shp
	(1044 kW) driving three-bladed Hamilton
	Standard 53C51-27 propellers

Performance	
Max. speed	251 kt (465 km/h) (OV-1D/SLAR)
	265 kt (491 km/h) (OV-1D/IR)
	265 kt (491 km/h) (RV-1D)
Take-off run	1,735 ft (529 m) (OV-1D/SLAR)
to 50 ft	1,750 ft (533 m) (OV-1D/IR)
(15 m)	1,610 ft (491 m) (RV-1D)
Landing run	1,137 ft (347 m) (OV-1D/SLAR)
from 50 ft	1,128 ft (344 m) (OV-1D) (IR)
(15 m)	1,720 ft (524 m) (RV-1D)
Service ceiling	25,000 ft (7620 m)
Range	892 nm (1651 km) (OV-1D/SLAR)
	938 nm (1737 km) (OV-1D/IR)
	877 nm (1652 km) (RV-1D)
Maximum	4.35 hours (OV-1D/SLAR)
endurance	4.54 hours (OV-1D/IR)
	4.59 hours (RV-1D)
Crew	pilot and observer

OV-1C/D and RV-1D underwing stores

In later years surviving OV-1Cs, along with all OV-1D and RV-1D models, were equipped with the full complement of six store stations. However, they were restricted to carrying mission-related equipment, including countermeasures stores and external fuel tanks.

Station 1/6
Aero 15C or 15D bomb rack
AN/ALQ-147A Hot Brick Infrared countermeasures pod (Stations 1 or 6) (OV-1C/D)
AN/ALQ-67 Fuze jammer countermeasures pod (Station 1) (OV-1C/D)
AN/ALQ-80 Noise countermeasures pod (Station 6) (OV-1C/D)

AN/ALQ-133 Emitter Locator (Stations 1/6) (RV-1D)

Station 3/4
Aero 65A bomb rack
150-US gal (568-litre) Aero 1C external fuel tank* (Stations 3 or 4)
AN/ALQ-147A (Sta. 3 or 4) (OV-1C/D)*
300-US gal (1136-litre) Aero 1D ferry fuel tank (Stations 3 or 4)
*The RV-1D was capable of carrying a single AN/ALQ-147 pod on station 3 or 4 in place of an external fuel tank

Station 5
Aero 15C or 15D bomb rack
LS-59A photo flasher (Station 5) (OV-1C/D)

Although the OV-1D was never armed in US Army service, two of the examples transferred to Argentina were tested with rocket pods and bombs. This example has two seven-round LAU-32 rocket pods alongside the standard 150-US gal tank.

RV-1D Quick Look II

The success of the RV-1C led the Army to order the modification of a number of ELINT aircraft. Beginning in 1974 Grumman modified 31 OV-1Bs to carry the AN/ALQ-133 Quick Look II airborne emitter locator identification system.

Like all previous modification programmes, this project was also carried out at Stuart. Approximately 20,000 man hours were required to convert an OV-1B into an RV-1D. The first RV-1D was accepted by the Army in February 1977 and the aircraft entered operational service in 1978.

As part of the modification programme, the KS-61 and KS-113A photographic systems and the AN/APS-94 SLAR were removed. The electronic intelligence (ELINT) system included a countermeasures receiving set comprising two pods carried on stations 1 and 6, a control indicator panel at the observer's station and associated electronics located in three avionics bays beneath the left wing. The receiving set was a tactical airborne emitter location system that provided real-time location of electromagnetic emitters. The RV-1D normally carried two 150-US gal (568-litre) fuel tanks on

68-15932 was the first RV-1D conversion. It incorporated several of the features adopted for the OV-1D, including beacon tracker, new communications antennas, enhanced cooling systems with larger intakes and a nose-mounted antenna for the ILS glideslope.

Above: The outward-facing antennas for the ALQ-133 Elint system were housed in pods under the wings, and are clearly visible in this view. The wartime mission of the RV-1D was to plot enemy radars in the battlefield, while in peacetime it monitored WarPac forces along the East/West German border.

Right: On the ground the antennas were usually covered for protection. This cover on a 2d MIB RV-1D sports a 'Quick Look' badge.

RV-1D mission equipment

AN/ALQ-133 Airborne Emitter Locator Identification System

Installed in the observer's position and known as Quick Look II, the electronic intelligence (ELINT) system consisted of the countermeasures receiving set and the AN/USQ-61 or 61A digital datalink. It weighed 1,124 lb (510 kg). The receiving set was a tactical airborne emitter location system that provided real-time location of electromagnetic emitters by collecting and processing specific data received from ground-based radio-frequency emitters detected along the aircraft's flight path. The collected data was downloaded from the receiving set when the RV-1D returned to its base. The system, however, was also capable of responding to control signals transmitted via the AN/USQ-61 or 61A digital datalink from the ground station. Data that was collected in response to signals from the ground station was capable of being transmitted back to the ground station via the datalink. The countermeasures receiving set was composed of two pods carried on stations 1 and 6, a control indicator panel at the observer's station, and associated electronics located in three avionics bays beneath the left wing.

stations 3 and 4 and, like the OV-1D, it was capable of being equipped with the louvred, scarfed shroud suppressor (LSSS) infrared signature suppression equipment.

OV-1B aircraft modified to RV-1D configuration comprised: 62-5891, 62-5897, 64-14238, 64-14239, 64-14242/64-14248, 64-14250, 64-14252/64-14256, 64-14258/ 64-14263, 64-14265, 6414267/ 64-14273.

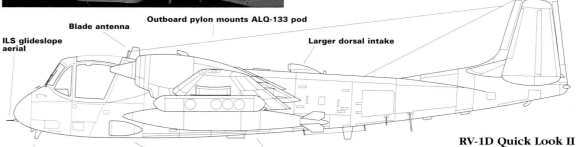

Blade antenna — Outboard pylon mounts ALQ-133 pod — ILS glideslope aerial — Larger dorsal intake — Blade antennas under nose — Large air intake — Two underwing pylons

RV-1D Quick Look II

OV-1E

Beginning in September 1986, a single OV-1D was equipped with updated cockpit systems and served as the

67-18922 served as the avionics prototype for what would have become the OV-1E. Notable is the undernose fairing associated with improved defences.

prototype under the block improvement plan. As part of the modification programme, serial 67-18922 was equipped with a digital instrumentation system comprising two multi-function displays (MFD), new radios, navigation equipment including a global positioning system, and new defensive avionics. During the same period, serial 68-15932 – another OV-1D – served as the

testbed for an uprated version of the AVCO Lycoming turboprop known as the T53-L-704, which produced 1,800 shp (1342 kW).

Had the Army followed through with the two projects, they would have been incorporated into a multi-stage improvement programme (MSIP) that would have resulted in the assignment of the formal designation OV-1E. Serial

67-18922, which served as the avionics test bed, is currently displayed at Hunter Army Air Field, Savannah, Georgia, in front of the 224th MIB (AE) headquarters building.

Suspended in an anechoic chamber, the avionics prototype undergoes systems compatibility trials . A new RWR 'pimple' was mounted at the tip of the nose.

Mohawk proposals

During 1961 Grumman proposed its model **134L**, which was more commonly referred to as the **AO-1EF Increased Capacity Mohawk**. The design retained the overall dimensions of the AO-1AF with the exception of its length, through the incorporation of a redesigned forward fuselage that added 28 in (71 cm) to the aircraft forward of the wing. The new fuselage provided a 300 per cent increase in fuselage volume that could simultaneously accommodate mission equipment and

dual controls. The ejection seats and frangible glass panels were replaced by fixed seats and an entry door was located aft of the propeller arc. In addition to offering a greater internal volume, the aircraft could carry a third crew member for specific missions or 800 lb (363 kg) of cargo if configured for the utility role. Production of this variant never advanced beyond a wooden mockup of the cockpit.

While Grumman produced new Mohawks it continued to look at

advanced concepts for improving the aircraft. A number of these projects revolved around providing the aircraft with new powerplants. These included advanced versions of the Lycoming T53, and a variant of the T55 turboshaft that equipped the Boeing Vertol CH-47 Chinook helicopter. Known by the model number **134N-3c**, the T55-powered Mohawk would also have been equipped with larger 11-ft 3-in (3.43-m) diameter propellers. The engines and propellers would have been connected via a cross shaft in a design that was similar to that used in today's MV-22

Osprey. The **134N-2a** combined the T53-L-7 with a single General Electric J85 turbojet installed in the aircraft's baggage compartment, while the **134N-4a** concept used two J85s to drive large ducted fans.

A multi-sensor version, proposed in 1962, was equipped with a 70-mm camera in the nose, a 5-in (12.7-cm) panoramic camera, lightweight SLAR, infrared mapper, a radar warning receiver and Doppler navigation system. Although this was not accepted for development, much of the design was incorporated in the later OV-1D variant.

OV-1 Mohawk operators

UNITED STATES ARMY

The first operational Mohawks were fielded by the 23d Special Warfare Aviation Detachment (SWAD) at Fort Rucker, Alabama, in July 1962. The aircraft were subsequently deployed with Aerial Surveillance Target Acquisition (ASTA) platoons assigned to divisional aviation battalions and artillery units assigned to US Army Europe, Continental Army Command, US Army Alaska, and units of the US Army Pacific within the Republics of Korea and Vietnam. Operations in Southeast Asia (SEA) began when the 23d Special Warfare Aviation Detachment (SWAD), relocated from Fort Rucker to Nha Trang, Republic of Vietnam (RVN) in September 1962.

When the 23d SWAD initially deployed to Vietnam it arrived with six modified JOV-1As that were equipped with cameras and the capability to carry machine-guns and rockets. In December 1964 the 4th Aerial Surveillance Target Acquisition (ASTA) Detachment joined the 23d SWAD when it arrived at Vung Tau, RVN, from Fort Bragg, North Carolina, with OV-1B and OV-1C models. Before the 4th ASTA could even fly its first mission it was consolidated with the 23d SWAD to form the 73d Aviation Company (Surveillance Airplane) or Surveillance Airplane Company (SAC) at Vung Tau in January 1965. During 1964 Mohawks were also deployed to the Korean peninsula and these aircraft would go on to patrol the demilitarised zone (DMZ) between North and South Korea for over three decades with a number of units.

In 1965 the 20th ASTA Detachment relocated from Fort Riley, Kansas, to Nha Trang, RVN, followed by the ASTA Platoon assigned to the 1st Infantry Division (ID), which arrived at Phu Loi from Fort Riley in October 1965. It was preceded by the 1st Cavalry Division's (CD) ASTA Platoon, which was attached to the 11th Aviation Company. The organisation previously operated as many as 24 Mohawks, from Lawson Army Air Field, Fort Benning, Georgia, while demonstrating the concept of using the Mohawk in the armed surveillance role. Throughout this period the unit operated as part of the 11th Air Assault Division's 226th Aerial Surveillance and Escort Battalion. The 11th arrived at An Khe, RVN, in July 1965. When the 226th was reorganised several of its OV-1As and OV-1Bs were assigned to the US Army Combat Developments Experimental Center (CDEC) at Fritzsche Army Air Field, Fort Ord, California, which continued the development of tactics that would be used in Southeast Asia. The 20th ASTA was eventually redesignated the 131st SAC in June 1966 while the units assigned to the 1st CD and 1st ID gave up their Mohawks and were inactivated. The 11th Aviation Company was also reorganised and its Mohawks were transferred to other units.

In Europe ASTA platoons were assigned to several divisional aviation battalions along with a smaller number of artillery units. The platoons were tasked with a variety of intelligence missions along the borders with the Warsaw Pact nations and were not equipped to carry under wing weapons.

Aerial Surveillance and Target Acquisition (ASTA) Platoons

1st AVN BN/1st Infantry Division	Marshall AAF, Fort Riley, Kansas; Phu Loi, Republic of Vietnam
2d AVN BN/2d Infantry Division	Lawson AAF, Fort Benning, Georgia; Republic of Korea
3d AVN BN/3d Infantry Division	Kitzengen AAF, Harvey Barracks, West Germany
4th AVN BN/4th Infantry Division	Polk AAF, Fort Polk, Louisiana
5th AVN BN/5th Infantry Division	Butts AAF, Fort Carson, Colorado
7th AVN BN/7th Infantry Division	Camp Casey, Republic of Korea
8th AVN BN/8th Infantry Division	Finthen AAF, West Germany
9th AVN BN/9th Infantry Division	Gray AAF, Fort Lewis, Washington
123d AVN BN/23d Infantry Division	Fort Stewart, Georgia
24th AVN BN /24th Infantry Division	Augsburg, West Germany
25th AVN BN/25th Infantry Division	Wheeler AAF, Schofield Barracks, Hawaii
501st AVN BN/1st Armored Division	Hood AAF, Fort Hood, Texas
502d AVN BN/2d Armored Cavalry Regiment	Nurnburg, West Germany
503d AVN BN/3d Armored Division	Hanau AAF, West Germany
504th AVN BN/4th Armored Division	Fuerth, West Germany
4th ASTA Det./82d AVN BN/82d Airborne Division	Simmons AAF, Fort Bragg, North Carolina; Vung Tau, Republic of Vietnam
101st AVN BN/101st Airborne Division	Campbell AAF, Fort Campbell, Kentucky
20th ASTA/14th AVN BN	Marshall AAF, Fort Riley, Kansas; Nha Trang, Republic of Vietnam; Phu Bai, Republic of Vietnam
11th Aviation Company (GS)/1st CAV DIV Airmobile)*	Lawson AAF, Fort Benning, Georgia; An Khe, Republic of Vietnam
110th Aviation Company/Southern Europe Task Force	Verona AAF, Italy
25th Artillery	Echterdingen AAF, Stuttgart, West Germany
26th Artillery	Darmstadt AHP, West Germany
55th Aviation Company/17th Combat Aviation Group	Seoul, Republic of Korea
12th Aviation Company/19th AVN BN	Bryant AHP, Fort Richardson, Alaska

*Previously designated the 11th Air Assault Division (Test) before deploying to Vietnam.

OV-1A of the 1st Infantry Division in revetments at Phu Loi, Vietnam.

116

A wholesale restructuring of the European-based ASTA platoons resulted in the activation of the 122d Aviation Company (Surveillance Airplane) at Fliegerhorst Army Air Field, Hanau, Germany, in 1965. Rather than being attached to a specific division, the unit supported US Army Europe, V Corps and VII Armored Corps.

Three additional surveillance airplane companies were deployed in Vietnam during 1967 when the 244th, 225th and 245th were transferred from Fort Lewis, Washington, during July, August and October. These units were, respectively, stationed at Can Tho, Phu Hiep and Da Nang. The five SACs based in the RVN were all assigned to the 1st Aviation Brigade, which reported directly to US Army Vietnam Headquarters (USARV-HQ). Five SACs were deployed to SEA and attached to each of four Corps areas that divided Vietnam. The fifth company was tasked with missions elsewhere in Southeast Asia.

Aviation Company (Surveillance Airplane)

73d Aviation Company (SA)	Vung Tau, RVN; Long Thanh North, Republic of Vietnam
131st Aviation Company (SA)	Phu Bai, Republic of Vietnam; Da Nang, Republic of Vietnam
225th Aviation Company (SA)	Gray AAF, Fort Lewis, Washington; Phu Hiep, Republic of Vietnam; Tuy Hoa, Republic of Vietnam
244th Aviation Company (SA)	Gray AAF, Fort Lewis, Washington; Can Tho, Republic of Vietnam
245th Aviation Company (SA)	Gray AAF, Fort Lewis, Washington; Da Nang, Republic of Vietnam

Mohawk operations in Southeast Asia were consolidated beginning in October 1970 when the 245th SAC was deactivated. The 225th followed in September 1971 and the 244th deactivated in December 1971, releasing its Mohawks to other units. The 131st SAC was deactivated in July 1971 and its assets were transferred to the 131st Military Intelligence Company (MI CO), which was activated in place of the SAC at Phu Bai. The 73d SAC remained in Vietnam until it, too, was deactivated in April 1972.

OV-1Bs of the 73rd Aviation Company (Surveillance) on the line at Vung Tau, Vietnam.

Operations in Southeast Asia ended in November 1972 when the 131st MI CO stood down. More than 200 Mohawks flew operationally in Southeast Asia and 63 were lost, including 27 that were brought down by enemy fire and 36 that were lost in operational incidents.

Post-Vietnam organisation

Following the US withdrawal from Southeast Asia Mohawk operations were consolidated within the CONUS while operations in Europe, Alaska and the Republic of Korea continued. The new operating sites included units at Fort Lewis, Washington, Fort Hood, Texas, Hunter Army Air Field, Georgia, along with Army National Guard units in Georgia and Oregon. Subsequently, a number of reorganisations took place within Army Aviation. As a result, the Mohawk fleet was grouped under the category of Special Electronic Mission Aircraft and, as such, many aircraft and units were placed under the cognisance of the US Army Security Agency (ASA). During this period Mohawks were assigned to Military Intelligence Detachment Aerial Surveillance (MIDAS) or Company (MICAS), Army Security Agency (ASA) Battalions, Aerial Surveillance Aviation Companies and so-called Aerial Exploitation Battalions (AEB).

Aviation Company (Surveillance Airplane)

122d Aviation Company (SA)/15th AVN BN (Combat)	Fliegerhorst AAF, Hanau, W. Germany
146th Aviation Company (SA)	Desederio AAF, Camp Humphreys, Rep. of Korea
159th Aviation Company (SA) (GA ARNG)	Dobbins AFB, Georgia
293d Aviation Company (SA)/55th AVN BN (CBT)	Robert Gray AAF, Fort Hood, Texas

In 1977 the US Army Intelligence and Security Command (INSCOM) was created and all Mohawk assets were subsequently consolidated within a new organisational unit known as a Military Intelligence Battalion (Aerial Exploitation), which were primarily organised as part of a respective US Army Corps or assigned directly to INSCOM. They were often referred to as Aerial Exploitation Battalions (AEB). Mohawks were fielded by a total of seven Military Intelligence Battalions, including two that were part of the Army National Guard and one assigned to the US Army Training and Doctrine Command (TRADOC).

OV-1D of the 146th MIB, based at Camp Humphreys in Korea.

RV-1D of the 2d MIB, one of two Mohawk-operating battalions in Germany.

Military Intelligence Detachment/Company Aerial Surveillance

73d MICAS/11th AVN GP	Fliegerhorst AAF, Hanau, West Germany; Echterdingen AAF Stuttgart, West Germany
131st MICAS/6th Cavalry Brigade	Robert Gray AAF, Fort Hood, Texas
158th MICAS/151st MIB (GA ARNG)	Winder-Barrow Airport, Georgia
159th MICAS/151st MIB (GA ARNG)	Dobbins AFB, Georgia
172d MIDAS	Wainwright AAF, Fort Wainwright, Alaska
184th MICAS/163d MIB	Fort Lewis, Washington
704th MIDAS/146th MIB	Desederio AAF, Camp Humphreys, Rep. of Korea
1042d MICAS/641st MIB (OR ARNG)	McNary Field, Salem, Oregon

OV-1C of the 151st Military Intelligence Battalion, Georgia National Guard.

1st Military Intelligence Battalion (Aerial Exploitation)
Assigned to V Corps at Campbell Barracks, Germany, the 1st MIB(AE) is based at Wiesbaden Army Air Field and reports to the 205th Military Intelligence Brigade (Communications-Electronic Warfare and Intelligence) (MIBDE [CEWI]). OV-1D and RV-1D variants were operated by A Company (A/1st MIB[AE]) until the company was inactivated in 1992.

2d Military Intelligence Battalion (Aerial Exploitation)
Assigned to VII Armored Corps at Ludwigsburg, Germany, the 2d MIB(AE) reported to the 207th Military Intelligence Brigade (Communications-Electronic Warfare and Intelligence) (MIBDE [CEWI]). The battalion was based alongside the brigade at Stuttgart Army Air Field in Echterdingen, Germany, and A Company operated the OV-1D and RV-1D until it was inactivated in 1991.

3d Military Intelligence Battalion (Aerial Exploitation)
Assigned directly to the Intelligence and Security Command (INSCOM), the 3d MIB is based at Desederio Army Air Field, Camp Humphreys, Republic of Korea. A/3d MIB(AE) operated the OV-1D and RV-1D until September 1996 when the last Mohawks were retired. The unit currently operates the Northrop Grumman RC-7B Airborne Reconnaissance Low-Multifunction (ARL-M) aircraft.

15th Military Intelligence Battalion (Aerial Exploitation)
Assigned to III Corps at Fort Hood, Texas, the 15th MIB(AE) reports to the 504th Military Intelligence Brigade (Communications-Electronic Warfare and Intelligence) (MIBDE [CEWI]). A/15th MIB(AE) retired its last OV-1Ds and RV-1Ds on 17 September 1993 and currently operates the TRW/IAI RQ-5A Hunter unmanned air vehicle.

224th Military Intelligence Battalion (Aerial Exploitation)
Assigned to the XVIII Airborne Corps, the 224th is based at Hunter Army Air Field in Savannah, Georgia, and reports to the 525th Military Intelligence Brigade (Communications-Electronic Warfare and Intelligence). A Company retired its last OV-1Ds and RV-1Ds in September 1996 and is currently inactive.

304th Military Intelligence Battalion (Training)
Assigned to the US Army Intelligence Center and School (USAICS) via the 111th Military Intelligence Brigade, B Company/304th Military Intelligence Battalion (Training) was based at Libby Army Air Field, Fort Huachuca in Arizona. The battalion conducted systems training for Mohawk observers until B Company retired its last OV-1Ds in September 1994.

151st Military Intelligence Battalion (Aerial Exploitation)
Activated on 1 October 1973, the Georgia Army National Guard's 151st Military Intelligence Battalion (Aerial Exploitation) was based at Dobbins AFB, Georgia. The battalion initially controlled the 158th Military Intelligence Company (MI Co) at Winder-Barrow Airport and the 159th MI Co at Dobbins AFB. Following a 1982 reorganisation only A Company/151st MIB(AE) at Dobbins remained. The unit operated the OV-1D until September 1991.

641st Military Intelligence Battalion (Aerial Exploitation)
Activated on 1 April 1973, the Oregon Army National Guard's 1042d Military Intelligence Company (Aerial Surveillance) (MI Co) was redesignated A Company as part of the 641st Military Intelligence Battalion (Aerial Exploitation) on 11 September 1982. The unit, based at McNary Field in Salem, operated the OV-1D until 1 October 1992.

OV-1D of the 3d MIB, which covered the Korean peninsula from Camp Humphreys.

OV-1D of the 15th MIB, CONUS-based at Fort Hood, Texas.

OV-1D of the 224th MIB, which supplied Mohawks for operations in Central America.

OV-1D of the 641st MIB, part of the Oregon National Guard organisation.

US Army Training units

Flight training for OV-1 aviators was carried out by the US Army Aviation Center's (USAAVNC) 1-223d Aviation at Cairns Army Air Field, Fort Rucker, Alabama. Observers received their systems training with the 304th Military Intelligence Battalion (Training), US Army Intelligence Center School (USAICS), at Libby Army Air Field, Fort Huachuca, Arizona. USAICS was established in 1971, however, it should be noted that the school was previously known as the United States Army Combat Surveillance and Electronic Warfare School (USACSEWS), the US Army Combat Surveillance School and Training Center (USACSS-TC) and the US Army Combat Surveillance School (USACSS) during the period that it trained Mohawk crews.

Most Army training aircraft feature liberal doses of orange paint to increase conspicuity in the busy Fort Rucker pattern. This was one of the OV-1Ds assigned to USAICS.

US Government and Test Agencies

National Aeronautics and Space Administration (NASA)

Several Mohawks were operated through the 1970s and 1980s in support of a variety of research projects. A single OV-1C (serial 67-18915) was operated by the Johnson Space Flight Research Center as NASA 928 beginning in November 1972. The aircraft, however, crashed on 25 April 1973 near League City, Texas.

During late 1972 and early 1973 NASA's Lewis Research Center (renamed NASA John H. Glenn Research Center at Lewis Field in 1999) bailed SLAR-equipped OV-1B serial 62-5903 from the Army, wearing the registration N512NA. NASA Lewis also used OV-1B 64-14244 (NASA 637) between December 1972 and April 1978 to analyse ice coverage and distribution over the Great Lakes as part of the Ice Warn project. The aircraft was subsequently transferred to the Langley Research Center, which operated the Mohawk as NASA 518 until March 1980. It was subsequently transferred to Grumman's Stuart facility and modified to RV-1D configuration.

Langley also operated OV-1B serial 62-5880 as NASA 512 from April 1980 through April 1989. Before entering service with Langley, however, Grumman's Stuart facility installed a pylon-mounted General Electric JT15D

This OV-1B operated from NASA Langley in full house colours. It was used for noise experiments, including those with a jet mounted under the wing.

NASA Lewis's OV-1B 64-14244 is seen over Lake Erie during the Project Ice Warn ice research programme in 1973.

turbofan engine under the right wing. Initially used for engine noise monitoring tests, the engine was later removed but the aircraft continued to support noise experiments.

Although assigned to the US Army Aviation Engineering Flight Activity (USAAEFA) at Edwards AFB, California, OV-1D serial 68-15932 was used in a joint NASA/US Army programme to test a NASA-designed stall-speed warning system in the early 1980s.

OV-1B	62-5880	NASA 512
OV-1B	62-5903	NASA 512
OV-1B	64-14244	NASA 637/518
OV-1C	67-18915	NASA 928
OV-1D	68-15932	

US Department of the Treasury (US Customs Service)

The US Customs Service acquired four OV-1Cs in 1973 for service in its drug interdiction programme. Prior to entering service the aircraft were modified by Lockheed Aircraft at its Marietta, Georgia, facility. As part of the modifications the aircraft were equipped with an elongated nose and a forward-looking infrared (FLIR) system built by Texas Instruments. They were initially flown in an Army scheme with military serials, but later received civil registrations and a custom paint scheme. They served with USCS in the surveillance role until retired in 1986.

OV-1C	60-3758	N6734
OV-1C	61-2699	N6740
OV-1C	62-5856	N6744
OV-1C	66-18896	N6745

Naval Air Test Center (NATC)/Test Pilot School (USNTPS)

The centre's weapon test division cleared all of the weapons carried by the Mohawk and operated a number of aircraft through the mid-1960s for follow-on flight tests at NAS Patuxent River, Maryland.

Beginning in 1965, the US Naval Test Pilot School, which trained US Army test pilots, received its first OV-1A. The school later operated the OV-1B and OV-1C models and the last of these was retired in September 1988 when they were replaced in the twin-engined handling role by Beech U-21As.

The Test Pilot School at Patuxent River operated Mohawks for over 20 years. These two examples are both OV-1Bs, one wearing a standard US Army training scheme (left).

US Environmental Protection Agency (USEPA)

During the mid-1970s single examples of the OV-1B and OV-1C were loaned to the Environmental Protection Agency (EPA), which used the aircraft for a variety of missions including environmental surveys of the areas around nuclear power plants.

| OV-1B | 64-14243 |
| OV-1C | 68-15953 |

US Department of Commerce (National Geodetic Survey)

In late 1971 the Army provided the National Ocean Service's National Geodetic Survey (NGS) with a SLAR-equipped OV-1B. The aircraft wore the civil registration N171 and was operated for several years in support of various NGS projects, such as studying state water resources and surveying the route of the trans-Alaskan pipeline.

| OV-1B | N171 |

US ARMY AVIATION ENGINEERING FLIGHT ACTIVITY (USAAEFA)

Over the Mohawk's 36-year career a variety of organisations undertook flight test activities with the aircraft. In later years the USAAEFA was the primary organisation involved in engineering activities. The unit operated a number of Mohawks that were dedicated to trials.

The unit, which was based at Edwards AFB, California, was later redesignated the US Army Airworthiness Qualification Test Directorate (AQTD). The unit retired its Mohawks before it relocated to Fort Rucker, Alabama, in October 1996. Assigned examples included:

OV-1C 61-2726
JOV-1C 60-3748
JOV-1D 62-5867
OV-1D 68-15932 (right)

ISRAEL DEFENCE FORCE/AIR FORCE

The US Army remained the sole operator until August 1974 when the Heyl Ha'Avir, or Israeli Air Force, acquired two OV-1Ds from the US Army. The aircraft were intended to fill an operational gap in real-time intelligence capabilities and were included in a package of equipment provided in return for Israeli concessions that allowed the Suez Canal to reopen. Known as the Atalef, or Bat, in Israeli service, the Mohawk's cameras, infrared sensors and SLAR were used to identify enemy force concentrations beyond the country's borders in the aftermath of the Yom Kippur War, the OV-1 serving as one of the IAF's primary intelligence platforms until 1982. The aircraft were operated by the Atalef Flight of a C-47 squadron based at Lod International Airport.

By the late 1970s and early 1980s unmanned aerial vehicles (UAV) began taking over the intelligence missions and the Mohawks were rendered redundant. While undergoing maintenance 4X-JRA was damaged by a fire in late 1980 or early 1981 and removed from service. The second aircraft was subsequently withdrawn from use in March 1982. Both aircraft were placed on display in Israel but were returned to the US in April 1984. They were subsequently refurbished by Grumman and returned to the US Army.

OV-1D 69-17021 4X-JRA/022
OV-1D 68-16993 4X-JRB/056

Israel's Atalefs were operated with olive drab upper surfaces and light grey undersides. They wore large national insignia and dual military/civil registrations. The pair was very active during the late 1970s, patrolling Israel's borders, partnered by Sigint-configured C-47s.

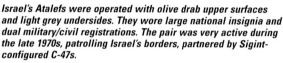

ARGENTINA

During 1992 the US approved a request from Argentina's Comando de Aviación de Ejército, or Army Aviation Command, to acquire a number of ex-US Army OV-1Ds. The initial two were handed over late that year after being accepted at Grumman's facility in Stuart, Florida. The aircraft were flown from Florida and arrived at Campo de Mayo Army Air Base, near Buenos Aires, on 24 December 1992. These aircraft were initially used to train Mohawk aircrews pending the delivery of the remainder of the fleet. Eleven additional aircraft were delivered by the end of 1993.

The Mohawks were initially assigned to Escuadrón de Aviación de Apoyo General 603 (Esc Av A Grl 603 or General Support Aviation Squadron), but in early 1994 the Escuadrón de Aviación de Exploración y Reconocimiento 601 (Esc Av Expl Rec or Exploration and Reconnaissance Aviation Squadron) was established. Both units were attached to Agrupación de Aviación de Ejército 601 (Agr Av Ej or Army Aviation Group). The last of 23 examples arrived in Argentina on 17 November 1994. A total of eight was equipped with dual controls and two of these aircraft were assigned to the Escuela de Aviación del Ejército (Ec Av Ej or Army Aviation School) at Campo de Mayo. The trainers were readily identified by the red external fuel tanks that they carried.

Equipment installed in the Argentine Mohawks comprises KA-60C and KA-76A cameras, AN/AAS-24 FLIR and the AN/APS-94F SLAR, which has received a small number of local upgrades, and the LS-59A strobe flash pods. The personnel that operate the sensors are assigned to the Compañía de Inteligencia Táctica (Tactical Intelligence Company), which reports to the Centro de Reunión de Inteligencia Militar (Military Intelligence Meeting Centre).

During 1995 two aircraft supported a flight test programme, at Base Aeronaval Comandante Espora (Comandante Espora Naval Aviation Base), and were equipped with a gunsight and six underwing hard points for carrying 2.75-in (70-mm) rockets in the seven-shot LAU-32 and 19-shot LAU-61 pods, and FAS 300 250-kg (551-lb) cluster bombs. These bombs, which were locally designed by CITEFA, are equipped with 510 anti-personnel and CAM 1 anti-tank grenades.

The army was prevented from accepting 11 additional aircraft offered by the US Army by a limited budget, while a lack of certain spares caused a rationalisation of the fleet in 1999. As a result, 11 aircraft were either scrapped as a source of spare parts or stored for future use, and the 10 remaining airframes are in service with Escuadrón de Aviación de Apoyo de Inteligencia 601 (Esc Av Apy Icia or Intelligence Support Aviation Squadron) and the Escuela de Aviación de Ejército. The squadron continues to operate the aircraft in the battlefield surveillance, tactical reconnaissance, and target acquisition roles.

OV-1D	62-5865	AE-027
OV-1D	62-5887	AE-030
OV-1D	67-18898	AE-035
OV-1D	67-18911	AE-022
OV-1D	67-18918	AE-032
OV-1D	67-18921	AE-028
OV-1D	68-15931	AE-023
OV-1D	68-15932	AE-021
OV-1D	68-15933	AE-034
OV-1D	68-15941	AE-024
OV-1D	68-15951	AE-025
OV-1D	68-15954	AE-037
OV-1D	68-15963	AE-031
OV-1D	68-16991	AE-026
OV-1D	68-16997	AE-039
OV-1D	69-17002	AE-038
OV-1D	69-17006	AE-040
OV-1D	69-17008	AE-020
OV-1D	69-17009	AE-029
OV-1D	69-17011	AE-041
OV-1D	69-17012	AE-033
OV-1D	69-17016	AE-042
OV-1D	69-17026	AE-036

Argentina's Meteors
Latin America's first jet fighter

On 11 July 1947 an English pilot named William Waterton climbed into the cockpit of a Gloster G.41G Meteor F.Mk IV, started its engines and headed down the makeshift runway of Edison Avenue in the port of Buenos Aires, bound for the Argentine air force base of El Palomar, thus completing the first flight of a jet fighter in Argentina. As the sole jet fighter type in Latin America the 'Glosters', as they became known in service, along with the Avro Lancaster and Lincoln bombers, gave the Argentine air force a strategic power that was unequalled in the region.

Above: Probably the most famous and undoubtedly the most colourful of all Argentine Meteors were those of the Escuadrón 46 display team. Painted in this striking red/white scheme, the team was active from 1962-64.

Below: Unlike later Meteor variants, the F.Mk 4 was not equipped with an ejection seat, making in-flight evacuation almost impossible. Only one successful 'bale-out' was achieved in its 22-year operational career with the Fuerza Aérea Argentina.

At the end of World War II, Argentina's aviation community benefited from the neutral stance it had maintained throughout most of the conflict. On one hand, there was a great immigration of German scientists, technicians and pilots who gave a substantial boost to the local aeronautic industry, which soon experienced its most fruitful period. On the other, there was the great debt that Great Britain owed Argentina from the war: Great Britain had been unable to pay for food and raw material with currency, so had offered manufactured products in lieu. Great Britain proposed to pay the debt with surplus military equipment of all kinds, even though this suggestion did not meet with the approval of the United States. Included in this material was a special lot of 15 Avro Lancaster heavy bombers, 30 Avro Lincolns, and 100 Gloster G.41G Meteor jet fighters. The latter were F.Mk IVs, featuring a slightly shorter wingspan than other versions.

At the beginning of 1947, the test pilot for the Fábrica Militar de Aviones (FMA), First Lieutenant Edmundo 'Pincho' Weiss, had been sent to the UK to evaluate the equipment, thereby becoming the first Argentine pilot to fly the Meteor. Later, another 11 pilots with experience in the Curtiss Hawk 75-O, the primary Argentine fighter of the time, were sent to England to begin their training as jet pilots. There, they had the chance to fly the Avro Anson, in order to get used to a two-engined aircraft; the de Havilland Dove, so they could learn landing techniques with a tricycle landing gear; and the Gloster Meteor Mk III. After each pilot had flown for about six hours, they took command of the six Argentine aircraft and flew 30 hours on them. On 6 June 1947, six Gloster G.41G Meteor F.Mk IVs were registered with the serial numbers I-007 to I-012, and subsequently used for training. After a training period of 60 hours, on 8 August they were discharged and replaced by another six aircraft.

The first 50 aircraft that ended up in Argentina had been built for RAF use; the second 50 were built especially for the Fuerza Aérea Argentina (FAA). They were initially allocated UK Class B registrations of G-5-101 to 200 for flight testing. The Argentine Meteors were

the first aircraft of this type to be exported anywhere in the world.

The first examples were delivered by ship to Buenos Aires, with I-001 to I-006 arriving on 8 July 1947. To get them to the base at El Palomar, a runway was improvised from Edison Avenue, in the port of Buenos Aires, from where the Meteors took off. On 19 July they participated in an aerial parade that stunned the watching crowd, which had never seen jet aircraft before. From 7 to 30 September, two Meteors (I-003 and I-006) were present at an aeronautical exhibition, together with the other new aircraft, held at Avenida 9 de Julio.

On 3 December 1947 the first flight unit equipped with the Gloster Meteor was created: Regimiento 4 de Caza Interceptora (R.4Caz), at Base Aérea Militar (BAM) Tandil in Buenos Aires province. Its aircraft entered service on 3 March 1948. The air base was improved in order to accommodate the new aircraft and, at the same time, the technicians and mechanics were mastering the characteristics of this new fighter. They faced a number of problems due to the complexity of this type compared to previous aircraft, and many incidents occurred during the first days of operation. The Meteor was a totally unknown concept in Latin America, although it was really quite a primitive jet aircraft.

The next aircraft to arrive in Argentina were slowly assembled at the Instituto Aerotécnico of the FMA at Córdoba, until all 100 were completed. During unloading, the entire fuselage of I-021 fell into the water and had to be replaced, although replacement did not come until 1949. In another incident, I-005 had an accident nine days after it began operations, which kept it out of service for more than three years. The first Argentine Meteor lost in flight was I-018, which exploded coming out of a barrel roll manoeuvre on 23 April 1948 during the evaluation period. This aircraft had been readied for an attempt on the world closed-circuit speed record, but this loss led to the abandonment of the project.

German assistance

Immediately after the arrival of the Meteors, pilot instruction began. The FAA utilised the invaluable assistance of legendary World War II fighter ace Adolf Galland, who taught the Argentine pilots contemporary air-to-air and air-to-ground techniques. Galland, working together with other renowned German wartime pilots, helped to modernise the combat tactics of the Argentine Air Force, which, like other Latin American air forces, was familiar only with pre-war tactics.

I-100 was the last of the 100 Meteor F.Mk 4s ordered in May 1947. The unit price per aircraft was quoted as £32,000, which included the training of some 12 Argentinian pilots and a cadre of groundcrew at Gloster's Moreton Valence headquarters in the UK.

On 15 March 1949, military air bases and regiments were converted to brigades – analogous to wings in the RAF – so R.4Caz and BAM Tandil became VI Brigade. On 7 June, sister unit the 6th Interceptor Regiment (R.6Caz), based at Tandil, received all the 'Glosters' with an odd serial number, while the even-numbered aircraft remained with R.4Caz.

On 9 January 1951, the Boletín Aeronáutico Confidencial 45 (Confidential Aeronautical Bulletin 45) posted the 112 Decree, under which the brigades became air brigades. The BAMs disappeared and became the base airfields for those units. Regiments were deleted, and in their place Grupos 2 and 3 de Caza (2 and 3 Fighter Groups) were created, based on the 4th and 6th Regiments. After these changes, both fighter groups, along with 55 Meteors, were sent to VII Brigada Aérea (VII Air Brigade) based at Morón, in the outskirts of Buenos Aires. The rest of the aircraft were stored at Tandil together with FMA IAe-22DL trainers, and five Avro Lincolns.

In addition to assembling the aircraft as they arrived, the FMA also assembled the jet engines and performed the Main Maintenance Cycles (ICMs). It also commenced a project to convert single-seat aircraft into two-seat form for use in observation missions. Subsequently the project was cancelled, but not before aircraft I-040, I-090 and I-095 had been modified to carry an extra seat behind the pilot, in place of communications equipment.

By 1951 unease with Perón's governance of the country was growing and for the first, but

Above: In 1947 the British government's mounting debt to Argentina forced it to renege on a 'gentlemen's agreement' with the USA not to supply arms to Perón's government due to its perceived pro-Nazi leanings during World War II. Along with 100 Meteors, the sale of 30 Avro Lincoln bombers was approved, the first 50 Meteors being re-directed from RAF orders to speed the delivery process. By 1948 a further order for 300 more Meteors and 30 Lincolns was expected, but growing reliance on the US eventually forced Britain to turn down these requests. The Meteors were delivered and initially operated in this bare metal finish, however, by the mid-1950s squadron commander's aircraft had begun to receive more colourful markings.

Right: Aircraft serial numbers were initially stencilled in black on the nose, but were soon transferred to the rear fuselage. In the late 1940s several examples were used as trials aircraft, including I-085 with a non-standard canopy.

Record-breakers and high-altitude research

On 12 April 1954, Captain Jorge Quagliani captured the South American altitude record in I-095 (below), modified with the wings of a Mk III aircraft and a reinforced pressurised cockpit, reaching 14000 m (45,932 ft). Earlier, in 1949, Lieutenant Mannuwal had also reached that altitude, but the achievement could not be confirmed at the time. In 1956 I-056 (right) was modified in a similar way and given this special colour scheme for use on high-altitude research flights, before crashing on approach to Morón in 1958.

I-090 with First Lt Osvaldo Rosito, and I-063 with Lt Ernesto Adradas, and their mission was to intercept aircraft flying over the city and to shoot down any aggressors. A few minutes later, two Meteors took off from Morón Air Base (one was I-032 flown by Capt. Jorge Mones Ruiz); they were tasked with the same mission, however, because these two pilots were sympathetic to the revolution, they merely overflew the city without intercepting aircraft. Another section took off soon after, commanded by First Lt Juan Carlos Carpio, but they too neglected to intercept attacking aircraft.

After bombing the Casa Rosada, the rebel Navy aircraft landed at Ezeiza airport, from where they continued operations.

After overflying the government building, which was smoking heavily, the first government Meteors continued their patrol. The weather was poor with limited visibility, making the search difficult. Lt García saw a section of Texans over the Aeroparque, about 5 km (3 miles) from the Casa Rosada, and ordered his compatriots follow him. The Meteors formed an echelon and approached the AT-6As flown by Teniente de Corbeta Máximo Rivero Kelly and Guardiamarina Armando Román, who were intending to land at the Aeroparque because of a low fuel state. Believing the Meteor pilots to be rebels – because the pilots of VII Brigada Aérea had stated that they supported the rebellion – the Texan pilots waggled their wings in salute, a gesture returned by the Meteors. When García was close enough, however, he opened fire on Rivero Kelly but with no effect. Rivero Kelly descended and flew close to the top of a train

not the last time, FAA Meteors were tasked with challenging revolutionary forces. Until the end of the 1951 a deployment of Meteors remained at Ezeiza airport, Buenos Aires, for the defence of the capital and the government.

Meanwhile, the pace of force-wide exercises increased. On 27 August 1952 Operativo Defensa began, wherein the Meteors intercepted the Avro Lincolns and Lancasters over Buenos Aires. These exercises were regularly repeated, often involving other Argentine cities.

Between 12 and 16 September 1953, 12 Meteors and four Lincolns visited Santiago, Chile on their first foray outside the country. Operating from Los Cerrillos airport, they conducted a formation flypast to mark Chile's Independence Day on 18 September.

Several Meteors were also allocated as trials aircraft. As part of the evaluations for the PT-1 air-to-ground missile project in 1953, I-087 received a belly-mounted stabiliser from the missile for aerodynamic testing. It was installed in place of the ventral fuel tank. Following the aerodynamic trials the Meteor was used as a chase aircraft during live-fire tests, until the project was cancelled in 1958.

Baptism of fire

On 16 June 1955, another attempt was made to overthrow President Juan Perón. The day brought the first air combat in the skies over Buenos Aires. At mid-day, four Beechcraft AT-11s and 15 North American AT-6A Texans from the Punta Indio naval aviation base bombed the Casa Rosada (headquarters of the government) in the belief that Perón was present. Having naval aircraft bomb Buenos Aires caused great alarm, and four Meteors were scrambled. They were I-039 flown by First Lt Juan García, I-077 with First Lt Mario Olezza,

Left: I-017 was one of a number of early Meteors used for unprepared runway trials in 1948. The Meteors at this stage were constantly being grounded owing to continual cracking of wing trailing-edge fillets. The problem reached its zenith in mid-1948 when only four hours per aircraft could be flown between component changes.

Below: During exercises the two fighter groups received different high-visibility markings to identify their role as 'attacking' or 'defensive' forces. This line-up in 1960 have already received the new C- serial prefix but have yet to be painted two-tone grey/green.

Above: Air traffic controllers in Morón's tower get a 'bird's eye' view of a home-based 'checker-tail' Meteor as it streaks low over its parked sister-ships in late 1959. Morón was the main operating base for the type from 1951-66.

Below: The overall silver finish, in which the Meteors were delivered, remained the standard scheme until the early 1960s. In the mid-1950s black anti-glare panels were added in front of the pilot's canopy.

Maintenance, repairs and overhauls were conducted at the Fábrica Militar de Aviones's facility at Córdoba. On emerging from overhaul in the early 1960s, the previously air defence-tasked Meteors in a silver finish (foreground) were re-painted in a two-tone grey/green scheme more suited to their new ground-attack role.

heading for Tigre, a suburb of Buenos Aires. Near San Isidro, en route to Tigre, the aircraft escaped into cloud and returned to Ezeiza.

Meanwhile, Román's Texan was attacked by the other Meteors, but Olezza and Rosito intentionally aimed away from the aircraft, because they were sympathetic to the revolution. However, the pilot of Meteor I-063 was loyal to the government and aimed directly at the AT-6A. The 20-mm bullets struck the right side of the aircraft and the wing burst into flames. The pilot escaped by parachute.

A few minutes later, Vicecomodoro Carlos Síster, chief of Escuadrón I, took off in I-052 to attack the naval aircraft at Ezeiza airport. Alerted to the presence of anti-aircraft artillery, he made his attack at low altitude and high speed. The rebel naval flyers thought he was on their side and were initially unconcerned until the Meteor began to fire its cannon at their aircraft sitting on the apron. After the first pass, Síster spied the returning Texan of Rivero Kelly and attacked it, but scored no hits. He made

another attack against two transports, an AT-11 and a Catalina, which responded with machine-gun fire. On the third pass, Síster's cannon locked so he returned to base. The result of the attack was damage to two civilian airliners (one SAS and one Aerolíneas Argentinas) and the AT-11 (Serial 0273/3-B-11).

Meanwhile, a group of revolutionary pilots had taken control of Morón Air Base. Initial missions involved attacking radio antennas and the Casa Rosada in support of the naval rebels. The first to take off was I-019, piloted by Lt Juan Boehler, who mounted an attack on the radio antennas. In mid-afternoon came the order to prepare a joint strike with naval aircraft on the Casa Rosada. The attack was started by a Texan, AT-11 and Catalina of the Navy, and as soon as their bombing ended, the Meteors arrived, flying low along Rivadavia Avenue which connected Morón Air Base to the government buildings. When over the Congress building, they descended even lower to fly between the buildings of the Avenida de Mayo, arriving at the

Plaza de Mayo where they opened fire on the Casa Rosada and the anti-aircraft artillery that had been deployed after the first attack. Lt Guillermo Palacio used his ventral auxiliary fuel tank as a makeshift napalm bomb, dropping it to destroy some parked cars. The attacks continued for some time, with the aircraft refuelling at Morón and returning. The radio antennas were attacked again, as were troops of Regimiento 3 de Infantería Motorizado who were pushing forward to Ezeiza.

By the end of the day, however, the revolutionary forces were facing defeat and the Meteors escaped to Uruguay along with the Navy's C-47 (0117/4-T-20). During this flight, I-058, I-064 and I-098 (flown by Lts Marelli and Jeannot, and Capt. Carús) attacked the Police Central Department and the Casa Rosada again. A total of six Meteors fled to Uruguay, however, I-064 ditched in a river near the coast because of fuel shortage and I-029 made a gear-up forced landing. Eventually, all the aircraft returned to Argentina.

Right: In September 1947 I-003 and I-006 were dismantled and towed through the streets of Buenos Aires to a major aeronautical exhibition, held at Avenida 9 de Julio. The outer wings were removed to allow the aircraft to pass various obstacles, including the railway bridges which crossed Libertado Avenue (seen here).

Below: Low flying was something of a forte of FAA's Meteor pilots, as demonstrated by C-048. Such bravado did, however, lead to several serious accidents.

Argentina's Meteors

The Revolución Libertadora – 1955

Following the June uprising there was, on 16 September 1955, a second attempt to overthrow Perón; it was much bigger, enjoyed great popular support and this time was successful. On this occasion, the Meteors fought on both sides. Morón remained loyal to the government, but at the FMA in Córdoba were three Meteors awaiting repairs (I-043, I-061 and I-079). These aircraft were hastily restored to flying condition by the rebels who had gained control of the FMA and the Escuela de Aviación Militar (EAM). Soon into action early on the first morning, the trio of Meteors proceeded to the rebel-aligned Army Artillery School which was under attack from the soldiers of the Infantry School located nearby. With their cannon armament yet to be refitted, the Meteors and some Percival Prentice, AT-11 and IAe-22DL trainers of the EAM overflew the Infantry School troops in an attempt to intimidate them.

For the remainder of the morning FMA technicians worked hard to fit each Meteor with two of the normal four 20-mm cannon and their sights. The FMA test pilots were First Lts Rogelio Balado and Alberto Herrero, commanded by Capt. Suárez. They were soon joined by First Lts Hellmuth Weber, Luis Morandini and Rossi.

The next Meteor action was on behalf of the government, when it was decided to attack the old destroyers ARA *Cervantes* (T-1) and ARA *La Rioja* (T-4) of the Fuerza Naval de Instrucción (Instruction Naval Force), which blockaded the Río de la Plata. Early in the morning, frigate captain Hugo Crexell received direct orders from Perón to co-operate with the air force in the attack on the ships, and subsequently made a reconnaissance flight over the river in a VII Brigada Aérea Meteor. At 09:00 four aircraft commanded by Vicecomodoro Síster took off to attack the destroyers. Crew on the ships saw the aircraft and readied the three 40-mm Bofors double cannon. At 09:30 the Meteors attacked *La Rioja*, firing their cannon. The ship opened fire, but its manual cannon were ineffective against the jet fighters.

The ship was hit and damaged by the four jets, two of which were themselves hit by artillery fire from the *Cervantes*, which was approaching *La Rioja* to assist. After this attack, three Meteors commanded by Vicecomodoro Pérez Laborda took off on a similar mission – this time attacking the *Cervantes*.

The next attack was executed against two of the Infantry Landing Ships (6 and 11), which were crossing the river from Martín García island where the rebel-controlled Escuela de Marinería was located. The ships, commanded by the icebreaker ARA *San Martín*, were en route to reinforce the base of Río Santiago near the city of La Plata. Four Meteors and five IAe-24 Calquins attacked the ships. Medium Landing Ship 1 set out from Río Santiago to support the vessels, but it lacked AAA and came under fire. The attack was repeated minutes later, leaving two men dead and a number injured.

The *Cervantes* and *La Rioja* were attacked for the third time at 11:00. As a result of the strikes a number of crew members were killed and injured. One of *La Rioja*'s guns was hit, but the uninjured men managed to resume firing after changing a barrel. As the destroyers made their way to the end of the river they were attacked again, near the Uruguayan coast. This time they suffered little damage, and the strikes ceased.

The next target was the Río Santiago base, attacked by an Avro Lincoln and Calquins and then by the Meteors. The attacks continued until nightfall.

By mid-day, the rebel Meteors had launched the first attacks against government troops trying to help the Infantry School, forcing the advance to end.

On 17 September, government Meteors saw action over Córdoba. Three aircraft had been readied at Las Higueras airport in the city of Río Cuarto to attack the airport of Pajas Blancas at Córdoba and took off after 17:00. The Meteors headed north at 700 km/h (435 mph), approaching at low altitude between the mountains. Their plan was to attack the airport from north to south, in line-astern. On the apron

Half a dozen Morón-based Meteors formate with an FAA C-47 shortly before the Revolucíon Libertadora. The next time the Meteors joined this type of aircraft the duty was to escort the triumphant revolutionary General Lonardi, in an Aerolíneas Argentinas example, to Buenos Aires to assume his role as head of state.

were three Lincolns, one being loaded with fuel and a full load of bombs. With Major Catalá leading the attack, and keeping a separation of 1000 m (3280 ft), the Meteors fired at the bombers, damaging one seriously (only one Lincoln managed to fly again during the revolution). The Meteors accelerated and returned to Río Cuarto, overflying the city of Córdoba, which was in the hands of the rebels. Major Aubone's Meteor was found to have been hit by several 20-mm bullets from Capt. Domínguez's aircraft, which had been flying too close during the attack.

While another strike was being prepared, the appearance of a column of rebel troops heading toward Río Cuarto forced the base's evacuation. The crews fled on an AT-11 (it was intercepted by a Calquin, but managed to escape with little damage).

During the night of 17 September, rebel Meteors flown by Balado and Weber attacked government troops who were approaching Córdoba from the east. Although the aircraft were damaged by AAA, they remained airworthy.

Combat continued the next day, first with the bombing of the EAM by two government Lincolns. In response First Lt Weber got airborne in a rebel Meteor as the attack was underway and intercepted the Lincolns, opening fire on the rearmost without hitting it. The Lincolns separated and Weber manoeuvred below one of the aircraft into a perfect firing position, however, his cannon failed to fire and he returned to base. Another Meteor, with Balado at the controls, got airborne just as the bombs fell onto the runway. He tried to convince one of the bomber crews to defect to the rebel side but they refused, after which he passed very close, menacing the crew and forcing them to return to Morón Air Base.

On 19 September, government troops began an assault on Córdoba from the northeast and initiated the occupation of the train station. Rebel aircraft responded, first the IAe-24 Calquin with napalm bombs, then the Meteors each fitted with two napalm bombs, intended for the Calquin, below their wings. This attack marked the first time such a weapon had been used for air-to-ground attack in Argentina. The artillery defence was massive, but the aircraft were able to strike, launching the bombs and also the ventral fuel tanks (as previously on 16 June). The cannon suffered problems in all the attacks due to poor preparation, forcing FMA technicians to fit one of the Meteors with two 20-mm cannon taken from a Lincoln.

By the late morning on 19 September, seeing that the situation was going against him, President Perón created a military government to negotiate a ceasefire with the rebels. Fighting stopped, but only long enough for both sides to reorganise their forces. Detecting troop movements to the southwest of Córdoba, the rebels decided to attack in the belief these government forces were violating the ceasefire. By this time, however, fuel supplies for the Meteors had run out, so normal automobile petrol was used in place of jet fuel (Rolls-Royce manuals stated it could be used for a short time in an emergency, but did not recommend entering combat with it). After mid-day the 'Glosters' took off, firing over the troops and stopping the advance. The first Meteor was Weber's, to be followed by Morandini, but because the latter's aircraft was not ready, Weber attacked again. Morandini finally got airborne when Weber returned. However, on his return to the FMA, one of the engines finally failed due to the use of petrol and I-079 stalled and crashed. Morandini was the last pilot – and the only Meteor pilot – to die during the revolution.

On 21 September, combat ceased for good: the revolutionary forces had won. The final Meteor action of the revolution came when two aircraft escorted the Aerolíneas Argentinas DC-3 of General Lonardi, leader of the revolution, on his trip to Buenos Aires to take over the government.

Above: I-043 was one of the trio of Meteors purloined by the rebels at Córdoba in the early stages of the Revolucíon Libertadora and was heavily involved in the ensuing combat. The rebel Meteors wore a cross within a 'V' with the legend 'Cristo vence' (Christ wins). Government-aligned Meteors wore a letter 'P' over a 'V' to denote that they were Perónistas.

Right: During a victory celebration, held on 22 September in Córdoba, the Meteors that had participated in the revolution flew with others from Buenos Aires flown by rebel pilots who had been unable to join in. Calquins, Lincolns, IAe-22DLs, Prentices and a prototype IA-33 Pulqui II jet fighter (foreground) also took part.

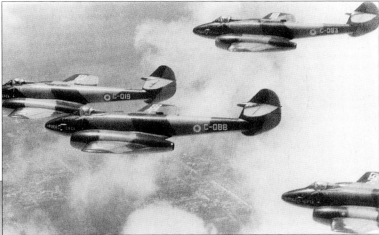

Above: C-002 was originally going to fly with the RAF as RA388 before its diversion to meet the Argentine order. The protrusion on the spine of the aircraft is a dorsal communications antenna and replaced the earlier 'pylon-type' antenna previously carried. The aircraft was one of about 20 which remained in operational service in early 1970, but by the end of the year all had been retired.

Above: A quartet of FAA Meteors formates over Buenos Aires province in the late 1960s. The lead aircraft, C-019, saw combat with rebel forces in the June 1955 anti-Perón uprising. In a strange coincidence all four aircraft (C-019, C-057, C-088 and C-093) survived until the Meteor's final days, and all are now preserved at various schools and institutions within Argentina.

Left: Seen at a ceremony marking the withdrawal of the Lincoln from service, a CB2 Meteor lines up alongside a newly delivered A-4B Skyhawk and an MS.760 Paris.

An unquiet peace

After the revolution, the FAA units resumed their peacetime activities, and 11 aircraft (I-006, 007, 019, 044, 051, 052, 068, 086, 089, 093 and 100) began operations from VI Brigada Aérea at Tandil, which continued until December 1955.

By 1957 the Meteor fleet had been reduced to about half of its original number, however, the delay in the arrival of the much-anticipated Sabre revived the Meteor's standing within the FAA. Along with forming various aerobatic teams to showcase the air force both at home and in neighbouring countries, the Meteors participated in numerous small- and larger-scale exercises.

In November 1959, the Meteors participated in a series of interception exercises with Avro Lincolns over Buenos Aires during Exercise Tigre, designed to test the FAA's ability to defend the capital. This was repeated in 1960 in Exercise Costanera.

On 21 September 1962 the Meteors saw combat action again, in yet another military uprising. This time they fired wing-mounted T-10 rockets against the Escuela de Suboficiales del Ejército (Army Sub-Officers Academy), whose members were entrenched at the Parque Chacabuco, in Buenos Aires. The Meteor's strikes proved conclusive and prompted the immediate surrender of the revolutionary forces.

A further uprising was quashed on 3 April 1963, when four 'Glosters' – in conjunction with two Lincolns, four Sabres, and four Morane-Saulnier Paris aircraft, as well as the Regimiento de Caballería de Tanques 8 – attacked Punta Indio naval aviation base in order to force the surrender of the Marines infantry that had revolted.

In 1963, due to the wide use of the Meteor in air-to-ground missions, its silver colour scheme was changed to a green-grey camouflage pattern, with the underside painted sky blue. This scheme remained until the end of the type's operational service.

Aerobatic display teams

The first FAA Meteor display team was formed in April 1951 especially for the visit of the prince of The Netherlands, who was in Argentina and was interested in the locally-built Pulqui II jet fighter prototype. Its success provoked the idea of creating a permanent aerobatic team with the Meteors, but a revolution attempt against President Juan Perón in that year resulted in the abandonment of these plans. The idea did not die, though, and in 1952 a team was formed, which debuted on 26 September 1952 at Aeroparque airport in Buenos Aires. The team gave other performances throughout 1953 and 1954, but were interrupted in 1955 by events that changed the history of Argentina – the Revolución Libertadora.

On 20 March 1957 a new aerobatic unit departed for a tour to Uruguay, during which it attempted to outperform the Meteor F.Mk 8 team of the Forca Aérea Brasileira. Shortly after, they faced their most important aerobatic challenge: on 17 November, during the XII Aeronautic Week, they appeared alongside the F-100D Super Sabres of the USAF's 'Thunderbirds' demonstration team.

On 21 September 1961, a five-Meteor team flew to Asunción in Paraguay to participate in independence anniversary celebrations. Then, one year later, the most famous of the FAA's Meteor teams, Escuadrilla 46, was created with eight red and white painted aircraft. The team continued until 1964, in that year performing in neighbouring Uruguay.

In spring 1966 a final Meteor team was formed at Morón, conducting a number of displays over the following 18 months.

Right: With a smoke generator fitted beneath each wing a Meteor belonging to the team formed in the early 1950s completes an impressively low diving pass at the Buenos Aires Aeroparque exhibition. In the foreground is a Bristol Freighter belonging to BA1, which was normally based at El Palomar.

Below: The most conspicuous of all FAA Meteor's were those of the Escuadrilla 46. Not an official unit, it was nonetheless the FAA's main publicity tool in the mid-1960s. The team comprised eight Meteors, fitted with smoke generators and adorned in a special red and white colour scheme. Five of the team's aircraft are seen here prior to being re-painted, but proudly displaying the team's unofficial badge beneath the cockpit.

Ground-attack Meteors

From 1960 the Meteors began to undertake fighter-bomber missions, for which rocket launching rails and bomb racks (built by the FMA) were fitted. In anticipation, in 1959 the role-identification had been changed from Interceptor to Fighter and, accordingly, serial numbers changed from I- to C- (for Caza, Fighter). On 12 November, Fighter Groups 2 and 3 became CB2 and CB3 (cazabombardeo, fighter bomber), officially assigned to VII Air Brigade from January 1960.

In 1960, the Meteors were deployed to Comodoro Rivadavia and then to Río Gallegos in Patagonia, where they took part in Operativo Sur, their first air-to-ground exercise using rockets and bombs. The aircraft also retained the four 20-mm Hispano cannon which were utilised for ground strafing as well as providing a degree of self-defence.

A Morón-based Meteor (above) demonstrates the range of bombs and rockets available for ground-attack. The cannon (foreground) were later modified to fire at 570 to 620 rounds per minute, with 180 rounds per gun. The most common rocket projectile used was the T-10 mounted on removable underwing rails (right), whereas bombs, ranging from 125 kg to 500 kg, but most typically 250-kg weapons (left), were mounted on wing hardpoints.

During the 1960s several of the aircraft had been modified with a non-standard weapons-fit including C-009 which was modified with two Colt-Browning 12.7-mm machine-guns, taken from a Sabre, in place of its four Hispano Suiza 20-mm cannon.

Meteor finale

As early as 1956 the Argentine government had signalled its intention to purchase 36 Canadair Sabre Mk 6s from Canada. However, a lack of finance saw this deal fall through and there was little choice but for the Meteors to continue in the air defence role.

Eventually, in 1960, 28 F-86F-30s (modified to F-40 standard) were purchased from the USA, allowing the remaining Meteors to be re-roled for ground-attack.

The beginning of the end for the Meteor in Argentine service coincided with the arrival in 1966 of the first batch of 25 US surplus A-4B Skyhawks. At this time CB2's Meteors moved to IV Brigada Aérea at Mendoza, together with the unit's Morane Saulnier MS.760 Paris. Then, from 1967, the last 20 Meteors were transferred to CB3 in VII Brigada Aérea, along with a few Beech 45 Mentors and a Bell UH-1H. In 1969, with the dissolution of CB3, the aircraft became part of Grupo Aéreo 7.

The arrival of a second batch of A-4Bs in 1970, along with the imminent arrival of the first Dassault Mirage IIIs, finally rendered the Meteor surplus to requirement after over two decades of frontline service.

The late 1960s had witnessed a number of 'lasts' for the Meteor. In 1966 the last Meteor aerobatic team was formed, giving its final performance the following year and, with aircraft being retired as they reached major overhaul, the final General Maintenance Service was carried out in 1969 at the FMA.

On 29 December 1970 the last dozen examples remaining in service (C-005, 027, 029, 037, 038, 051, 057, 071, 088, 093, 094 and 099) flew over Buenos Aires and that afternoon they all (except C-051 and C-088) flew over the city's western suburbs, as a final farewell.

So ended the operational career of the first Latin American jet fighter. Many Argentine Meteors ended their 'lives' under the scrapman's torch, but around 20 airframes survive today; they are in different museums, monuments, and technical schools, where a number are still used as instructional airframes.

Juan Carlos Cicalesi and Santiago Rivas

The introduction of the North American F-86F Sabre in the early 1960s relieved the Meteors of their air defence duties. The arrival of the Sabres also provided FMA technicians with an opportunity to extend the operational radius of the Meteors by re-plumbing and locally strengthening the wing to take the Sabre's 120-US gal (454-litre) underwing external fuel tanks. It is not known how many Meteors were capable of being fitted with the tanks but C-005 (left) was used as a trials aircraft and for a period had a shark's mouth adorning the forward section of each external tank. C-005 was one of the 12 aircraft which participated in the final 12-aircraft flypast over Buenos Aires on 29 December 1970, and is now preserved at the Museo Fortín Independencia located in Tandil City.

Argentine air force Meteor F.Mk IVs

Serial	Prev. Ident.	Entered Service	Struck off	Comments
I-001	RA384	3/3/48	-/66	Damaged at Mar del Plata 5/11/66.
I-002	RA386	3/3/48	30/12/70	Damaged at Morón 1/9/64 as u/c retracted involuntarily
I-003	RA388	3/3/48	3/58	W/o at Gen. Rodriguez 18/3/58
I-004	RA389	3/3/48	4/63	W/o in mid-air collision with C-088 over Matheu 9/4/63
I-005	RA390	3/3/48	30/12/70	Light damage in acc. at Tandil 12/3/48. Preserved at Museo Fortín Independencia
I-006	RA391	3/3/48	10/58	W/o at Morón 14/10/58 during aerobatics
I-007	RA370	2/10/48	6/58	W/o at Gen. Rodriguez 16/6/58
I-008	RA385	2/10/48	12/58	W/o at Morón 21/11/58
I-009	RA392	2/10/48	28/8/69	Damaged at Morón 2/9/60
I-010	RA393	3/3/48	12/5/70	Damaged at Morón 3/4/54 and 14/8/64. Preserved at Area de Material Quilmes
I-011	RA395	2/10/48	12/5/70	Collided with C-013 at Morón 10/1/66
I-012	RA396	2/10/48	8/53	W/o at Morón 8/53
I-013	EE570	2/10/48	6/3/70	Damaged at Morón 18/8/60 and 10/1/66
I-014	EE575	2/10/28	12/5/70	Scrapped 1970
I-015	EE551	2/10/48	6/62	Damaged at Córdoba 29/6/60. W/o after overshooting runway at Morón 16/6/62
I-016	EE569	3/3/48	4/51	W/o in accident at Tandil 10/4/51
I-017	EE554	2/10/48	1/56	W/o in accident at Morón 3/1/56
I-018	EE571	-	4/48	Exploded in mid-air during aerobatics at Morón prior to service entry
I-019	EE553	2/10/48	6/3/70	Preserved at Santa Rosa de la Pampa as I-021
I-020	EE546	2/10/48	12/5/70	Gear-up landing at Morón 19/1/61
I-021	EE544	2/10/48	2/52	W/o in accident at Morón 2/2/52
I-022	EE552	2/10/48	11/62	W/o at Mendoza after nosewheel failure 20/11/62
I-023	EE576	2/10/48	2/56	W/o in accident at Castelar 26/1/56
I-024	EE548	2/10/48	1/54	W/o after impacting with water near Isla Martín García 16/1/54
I-025	EE532	2/10/48	10/62	Damaged in heavy landing at Morón 10/62. Now at Escuela de Aviación Militar, Córdoba
I-026	EE572	2/10/48	10/56	W/o at Tandil 28/9/56
I-027	EE527	2/10/48	30/12/70	Collided with towed target 21/12/66. Now used for training at Córdoba University
I-028	EE535	2/10/48	4/54	W/o at Lujan 8/4/54
I-029	EE537	2/10/48	30/12/70	Preserved at Estancia La Romana
I-030	EE542	2/10/48	10/62	W/o at Morón in gear-up landing 24/10/62
I-031	EE588	2/10/48	25/11/69	Preserved at the Liceo Aeronáutico Militar
I-032	EE581	2/10/48	3/67	W/o in accident at Morón 19/3/67
I-033	EE582	2/10/48	11/54	W/o in accident at Morón 11/54
I-034	EE574	2/10/48	3/66	W/o 16/3/66
I-035	EE580	2/10/48	7/60	W/o in mid-air collision with C-096 over Pontevedra 12/7/60
I-036	EE577	2/10/48	5/68	Destroyed in fire at Mendoza 2/5/68
I-037	EE583	2/10/48	30/12/70	Scrapped 1971
I-038	EE587	2/10/48	30/12/70	Preserved at Junín, Buenos Aires
I-039	EE585	2/10/48	5/60	W/o in heavy landing at Morón 18/5/60
I-040	EE589	2/10/48	11/52	W/o in accident at Morón 11/52
I-041	EE586	2/10/48	12/5/70	Preserved at Museo Nacional de Aeronáutica
I-042	EE526	2/10/48	-/52	W/o in mid-air collision with I-078 over Los Cardales -/52
I-043	EE540	2/10/48	5/56	W/o in accident at Córdoba 5/56
I-044	EE534	2/10/48	25/11/66	Scrapped
I-045	EE547	2/10/48	12/55	W/o in accident at Morón 11/55
I-046	EE543	2/10/48	1/50	W/o in accident at Tandil 16/1/50
I-047	EE533	2/10/48	4/51	W/o in mid-air collision with I-097 over Tandil 21/4/51
I-048	EE539	2/10/48	1/66	W/o after landing short at Morón 21/1/66
I-049	EE536	2/10/48	5/60	W/o after landing short at Morón 29/4/60
I-050	EE541	2/10/48	8/55	W/o in accident at Morón 8/55
I-051	G-5-151	6/7/49	12/5/70	Preserved as C-002 at Baradero Airport
I-052	G-5-152	6/7/49	5/56	W/o in accident at Morón 5/56
I-053	G-5-153	20/12/48	6/62	W/o after heavy landing at Morón 12/6/62
I-054	G-5-154	20/12/48	8/57	W/o in accident at Morón 21/8/57
I-055	G-5-155	20/12/48	4/50	Exploded in mid-air over Tandil 22/4/50
I-056	G-5-156	20/12/48	6/58	Crashed on approach to Morón 22/6/58
I-057	G-5-157	20/12/48	30/12/70	Preserved at Loreto, Santiago del Estero
I-058	G-5-158	20/12/48	10/55	W/o in accident at Morón 10/55
I-059	G-5-159	20/12/48	5/49	W/o in accident at Tandil 24/5/49
I-060	G-5-160	20/12/48	8/54	W/o in accident at Córdoba 12/8/54
I-061	G-5-161	20/12/48	12/57	Crashed at Aeroparque exhibition 19/12/57
I-062	G-5-162	20/12/48	4/61	W/o in gear-up landing at Morón 6/4/61
I-063	G-5-163	11/5/49	5/66	W/o after heavy landing at Morón 5/5/66
I-064	G-5-164	6/7/49	6/55	W/o after heavy landing at Colonia, Uruguay 16/6/55
I-065	G-5-165	20/12/48	3/66	Wfu 16/3/66
I-066	G-5-166	20/12/48	-/69	Wfu -/69
I-067	G-5-167	11/5/49	11/51	W/o in accident at Ezeiza 19/11/51
I-068	G-5-168	20/12/48	8/63	W/o in landing accident at Morón 20/8/63
I-069	G-5-169	11/5/49	9/60	W/o in accident at Morón 30/8/60
I-070	G-5-170	11/5/49	6/49	W/o at Tandil after mid-air collision with I-092 27/6/49
I-071	G-5-171	6/7/49	30/12/70	Preserved for training at Escuela Técnica Jorge Newbery, Buenos Aires
I-072	G-5-172	11/5/49	10/62	W/o in accident at Morón 18/10/62
I-073	G-5-173	6/7/49	30/12/70	Preserved at Mar del Plata air base
I-074	G-5-174	6/7/49	7/56	W/o in mid-air collision with I-098 over Ituzaingo 16/7/56
I-075	G-5-175	11/5/49	6/58	W/o in accident at Córdoba 6/58
I-076	G-5-176	6/7/49	29/1/70	Damaged in collision with C-063 on t/o at Morón 31/8/64
I-077	G-5-177	6/7/49	6/56	W/o at Ezeiza 6/56
I-078	G-5-178	6/7/49	11/52	W/o in mid-air collision with I-042 over Los Cardales 17/11/52
I-079	G-5-179	6/7/49	8/55	W/o in mid-air fire during the Revolución Libertadora 19/9/55
I-080	G-5-180	6/7/49	9/58	W/o after heavy landing at Morón 19/9/58
I-081	G-5-181	11/5/49	12/5/70	Damaged in heavy landing at Morón 3/63
I-082	G-5-182	11/5/49	11/52	W/o in accident at El Palomar 15/11/52
I-083	G-5-183	11/5/49	10/67	W/o in accident at Resistencia 12/10/67
I-084	G-5-184	6/7/49	12/5/70	Scrapped
I-085	G-5-185	6/7/49	4/60	W/o after pilot baled-out over Garin 6/4/60
I-086	G-5-186	11/5/49	3/68	Wfu 5/3/68
I-087	G-5-187	6/7/49	3/58	Crashed at Castelar 10/3/58
I-088	G-5-188	11/5/49	30/12/70	Damaged in mid-air collision with C-004 9/4/63. Preserved at Aero Club Chivilcoy
I-089	G-5-189	6/7/49	11/64	W/o at Morón 6/11/64
I-090	G-5-190	11/5/49	16/3/66	Preserved at Salta City
I-091	G-5-191	11/5/49	10/64	Crashed into a ship at Entre Ríos 19/10/64
I-092	G-5-192	6/7/49	11/51	W/o in accident at Ezeiza 15/11/51
I-093	G-5-193	11/5/49	30/12/70	Preserved at VII Brigada Aérea, José C.Paz
I-094	G-5-194	6/7/49	30/12/70	Preserved at Grupo 1 de Vigilancia Aérea, Merlo, near Buenos Aires
I-095	G-5-195	11/5/49	30/12/70	Preserved at Air Force HQ, Buenos Aires
I-096	G-5-196	11/5/49	7/60	W/o in mid-air collision with C-035 over Pontevedra 12/7/60
I-097	G-5-197	6/7/49	4/51	W/o in mid-air collision with I-047 over Tandil 21/4/51
I-098	G-5-198	11/5/49	7/56	W/o in mid-air collision with I-074 over Morón 16/7/56
I-099	G-5-199	11/5/49	30/12/70	Preserved at Neuquén Airport
I-100	G-5-200	6/7/49	10/62	W/o in landing at accident at Córdoba 10/62

*Serial prefixes for surviving Meteors changed from I- to C- in 1959. Serials I-001 to I-050 were also allocated the consecutive UK Class B registrations G-5-101 to G-5-150.

C-051 was the first of the second batch of F.Mk 4s built specifically by Gloster for Argentina and is seen here towards the end of its service life. On retirement it was allocated for preservation and is currently on display at Baradero airport painted as C-002. Of the 100 Meteors delivered to Argentina, some 67 were lost in accidents during its 22-year career, including 10 lost in mid-air collisions and two lost to ground fires. Of the remaining 33 aircraft, 19 survive today as whole or near-whole airframes – the other 14 succumbed to the scrapman's torch.

P-47 Thunderbolt

Part 2: Final developments and combat in the Mediterranean, Far East and Pacific

Enlarged outer wing panels with squared-off wing tips and a pronounced dorsal fin forward of the vertical tail were the distinguishing features of the last Thunderbolt variant, the P-47N. The P-47's outline remained remarkably unchanged throughout its five-year production life, save for the obvious revision brought about by the change to a 'bubble' canopy during P-47D production.

Thunderbolt development culminated in the P-47N, a much improved aircraft with extra fuel capacity in redesigned wings. 7,750 were on order on VJ-Day; 1,816 were delivered, the last in October 1945.

Below: The RAF used the Thunderbolt widely in the Burma/India theatre.

Although it was overshadowed and replaced by North American's cheaper, longer-ranged P-51 Mustang in the USAAF's Eighth Air Force, Republic's P-47 Thunderbolt was probably a better all-round fighter, having superior agility, better flying characteristics, and better air-to-ground capabilities. These attributes led to a massive production run, exceeding that of any other US fighter, and the aircraft saw extensive operational service. The type was first pressed into service as a long-range escort fighter and fighter-bomber in the European Theatre of Operations, serving with the Eighth and Ninth Air Forces. This article describes the aircraft's service in large numbers in the Mediterranean and its role as the vital backbone of the air war in the Far East, followed by its brief but useful post-war career.

Rejected by the 'Mighty Eighth'

In the high-profile Eighth Air Force, all but one P-47-equipped Fighter Group transitioned to the Mustang between November 1943 and November 1944. The superseded P-47 had been constrained in the crucial long-range escort role by its relatively short range. Although the aircraft had respectable internal fuel tankage (305 US gal; 1325 litres), the big 18-cylinder Pratt & Whitney R-2800 radial engine drank fuel at a prodigious rate. This limited the aircraft's usefulness to the 'Mighty Eighth', which was beginning to send out its bombers to farther-flung targets beyond the range of the Thunderbolt.

The P-47 had a cramped and relatively uncomfortable cockpit, with inadequate thigh support and rudder pedal adjustment for taller pilots. The cockpit was also poorly laid out, with too much room for error in selecting oil cooler flaps or intercooler flaps. The engine operated very badly at incorrect settings, also making the aircraft poorly suited for long-duration escort missions.

The P-51 Mustang, by comparison, could accompany the bombers all the way to Berlin without difficulty, and had an extremely comfortable cockpit. The Mustang pilot enjoyed automatic control of the oil and coolant radiator flaps, which significantly reduced pilot workload.

It was this long-range capability which led to the rapid rise to dominance of the P-51 Mustang within the Eighth Air Force, and which quickly obscured the importance of the P-47. The latter was relegated to the fighter-bomber role and to service in other theatres. The tactical specialists of the less glamorous (but arguably more important) Ninth Air Force quickly became the major operator of the P-47, operating 16 Thunderbolt groups mainly in the fighter-bomber role.

This role switch has sometimes led to the assumption that

An Air Service Command test pilot prepares to take a newly reassembled P-47D aloft after reassembly on a North African airstrip, almost certainly Solimon, in Tunisia. These aircraft were destined for the Twelfth Air Force's 325th Fighter Group, which became the first group in the MTO to convert to the P-47 in September 1943, two months before the 325th joined the newly formed Fifteenth Air Force. Operations – bomber escort missions alongside 15th AF P-38s – began in December, though by then the group had moved to Foggia, Italy. When the 325th FG converted to P-51s in mid-1944, its P-47s were then briefly used by the 332nd FG.

the P-47 was an inferior air-to-air dogfighter, but this was never the case. On paper, the Thunderbolt did have a marginally worse turn performance than the P-51, and did bleed energy more rapidly in a maximum-rate turn. In its favour, the Thunderbolt had much lower stick forces (about 7.7 lb/g, compared to the P-51's 20 lb/g), which made the aircraft less tiring to manoeuvre and facilitated accurate gun tracking. The P-47 also had a better rate of roll than the Mustang, so although it turned more slowly in a sustained turn, it could change direction more quickly; its instantaneous turn rate was no less impressive. The P-51 could not accelerate or dive as quickly as the Thunderbolt and had unpredictable and dangerous stall characteristics, with little warning and a vicious departure. This meant that the P-51 pilot could easily 'overcook' a combat manoeuvre and find his aircraft rolling rapidly through as much as 270°, losing perhaps 500 ft (152 m) and experiencing a violent aileron snatch that could tear the stick out of his hands.

By contrast, the Thunderbolt was generally easy and pleasant to manoeuvre, and was stable enough to allow its pilot to place the pipper on the target and keep it there throughout energetic manoeuvres. The P-47 pilot could safely fly in the pre-stall buffet, maximising his turn performance, without risking the type of departure that ruined many a Mustang pilots' day.

Mediterranean service

Fifteenth Air Force's sole Thunderbolt-equipped group, the 325th, introduced the type to the Mediterranean theatre in December 1943, to serve alongside three P-38 groups in the bomber escort role. By the end of the year the tactical Twelfth Air Force was converting the first of an eventual total of six groups to the P-47. While the 325th FG's use of the Thunderbolt lasted a brief five months, the Twelfth's groups relied on the P-47 until VE-Day.

Above: The 325th FG's first Thunderbolts were P-47D-6s and D-10s (including 42-75001 of the 318th FS); by January 1944 these had been passed to the 57th and 79th FGs and replaced with D-15s with improved range.

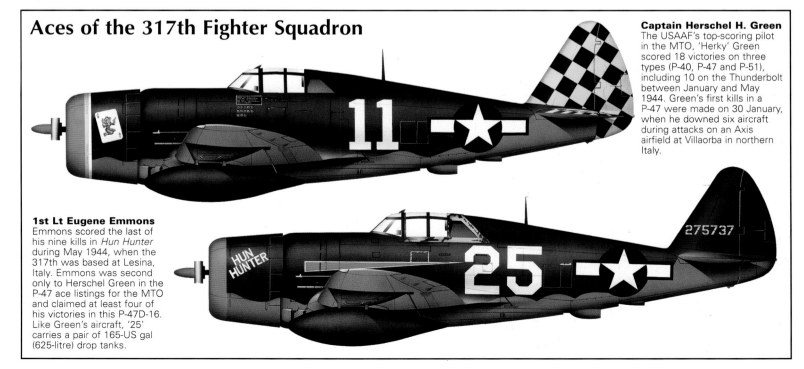

Aces of the 317th Fighter Squadron

Captain Herschel H. Green
The USAAF's top-scoring pilot in the MTO, 'Herky' Green scored 18 victories on three types (P-40, P-47 and P-51), including 10 on the Thunderbolt between January and May 1944. Green's first kills in a P-47 were made on 30 January, when he downed six aircraft during attacks on an Axis airfield at Villaorba in northern Italy.

1st Lt Eugene Emmons
Emmons scored the last of his nine kills in *Hun Hunter* during May 1944, when the 317th was based at Lesina, Italy. Emmons was second only to Herschel Green in the P-47 ace listings for the MTO and claimed at least four of his victories in this P-47D-16. Like Green's aircraft, '25' carries a pair of 165-US gal (625-litre) drop tanks.

The P-47 could carry a pair of 1,000-lb bombs underwing, but they made handling tricky on take-off and in formation-keeping, and 500-lb bombs were usually used instead. The aircraft also used rockets and napalm tanks in its new-found air-to-ground role, and proved correspondingly popular with the squadrons of the Ninth Air Force.

In other theatres, the Thunderbolt's career was only just beginning, and during late 1943 it began to equip numerous fighter groups in the Mediterranean and Pacific.

US operations in the Mediterranean

The Thunderbolt served extensively in the Mediterranean with two numbered USAAF air forces. As a strategic air force, the 15th Air Force was in some respects broadly equivalent to the 8th, and its fighters were heavily tasked with bomber escort duties. Similarly, the 15th Air Force kept its P-47s for a relatively brief period, replacing them with P-51 Mustangs.

The 15th Air Force formed on 1 November 1943 under General Jimmy Doolittle (soon after the capitulation of Italy), and took over the long-range assets of the 12th Air Force, including six Heavy Bomber Groups, three P-38 Fighter Groups and the 325th Fighter Group. The 325th Fighter Group transitioned to the P-47 from September 1943 but did not begin flying operations with these aircraft until it joined the new 15th Air Force, from December 1943. The unit flew the P-47 until May 1944, notching up an impressive combat record.

Six of the 325th FG's pilots scored more than five kills each in the P-47. Major Herschel Green of the 317th was the most successful, adding 10 kills to the three he had scored while flying the Curtiss P-40. (Green later added five more while flying the P-51D.) Other pilots opened their scoring on the P-47 and attained ace status while flying the P-51.

The 'Checkertails' traded their Thunderbolts for Mustangs after five months in action. Its aircraft were then passed to the African-American-manned 332nd Fighter Group (the famous 'Tuskegee Airmen'), which joined the 15th Air Force and flew the new type for fewer than two months, before re-equipping with P-51 Mustangs. During this brief interlude,

Jenny 'A' (named for a particularly conscientious Republic factory worker) is a P-47D-15-RE of the 319th FS, pictured at Lesina, the unit's base for 12 months from March 1944. The 325th FG's famous black/yellow checkerboard tail marking was not applied to its Thunderbolts until the spring of 1944. P-47D-15s had 70 US gal (265 litres) of additional internal fuel tankage and were equipped with underwing pylons at the factory. The group's bomber escort role necessitated the carriage of 165-US gal (625-litre) drop tanks, as seen on this aircraft. They were the largest underwing tanks carried by the 'Jug'.

The 57th Fighter Group was the first 12th AF group to convert to the P-47, in December 1943, and the first to pioneer the use of the P-47 as a low-level strafer and dive-bomber. This 64th FS aircraft, with a scorpion emblem on its cowling and pyramid – a reference to its earlier service in North Africa – is seen at Grosseto during 1945.

An evaluation of the Mustang, Thunderbolt, Grumman F6F Hellcat and Vought F4U Corsair by modern test pilots during the 1990s judged the P-51 Mustang to be "totally unsuited to the ACM [air combat manoeuvring] environment" but conceded that it was "well suited to long-range escort missions, intercepting and defending against non-manoeuvring targets". Fortunately, by the time the Mustang replaced the P-47 in the long-range escort role, the main German home defence fighter was the heavyweight and rather sluggish Messerschmitt Bf 109G, with which even the P-51 could deal without difficulty.

With this in mind, the switch of most P-47s to tactical and ground attack duties – where they would come up against more agile enemy fighters, including the Fw 190 – can be seen to have been a wise step.

The Focke-Wulf Fw 190 was a remarkably nimble fighter, and gave even the Spitfire a hard time. Unsurprisingly, the P-47C or D stood little chance in a low-level turning fight against the Fw 190. Remarkably, though, the P-47 could outrun an Fw 190 in level flight, and would (eventually) catch up with a diving Fw 190. Even more surprising, a P-47 could outmanoeuvre an Fw 190 at airspeeds above 250 mph (402 km/h) at all heights, and enjoyed an advantage above 15,000 ft (4572 m).

The Thunderbolt's final advantage over the P-51, and a key component in its success as a fighter-bomber, lay in its armament. Although never armed with 20-mm cannon (found to be the optimum fighter weapon), it did pack a heavier punch than other US fighters of the period; some P-47 variants mounted as many as eight 0.50-in machine-guns. (Early Mustangs, by contrast, had only four machine-guns). This was particularly useful for strafing ground targets, for which the Thunderbolt's accurate air-to-ground target-tracking made it well suited.

Gun camera footage from a following aircraft shows a Twelfth Air Force P-47 in its element, strafing motorised transport on a road south of the Brenner Pass, near Verona.

Converting to the P-47 shortly after the 57th Fighter Group, the 79th Fighter Group used codes with an 'X' prefix – these are 86th Fighter Squadron aircraft. Pale blue tailfins with yellow lightning bolts, as applied to aircraft 'X51', were introduced during the spring of 1945.

the 332nd made good use of its Thunderbolts, downing five enemy aircraft on 7 June during its first mission with the 15th AF (and its first with the P-47). On 25 June, off Trieste, two pilots (Lieutenants Wendell Pruitt and Lee Archer, the latter subsequently becoming an ace) attacked and sank a German destroyer with well-aimed machine-gun fire.

If the 15th Air Force was the MTO counterpart to the Eighth, then the tactical 12th Air Force was the Mediterranean equivalent to the Ninth. Like the Ninth, the 12th Air Force came late to the Thunderbolt, but then adopted the type in large numbers and used it to devastating effect, primarily in the fighter-bomber role. These Thunderbolts scored relatively few air-to-air victories (129 in total, between the six groups), and the units produced no aces, so they have been largely sidelined in the history books.

The first of the 12th Air Force fighter groups to re-equip with the P-47 was the 57th FG, which traded its P-40s for Thunderbolts in December 1943. It was joined by the 79th FG in February 1944, the 27th FG in June 1944, the 86th and 324th FGs in July 1944, and the 350th FG in August 1944. These units supported the Allied push into Italy (the supposed "soft underbelly of the Axis") and Operation Dragoon, the invasion of southern France in August 1944. From the autumn of 1944, some units supported partisan operations in Yugoslavia, while others turned their attention to targets in southern Germany. During the closing stages of the war, 12th Air Force Thunderbolts introduced a range of new weapons. The 79th FG used aircraft armed with one 500-lb and two 1,000-lb bombs during attacks against Cassino, while tube-launched air-to-ground rockets (three per side) were used as it supported the British Eighth Army's push into the Reich. A 12th Air Force P-47 pilot, Lieutenant Raymond Knight of the 346th FS, 350th FG, won the Medal of Honor after leading two daring low-level raids against German airfields. The medal was awarded posthumously, as Knight's battle-damaged aircraft crashed into a mountain as he tried to regain his base. The 12th Air Force

Thunderbolts took a heavy toll of enemy trains, artillery guns, motor- and horse-drawn vehicles.

Non-US operators in the MTO

Two foreign users of the P-47 also used the type operationally in the Mediterranean theatre. The first were the Free French, who began receiving the first of 446 Thunderbolts in March 1944. These aircraft were used to re-equip units in North Africa that had previously used Hawker Hurricanes. By May 1944 Escadre de Chasse 4 was operating from Corsica with two P-47-equipped Groupes de Chasse (GC II/3 'Dauphiné' and GC II/5 'La Fayette'), and a third Groupe de Chasse (GC I/4 'Navarre') was added before the invasion of southern France. GC II/3 was initially attached to the 57th FG. The Thunderbolt units subsequently moved to airfields in southern France and supported US and Free French units as they advanced on Germany. A second Thunderbolt escadre was formed late

The emblem on the cowling of this aircraft is that of the 86th Fighter Squadron, the first unit in the 79th Fighter Group to fly a combat mission after it re-equipped with Thunderbolts, on 9 March 1944. The 79th was an early exponent of the use of M10 4.5-in rockets, as seen on this aircraft, from October 1944.

Republic's 10,000th P-47 was one of the ubiquitous P-47D-30s, 2,600 of which were completed. These introduced such features as blunt-nosed ailerons, to improve handling at high speeds, and a set of dive flaps positioned at 30 per cent of chord from the leading edge. Delivered to the 87th FS/79th FG in Italy, 10 Grand Thunderbolt 44-20441 flew its first mission on 19 January 1945, flown by the group's CO, Colonel John Martin. Later in the month Martin, himself a veteran of over 200 missions, flew the group's 30,000th combat sortie in this machine.

The aircraft nearest the camera in this view of a quartet of 345th Fighter Squadron/350th Fighter Group machines, has an impressive tally of 70 bombing missions to its name. Each carries a 75-US gal (284-litre) drop tank and a pair of 500-lb bombs. While '5C6' is an older P-47D-28, the middle pair of Thunderbolts are P-47D-30s, presumably recently delivered, given their lack of markings. The 350th FG used an aircraft code system whereby the first digit was the last digit in the squadron's number, the letter indicated the flight to which the aircraft was assigned and the second digit was the aircraft's number in the flight.

Above: The 350th FG was a relatively late convert to the P-47, making the transition during August/September 1944. Among its three squadrons was the 347th FS, known as the 'Screaming Red Ass Squadron'. This 347th aircraft took a hit from flak during a low-level mission but managed to get down safely before the engine seized solid. Seeing out of the oil-covered cockpit must have been a struggle for the pilot.

Right: Disney's cartoon character 'Goofy', flying a P-47 through flak bursts, was the unofficial emblem of the 346th Fighter Squadron. The unit later employed a black/white checker pattern on the rudder. It finished the war with 11 kills to its name.

The other operator of the Thunderbolt in the Mediterranean was 1° Grupo Aviação de Caça (1° GavCa) of the Força Aérea Brasileira. During World War I, Brazil had been the only Latin American nation to join the Allies, and (despite strong neutral and even pro-German forces) did so again during World War II, prompted by German U-boat activity in Brazilian waters and by the desire to gain greater international status after the war. Brazil therefore declared war on Germany and Italy on 22 August 1942.

Rather than fighting a simple anti-submarine war in the South Atlantic, and participating in regional defence, Brazil made a very real contribution to the Allied effort in Europe, sending a division-sized Brazilian army unit to Italy as part of the US IV Corps.

It was decided at an early stage to select, train and deploy a Brazilian fighter unit in Europe alongside the army division, and the 1° GavCa was formed by Presidential decree on 18 December 1943. Manned by volunteers, the unit trained with the USAAF's 30th FS in the Canal Zone, with each pilot flying 60 hours in the P-40 before flying normal operations on the type, until the end of June 1944. The unit then moved to Suffolk Field, Long Island, where the pilots and ground crew converted to the Thunderbolt, each pilot on average getting 80 hours on the P-47G. The majority of the Brazilians already had extensive flying experience, so the unit compared well with the best American fighter units, whose new pilots often had as few as 10 fighter hours upon joining their first squadron.

The 1° Grupo Aviação de Caça sailed for the Mediterranean in September 1944, and joined the 350th Fighter Group on 7 October. Brazilian aircrew began flying operational missions one week later, and the unit flew its first independent operational mission on 11 November 1944, using the callsign 'Jambock'.

Although the pilots of the 1° GavCa had trained in escort/fighter tactics, the Brazilian Thunderbolts were used primarily in the fighter-bomber role, to which the Brazilians quickly adapted with great success. Their aircraft wore green and yellow rudder stripes, with the Brazilian national insignia superimposed on the standard USAAF star and bar, although later the bars were often overpainted. The Brazilians lost 17 men in combat, and usually maintained a strength of 26 aircraft, drawn from a pool of some 88 P-47s. General Arnold later said that the Brazilians "flew them [the P-47s] very, very well", and the only criticism of the Brazilians offered by Major William Marshall, a USAAF pilot with the 350th FG, was that "they stayed with a target too long... perhaps they were braver than I was!". Marshall also observed that they "probably contributed as much 'kill' to the Germans in the Brenner Pass and all through the Po Valley as any other squadron".

in 1944, with three more groupes de chasse (GC I/5 'Champagne', GC III/3 'Ardennes' and GC III/6 'Roussillon'). By the end of the war, only the Spitfire Mk IX outnumbered the P-47 in French service. Thunderbolt units then formed part of the occupation forces in Germany until late 1949, when they returned to re-equip with jet fighters.

First Free-French P-47s – GC II/5 'La Fayette'

P-47D-28
Details of the pilot of this aircraft, based at Amberieu, near Lyon, during the autumn of 1944, are unknown, but it is representative of a P-47 GC II/5, one of the six Armée de l'Air units raised as part of 1st TACAF in 1944/45. Few victories were scored by French P-47 pilots, given the lack of opportunity for air combat; kill markings displayed are likely to be those of a veteran of the Battle of France in 1940.

At the end of the war, 26 surviving Brazilian aircraft then in use were shipped home. Nineteen reserve aircraft in the depot at Naples were dispersed to USAAF units and replaced by 19 virtually new aircraft delivered to Brazil from Kelly Field, Texas. Post-war attrition was more than countered by the delivery of 25 additional aircraft in 1947 and 25 more in 1953. The Brazilian Thunderbolts were relegated to the fighter training role from 1954, and finally were deactivated in 1957-58.

Pacific action

The P-47 entered operational service with the Eighth Air Force in England in April 1943, and finally ended up in the Pacific in December that year, when a batch of P-47s reached the 5th Air Force in Australia. Originally based in the Philippines (forming from the Philippine Department Air Force in February 1942), the handful of personnel who escaped the Japanese invasion that spring formed the cadre of a new 5th Air Force in northern Australia, which was sent forward to Java to delay or halt the advancing Japanese. Pushed back to Australia following Japan's seizure of the Dutch East Indies, the command was tasked with defending Australia and the remaining Allied-held areas within New Guinea against enemy air attack.

When the first Thunderbolts arrived, General George C. Kenney was furious, believing that the aircraft had neither the range to cover the massive distances involved nor the manoeuvrability to deal with Japan's agile lightweight fighters. He had hoped to receive more P-38 Lightnings.

Fortunately, the Thunderbolt's combat record in the southwest Pacific was much better than Kenney could have predicted. Together, the three groups (and two squadrons) claimed 511 enemy aircraft and produced a string of aces, including the 348th FG's Colonel Neel Kearby, the highest-scoring P-47 ace in the Pacific with 22 victories, and Major William Dunham from the same unit, whose 16 kills included 15 scored in the P-47.

The 348th FG was the first P-47 unit in the Pacific, and became the most successful. The group arrived in-theatre on 23 June 1943, and for the rest of the war moved up

through New Guinea and on to Leyte, the Philippines and finally to Ie Shima. The Thunderbolts flew intensive ground attack and close air support missions, but often encountered enemy aircraft, and rapidly built up an impressive kill tally (this eventually reached 326 aircraft, many downed by the group's 20 aces). Like RAAF and RNZAF aircraft in the theatre, the 348th FG's P-47s wore white tail surfaces and wing leading edges on their drab green camouflage, eventually gaining red candy stripes on the rudders when natural metal aircraft were received.

In October 1943 the 36th FS of the 8th FG briefly converted to the P-47, while the group's other squadrons received P-38s. This situation was clearly unsustainable, and the 36th FS soon received P-38s, bringing it into line with the 8th FG's other squadrons. It was a similar story for the 49th FG's 9th FS, which used P-47s only from November 1943 until April 1944, before switching back to the P-38s that equipped the rest of the group. The 35th FG converted to P-47s in November 1943 and used them until March 1945, when the group finally converted to the Mustang. The 35th flew some very long-range missions, using fuel conservation techniques developed and passed on by one Colonel Charles A. Lindbergh. The group's aircraft flew some missions that lasted up to eight hours,

Bottom: With the conclusion of the North African campaign, 4ème Escadre de Chasse was formed on Corsica. These 'razorback' P-47Ds, pictured on the island during the summer of 1944, are former 324th Fighter Group machines, 'handed down' by the 12th Air Force unit as it re-equipped with newer examples.

Below: 3ème Escadre de Chasse was formed in late 1944, with GC I/5 'Champagne' among its three constituent 'groupes de chasse'. This P-47D-27 of GC I/5, pictured during 1945, displays typical Armée de l'Air markings, including a blue/white/red rudder and alphanumeric codes.

In combat from November 1944, Brazil's 1° GAvCa joined 12th Air Force's 350th FG in providing air support for US IV Corps, which incorporated a division-sized Brazilian army unit. These aircraft, pictured at the unit's Tarquinia base, carry M10 'bazooka' rockets and napalm for just such a mission.

Right: Flown almost exclusively by 1° GAvCa's commanding officer, Lt Col Nero Moura, P-47D-25 42-26450 was technically the FAB's first operational Thunderbolt; having survived the war it was shipped to Brazil, becoming 4104 in the post-war FAB.

Early attempts to improve the P-47's range included a 200-US gal (757-litre) drop tank, seen fitted to a 348th FG P-47D-2 at Ward Field, Port Moresby, in September 1943. Another example of the tank, developed by Fifth Air Force Service Command in Brisbane, Australia (and consequently known as the 'Brisbane' tank), is visible at the extreme right.

and on 7 August 1944 Major Cella led 26 Thunderbolts on a 900-mile (1450-km) mission.

The final Thunderbolt unit to arrive in-theatre was actually the first to have converted to type: the 58th FG had received its P-47s in October 1942 in Virginia, but moved to Australia only in December 1943, after which it followed a similar pattern to the 5th Air Force's other 'long-term' P-47 groups.

This group flew a very unusual operation on 26 December 1944, for which it won a Distinguished Unit Citation. Colonel Atkinson led 32 Thunderbolts in a daring low-level strafing attack against a Japanese battleship, heavy cruiser and six destroyers that were about to bombard Mindoro. The force took off at 20.15 in the midst of an air attack by Japanese aircraft, during which two aircraft crashed and one returned with engine trouble. The remaining 29 P-47s then attacked the enemy warships from all directions, turning on landing lights to avoid collision after each pass, then turning them off as they attacked once more. The pilots attacked again and again until they ran out of ammunition, averaging nine firing passes each. Nine

aircraft were lost, but six of the downed pilots were rescued. Atkinson won a DSC, two majors each won a Silver Star, and 23 pilots were awarded DFCs.

The 'Pineapple Air Force'

Sometimes known as the 'Pineapple Air Force', the 7th Air Force was primarily responsible for the defence of Hawaii. Although four 7th Air Force groups eventually operated P-47s, three never saw combat with the Thunderbolt. The 15th FG flew P-47Ds in defence of Hawaii between June and September 1944, but converted to the Mustang before being sent forward to Saipan. Its place was taken eventually by the 508th FG, but this remained a training and ferry unit, and its P-47s never saw action. The 18th FG never became operational.

The remaining 7th Air Force P-47 group (the 318th FG) converted to the type in November-December 1943, and entered the fray in mid-1944, being shipped to Saipan aboard two aircraft-carriers and then catapult-launched from these ships for the brief hop to its new island home. When the unit arrived, parts of Saipan were still under Japanese control, and pilots were frequently in danger from snipers as they walked to their aircraft. On at least one occasion a Japanese soldier infiltrated the flight line, punctured a P-47's tanks with his bayonet, and set it alight. The group then moved to Tinian, and continued to support the

When Fifth Air Force began to receive P-47s equipped with wing racks (from P-47D-15 onwards; a D-16 is pictured), it was able to utilise 165-US gal (625-litre) drop tanks of the type used by the P-38 and also adopted by Thunderbolt units in the MTO. This February 1944 photograph shows a tank drop during separation testing.

Thunderbolts in the southwest Pacific

The first Thunderbolts to reach the PTO were those delivered to New Guinea in response to a Fifth Air Force request for aircraft to escort its bomber force, as it attacked Japanese airfields in the push northwards. The Fifth Air Force would have much preferred more long-range P-38 Lightnings and immediately took a dislike to the P-47 for its short range and lacklustre performance at low altitude. However, under leaders such as the 348th's Neel Kearby, the P-47's qualities were eventually recognised and new tactics developed to exploit the aircraft's high dive speed in the ground attack role.

5th Air Force colours, late 1944

39th FS, 35th FG
Lieutenant LeRoy V. Grossheusch was among the top scorers in the southwest Pacific, finishing the war having claimed eight victories, all but one while flying P-47s, including this aircraft. By the summer of 1944, new Thunderbolts were arriving in the SWPA in natural metal and surviving early examples, like this D-4 (based at Morotai in September 1944), were stripped of their olive drab. Light blue was the 39th's squadron colour.

40th FS, 35th FG
Captain Alvaro Jay Hunter claimed five air-to-air victories, the last three while flying P-47D-28s like this aircraft, his mount during his final months with the 40th in late 1944. Carrying red 40th FS squadron colours and a black theatre ID band, *My Baby* (named on the right side of the aircraft only) lacks the 35th FG's usual pre-war style 'USAAC' rudder striping. The 40th FS carried individual aircraft numbers in the range 40-69.

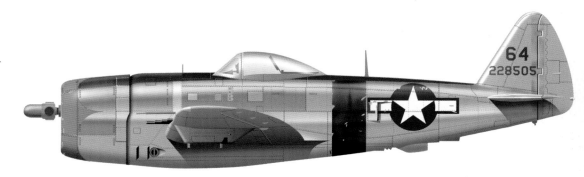

340th FS, 348th FG
Josie was the mount of Lieutenant Mike Dikovitsky, a five-kill ace based on Leyte with the 340th FS in December 1944. The 348th FG adopted similar rudder striping to that employed by the 35th, though the tails of its aircraft also carried a vertical stripe in the squadron colour over which was painted the aircraft's individual aircraft number in white. The group's aircraft also carried a variation on theatre ID striping, with a pair of black bands on the fuselage and three on the wings.

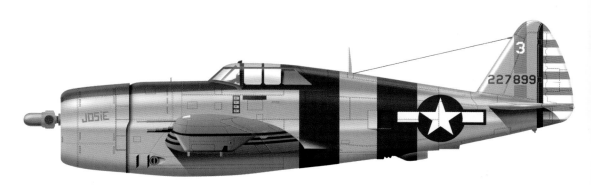

342nd FS, 348th FG
With seven victories to his name by the end of the war, Lieutenant Marvin E. Grant was another of the top-scoring pilots of the SWPA. All were claimed by mid-June 1944 and while flying early P-47Ds; P-47D-23 *Sylvia/Racine Belle* was his mount in late 1944/early 1945 when the 342nd was based on Leyte. The aircraft carries some of the 348th's usual markings, a tail stripe in the 342nd FS's light blue and an unusual red/white fuselage stripe, also a 342nd marking, introduced in late 1944.

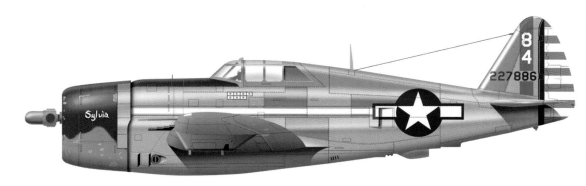

460th FS, 348th FG
A colourful tally board adorns *Bonnie*, a P-47D-23 flown by Major William D. Dunham, CO of the 460th FS and, by the end of the war, the second highest-scoring ace in the southwest Pacific, with 16 kills (the last of which did not come until August 1945; by then the 348th was flying Mustangs). This aircraft carries standard 348th FG markings of the period with black tail striping, as used by the 460th FS, which joined the group in September 1944 (using individual aircraft numbers in the range 91-120).

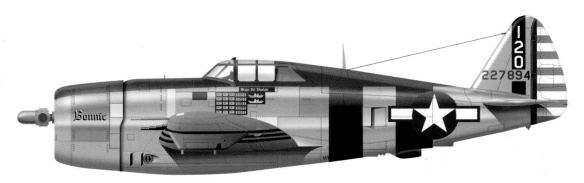

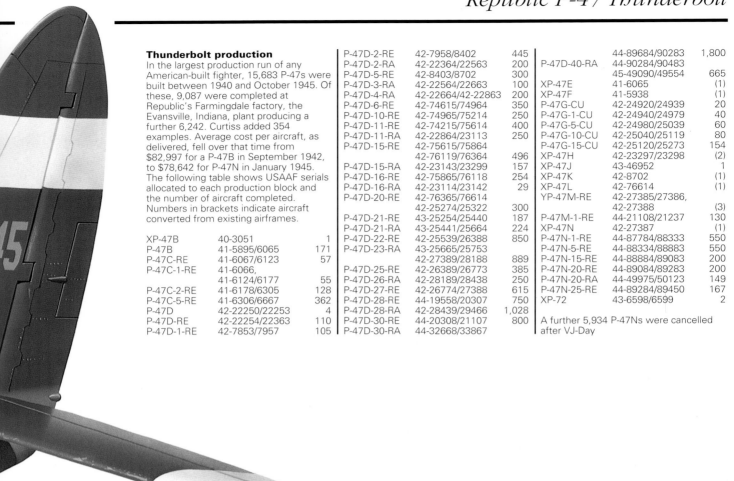

Thunderbolt production

In the largest production run of any American-built fighter, 15,683 P-47s were built between 1940 and October 1945. Of these, 9,087 were completed at Republic's Farmingdale factory, the Evansville, Indiana, plant producing a further 6,242. Curtiss added 354 examples. Average cost per aircraft, as delivered, fell over that time from $82,997 for a P-47B in September 1942, to $78,642 for P-47N in January 1945. The following table shows USAAF serials allocated to each production block and the number of aircraft completed. Numbers in brackets indicate aircraft converted from existing airframes.

Designation	Serials	Number
XP-47B	40-3051	1
P-47B	41-5895/6065	171
P-47C-RE	41-6067/6123	57
P-47C-1-RE	41-6066, 41-6124/6177	55
P-47C-2-RE	41-6178/6305	128
P-47C-5-RE	41-6306/6667	362
P-47D	42-22250/22253	4
P-47D-RE	42-22254/22363	110
P-47D-1-RE	42-7853/7957	105
P-47D-2-RE	42-7958/8402	445
P-47D-2-RA	42-22364/22563	200
P-47D-5-RE	42-8403/8702	300
P-47D-3-RA	42-22564/22663	100
P-47D-4-RA	42-22664/42-22863	200
P-47D-6-RE	42-74615/74964	350
P-47D-10-RE	42-74965/75214	250
P-47D-11-RE	42-74215/75614	400
P-47D-11-RA	42-22864/23113	250
P-47D-15-RE	42-75615/75864	
	42-76119/76364	496
P-47D-15-RA	42-23143/23299	157
P-47D-16-RE	42-75865/76118	254
P-47D-16-RA	42-23114/23142	29
P-47D-20-RE	42-76365/76614	
	42-25274/25322	300
P-47D-21-RE	43-25254/25440	187
P-47D-21-RA	43-25441/25664	224
P-47D-22-RE	42-25539/26388	850
P-47D-23-RA	42-25665/25753	
	42-27389/28188	889
P-47D-25-RE	42-26389/26773	385
P-47D-26-RE	42-28189/28438	250
P-47D-27-RE	42-26774/27388	615
P-47D-28-RE	44-19558/20307	750
P-47D-28-RA	42-28439/29466	1,028
P-47D-30-RE	44-20308/21107	800
P-47D-30-RA	44-32668/33867	
	44-89684/90283	1,800
P-47D-40-RA	44-90284/90483	
	45-49090/49554	665
XP-47E	41-6065	(1)
XP-47F	41-5938	(1)
P-47G-CU	42-24920/24939	20
P-47G-1-CU	42-24940/24979	40
P-47G-5-CU	42-24980/25039	60
P-47G-10-CU	42-25040/25119	80
P-47G-15-CU	42-25120/25273	154
XP-47H	42-23297/23298	(2)
XP-47J	43-46952	1
XP-47K	42-8702	(1)
XP-47L	42-76614	(1)
YP-47M-RE	42-27385/27386, 42-27388	(3)
P-47M-1-RE	44-21108/21237	130
XP-47N	42-27387	(1)
P-47N-1-RE	44-87784/88333	550
P-47N-5-RE	44-88334/88883	550
P-47N-15-RE	44-88884/89083	200
P-47N-20-RE	44-89084/89283	200
P-47N-20-RA	44-49975/50123	149
P-47N-25-RE	44-89284/89450	167
XP-72	43-6598/6599	2

A further 5,934 P-47Ns were cancelled after VJ-Day

Thunderbolt specifications

	P-47B	P-47C-5	P-47D-15	P-47D-30	P-47M-1	P-47N-1
nt:						
e:	R-2800-21	R-2800-21	R-2800-63	R-2800-59	R-2800-57	R-2800-57
g:	2,000 hp (1491 kW)	2,000 hp (1491 kW)	2,000 hp (1491 kW)	2,000 hp (1491 kW)	2,100 hp (1566 kW)	2,100 hp (1566 kW)
g (w/m inj.):	–	–	2,300 hp (1715 kW)	2,430 hp (1812 kW)	2,800 hp (2088 kW)	2,800 hp (2088 kW)
h:	40 ft 9 9/16 in (12.43 m)	40 ft 9 9/16 in (12.43 m)	40 ft 9 9/16 in (12.43 m)	40 ft 9 9/16 in (12.43 m)	40 ft 9 9/16 in (12.43 m)	42 ft 6 5/16 in (12.96 m)
	35 ft 4 3/16 in (10.77 m)	36 ft 1 3/16 in (11 m)	36 ft 1 3/16 in (11 m)	36 ft 1 3/4 in (11.02 m)	36 ft 1 3/4 in (11.02 m)	36 ft 1 3/4 in (11.02 m)
t:	14 ft 2 in (4.32 m)	14 ft 3 5/16 in (4.35 m)	14 ft 3 5/16 in (4.35 m)	14 ft 8 1/16 in (4.47 m)	14 ft 8 1/16 in (4.47 m)	14 ft 6 in (4.42 m)
area:	300 sq ft (27.87 m²)	300 sq ft (27.87 m²)	300 sq ft (27.87 m²)	300 sq ft (27.87 m²)	300 sq ft (27.87 m²)	322.2 sq ft (29.93 m²)
y weight:	9,346 lb (4239 kg)	9,900 lb (4491 kg)	9,900 lb (4491 kg)	10,000 lb (4536 kg)	10,340 lb (4690 kg)	10,988 lb (4984 kg)
weight:	12,245 lb (5554 kg)	13,500 lb (6123 kg)	13,500 lb (6123 kg)	14,500 lb (6577 kg)	15,000 lb (6804 kg)	13,823 lb (6284 kg)
t/o weight:	13,360 lb (6060 kg)	14,925 lb (6770 kg)	15,000 lb (6804 kg)	17,500 lb (7938 kg)	18,000 lb (8165 kg)	21,200 lb (9616 kg)
num speed:	429 mph at 27,800 ft (690 km/h at 8473 m)	420 mph at 30,000 ft (676 km/h at 9144 m)	433 mph at 30,000 ft (697 km/h at 9144 m)	423 mph at 30,000 ft (681 km/h at 9144 m)	473 mph at 32,000 ft (761 km/h at 9754 m)	467 mph at 32,000 ft (752 km/h at 9754 m)
ng speed:	100 mph (161 km/h)	104 mph (167 km/h)	104 mph (167 km/h)	105 mph (169 km/h)	99 mph (159 km/h)	98 mph (158 km/h)
speed:	6.7 min to 15,000 ft (4572 m)	7.2 min to 15,000 ft (4572 m)	7.2 min to 15,000 ft (4572 m)	6.2 min to 15,000 ft (4572 m)	5 min to 15,000 ft (4572 m)	9 min to 15,000 ft (4572 m)
ce ceiling:	42,000 ft (12802 m)	42,000 ft (12802 m)	42,000 ft (12802 m)	42,000 ft (12802 m)	41,000 ft (12497 m)	43,000 ft (13106 m)
al fuel:	305 US gal (1155 litres)	305 US gal (1155 litres)	305 US gal (1155 litres)	370 US gal (1401 litres)	370 US gal (1401 litres)	556 US gal (2105 litres)
al fuel:	–	200 US gal (757 litres)	200 US gal (757 litres)	410 US gal (1552 litres)	410 US gal (1552 litres)	700 US gal (2650 litres)
num range:	835 miles at 10,000 ft (1344 km at 3048 m)	835 miles at 10,000 ft (1344 km at 3048 m)	835 miles at 10,000 ft (1344 km at 3048 m)	1,030 miles at 10,000 ft (1658 km at 3048 m)	–	2,000 miles at 25,000 ft (3219 km at 7620 m)
al range:	550 miles at 25,000 ft (885 km at 7620 m)	400 miles at 25,000 ft (644 km at 7620 m)	400 miles at 25,000 ft (644 km at 7620 m)	590 miles at 25,000 ft (950 km at 7620 m)	530 miles at 26,000 ft (853 km at 7925 m)	800 miles at 25,000 ft (1287 km at 7620 m)
ment:	eight 0.5-in m/gs with 500 rpg	eight 0.5-in m/gs with 300-425 rpg	eight 0.5-in m/gs with 267/425 rpg bomb load up to 2,500 lb (1134 kg)	eight 0.5-in m/gs with 267/425 rpg bomb load up to 2,500 lb (1134 kg)	eight 0.5-in m/gs with 267 rpg bomb load up to 2,500 lb (1134 kg)	eight 0.5-in m/gs with 267/500 rpg bomb load up to 3,000 lb (1361 kg)

2794

QP-★-D

Iain Wyllie

Thu
Vari
Eng
Rati
Rati
Spa
Leng
Heig
Wing
Emp
Gros
Max
Max

Lan
Clim

Serv
Inter
Exte
Max

Nor

Arm

Initial sorties
Initial sorties were fighter sweeps over occupied Europe, the
Thunderbolt's lack of range proving a headache for VIII Fighter
Command given that bomber escort was the type's stated role.
Escort missions were flown from mid-1943 (out of Manston, Kent,
in order that fuel tanks could be topped up before the fighters
crossed the Channel) and during the summer of 1943 the 4th FG,
by then equipped with P-47C-5s and P-47D-1s, flew the first
missions equipped with early belly-mounted drop tanks. These
provided sufficient range to allow the 4th's Thunderbolts to
become the first Eighth Air Force fighter group to venture over
Germany, on 28 July.

TP-47G

Left: The Eighth Air Force is known to have converted three Thunderbolts (a P-47C and a pair of P-47Ds) as two-seat liaison, training and radar interception trials aircraft. Curtiss, having gained experience in producing the two-seat TP-40N, developed the dual-control TP-47G-16-CU for the gunnery training role. Two P-47G-15-CUs (42-25266 and -26267) were converted by reducing the size of the aircraft's main fuel tank and constructing a second cockpit above it, ahead of the existing cockpit. Fully armed, these 'Doublebolts' would have carried an instructor in the rear seat, had the project proceeded past the trials stage.

XP-47H

Right and below: Notable as the only Thunderbolts powered by a liquid-cooled engine, the XP-47Hs (converted from P-47D-15-RAs 42-23297 and 42-23298, the first making its maiden flight on 26 July 1945) were intended purely as testbeds for Chrysler's XI-2220 inverted-Vee, 16-cylinder powerplant, developed to fill a perceived requirement for new, high-powered, liquid-cooled fighter engines. Rated at 2,500 hp (1864 kW), the XI-2220 was unique in that power from its crankshaft was transferred to the propeller shaft via gearing located between the fourth and fifth pair of cylinders. It was the failure, on the aircraft's 27th flight, of the long prop shaft that brought about the programme's demise in November 1945; by then the USAAF had little requirement for a new piston engine.

XP-47K

Completed on 3 July 1943 by the conversion of the last P-47D-5-RE, the XP-47K (42-8702, left) was the first Thunderbolt with a bubble canopy. In the event, the production version of this aircraft retained the P-47D designation and the XP-47K was reworked as a testbed for the long-range wing with integral fuel cells intended for the P-47N.

XP-47L

Another new variant that entered production simply as a P-47D variant, the XP-47L (42-76614) was a conversion of the last production P-47D-20-RE with extra internal fuel tankage (totalling 370 US gal/1401 litres compared with the then-standard 305 US gal/1155 litres). These and other detail changes were subsequently introduced in the P-47D-25-RE. 42-76614 went on to test the R-2800 'C' series engine which later powered the P-47M.

XP-47J

Above and right: Projected in 1942 and cleared for production, in prototype form, in June 1943, the XP-47J was an attempt to produce a lightweight Thunderbolt powered by the new R-2800 'C' series engine, rated at 2,800 hp (2088 kW). This power increase had been accomplished without an increase in engine weight; Republic also made structural refinements to the aircraft, reduced its armament to six '50-calibers', fitted smaller fuel tanks, deleted its external stores capability and simplified its radio equipment. After a maiden flight on 26 November 1943, test flying proceeded into the following spring and by mid-1944 the XP-47J (43-46952) was approaching speeds of 500 mph (805 km/h). After a larger propeller and a more powerful CH-5 turbo-supercharger were fitted Republic claimed a top speed of 504 mph (811 km/h) at 34,450 ft (10500 m) – a record for a wartime piston-engined fighter (though doubts were cast as to the accuracy of this claim by the USAAF, who could only manage 493 mph (793 km/h) during subsequent testing at Wright Field). Though a promising aircraft, the P-47J failed to reach production. Not only was it little faster in 'combat trim' than the then production P-47, it would have necessitated considerable retooling (and inevitable delay) at Republic's factories. Perhaps the greatest threat to the P-47 however, was the XP-72 – an altogether more advanced machine.

In concept the XP-72 was a development of the XP-47J, but benefited from a new engine – the 3,000-hp (2237-kW) Pratt & Whitney R-4360-13 Wasp Major radial, nicknamed the 'corn cob' after its 28-cylinder, four-row layout. The first of two prototypes (43-6598, left) flew on 2 February 1944 and was equipped with a large four-bladed propeller; a second example (43-6599, below left) flew in July 1944, with a six-bladed contra-rotating propeller of smaller diameter (and was wrecked in a take-off accident early on in flight testing). Although the XP-72 was some 1,500 lb (680 kg) heavier than the P-47D, the aircraft had a top speed of 490 mph (789 km/h) and the potential to better 500 mph (805 km/h) once the R-4360 was developing full power. Attracted by the aircraft's superficial similarity to the P-47, the USAAF ordered 100 production examples (though it is difficult to see how retooling for P-72 production would have been any easier than for the P-47J), but these were soon cancelled as the war ended and the 'writing was on the wall' for the piston-engined fighter.

XP-69 – high-powered radial-engined project

At the time of the XP-72's conception, shortly after the Thunderbolt's first flight in 1941, Alexander Kartveli proposed two new fighters equipped with high-powered radial engines. The more conventional of the two, the Pratt & Whitney R-4360-powered XP-72, was selected by the USAAF for further development; the other machine – the Republic Model AP-18 (XP-69) – was comparatively unconventional by the standards of the day and did not proceed past the mock-up stage. Similar in layout to the Bell P-39/P-63 fighters, the XP-69 was powered by a liquid-cooled, six-row, 42-cylinder Wright R-2160-3 Tornado radial rated at 2,350-hp (1752-kW). This was in the fuselage, aft of the pilot, and drove a propeller in the nose via a long prop shaft. A pressurised cabin was envisaged for the high-altitude fighter, along with armament comprising a pair of 37-mm cannon and four 0.5-in machine-guns. Wingspan was given as 51 ft 8 in (15.77 m); length was 51 ft 6 in (15.70 m).

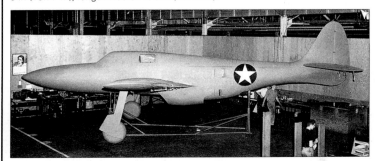

Fifth Air Force in New Guinea

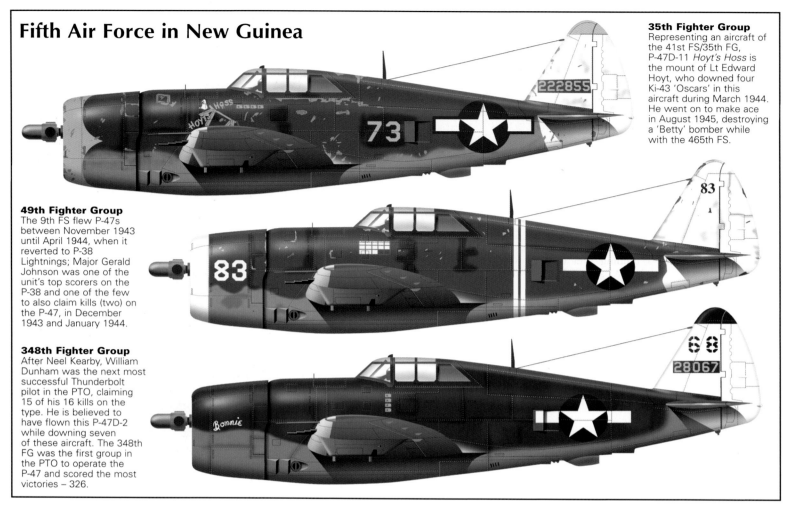

35th Fighter Group
Representing an aircraft of the 41st FS/35th FG, P-47D-11 *Hoyt's Hoss* is the mount of Lt Edward Hoyt, who downed four Ki-43 'Oscars' in this aircraft during March 1944. He went on to make ace in August 1945, destroying a 'Betty' bomber while with the 465th FS.

49th Fighter Group
The 9th FS flew P-47s between November 1943 until April 1944, when it reverted to P-38 Lightnings; Major Gerald Johnson was one of the unit's top scorers on the P-38 and one of the few to also claim kills (two) on the P-47, in December 1943 and January 1944.

348th Fighter Group
After Neel Kearby, William Dunham was the next most successful Thunderbolt pilot in the PTO, claiming 15 of his 16 kills on the type. He is believed to have flown this P-47D-2 while downing seven of these aircraft. The 348th FG was the first group in the PTO to operate the P-47 and scored the most victories – 326.

Thunderbolt units, though the 33rd FG soon returned to India and the 10th Air Force.

This left the 14th Air Force with the 81st Fighter Group, which had flown with the 12th Air Force in North Africa before moving to India in February-March 1944, where it converted to the P-47. The group moved to Kwanghan and a number of dispersed airfields in China after a period of training, and began flying operations in June 1944. The unit operated principally in the fighter-bomber role until it disbanded in December 1945.

RAF Thunderbolts

Prior to Pearl Harbor (7 December 1941), it had been expected that the P-47 would begin its operational career in Royal Air Force hands, as the Air Ministry expressed an interest in acquiring 100 P-47Bs in place of some of the Curtiss P-40 Kittyhawks then on order. The need to meet USAAF requirements reduced the number of aircraft that would be available for RAF use, while deficiencies with the P-47B led the USAAF to inform the British Air Ministry in September 1941 that "more bugs needed to be wrung out of the design" before it was fit for combat use. The RAF remained interested in the Thunderbolt and eventually acquired large numbers of P-47Ds (primarily for use in South East Asia Command), but the Desert Air Force fought on without P-47s.

In the CBI, the most important Thunderbolt operator in terms of numbers was the Royal Air Force. Apart from a

By the end of the war the only fighter group still operating P-47s in the SWPA was the 58th Fighter Group, by then operating in a tactical role. Comprising the 69th, 310th and 311th Fighter Squadrons, the group arrived in New Guinea in late 1943 and saw its first action the following February. Bomber escort and strafing missions were typical of the 58th's stock-in-trade, along with convoy escort work over the Bismarck Sea to the Admiralty Islands. By November 1944 the group had reached the Philippines and six months later had moved northward, with the progress of the war, to Okinawa. Pictured is a 311th FS aircraft over the Philippines in January 1945.

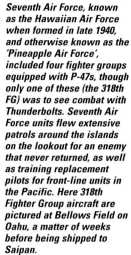

Seventh Air Force, known
as the Hawaiian Air Force
when formed in late 1940,
and otherwise known as the
'Pineapple Air Force',
included four fighter groups
equipped with P-47s, though
only one of these (the 318th
FG) was to see combat with
Thunderbolts. Seventh Air
Force units flew extensive
patrols around the islands
on the lookout for an enemy
that never returned, as well
as training replacement
pilots for front-line units in
the Pacific. Here 318th
Fighter Group aircraft are
pictured at Bellows Field on
Oahu, a matter of weeks
before being shipped to
Saipan.

Left: Coded '98' this
somewhat weatherbeaten
P-47D-11, pictured over
Kahuhu Airfield on Oahu, is
probably a 45th FS/15th FG
aircraft. As well as
patrolling the skies over the
Hawaiian islands, Seventh
Air Force pilots ferried
aircraft repaired in Hawaii
to forward areas.

handful of aircraft retained in the UK for evaluation, test and trials duties, and a training unit (No. 73 OTU) in Egypt's canal zone, all of the RAF's Thunderbolts were used by units in the Far East.

After failing to take delivery of the P-47Bs that it had once requested, the RAF did eventually receive a total of 826 P-47Ds of various sub-types. In RAF service, the 'razor-backed' P-47D-15-RE, D-21-RE and D-22-RE aircraft were designated Thunderbolt Mk Is (FL731-FL850 and HB962-HD181), while P-47D-25/-30-RE and P-47D-30/-40-RA aircraft with the later 'bubble canopy' were known as Thunderbolt Mk IIs (HD182-HD301, KJ128-KJ367, KL168-KL347, and KL838-KL976). A few aircraft in the last two batches were equipped with the dorsal fin strake.

The Thunderbolt was a very strange and unfamiliar aircraft to RAF pilots, who were used to the snug and claustrophobic cockpits of the diminutive Spitfire and Hurricane. The type dwarfed even its closest RAF equivalents, the Typhoon and Tempest. Its sluggish (by British standards) handling led many to conclude that the type was barely fit to fulfil any operational role.

In Air HQ India, however, RAF staff officers could see a requirement for eight squadrons of Thunderbolts to replace shorter-ranged and increasingly elderly Hurricanes with No. 221 Group on the Central Imphal Front and No. 224 Group on the Arakan Front. On 8 November 1943 the Air Ministry tersely reminded AHQ India that the Thunderbolt was "a high-altitude long-range fighter, relatively unmanoeuvrable and with slow initial rate of climb… therefore unsuitable for the ground attack role". Although Whitehall

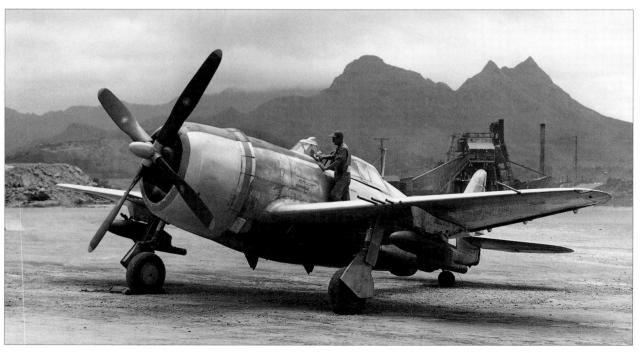

may not have been taking note of the success of the USAAF's P-47s in the ETO, AHQ India was convinced of the type's usefulness, not least because its air-cooled radial engine promised greater reliability than engines like the Merlin in harsh tropical conditions.

It is perhaps just as well that AHQ India was so enthusiastic about the P-47, since an RAF request for 120 P-47s and 900 P-51s was turned down. A counter-offer was made of only 300 Mustangs, but 420 Thunderbolts.

Two P-47D-15-REs were shipped directly to No. 301 MU at Marachi for evaluation in February 1944, and conversion training began in May at C&CU Mauripur and No. 1670 CU at Yelahanka. The first front-line units were Nos 146 and 261 Squadrons in India, and No. 135 Squadron at Minneriya

A typically dramatic
volcanic landscape on Oahu
provides a backdrop for this
view of a 15th Fighter Group
P-47D at Bellows Field
during 1944. Training was
an important role for
Seventh Air Force units; this
aircraft carries smoke
canisters for use on a
training exercise. The 15th
Fighter Group finally
received P-51 Mustangs in
late 1944 and was shipped
to Saipan in early 1945.

bomber sorties, causing colossal damage to Japanese supply lines and front-line positions, as they supported the Allied advance that steadily drove the Japanese from Burma (now Myanmar). They also flew occasional bomber escort missions, and during these and the low-level ground attack missions a handful of air-to-air victories were claimed, their relative paucity being a token of the unwillingness of the Japanese to engage in air combat. The low-level ground attack role was inevitably costly, and losses were steady, though the P-47 demonstrated an extraordinary ability to return to base with major damage (where the aircraft would usually be scrapped rather than repaired). Even so, the stock of P-47s in the Indian MUs was sufficient to convert additional Hurricane units, and by the end of the war the RAF had 16 Thunderbolt squadrons.

A reserve of aircraft was built up for Operation Zipper (a planned landing in Malaya) and Operation Tiderace (the subsequent drive to Singapore), but these operations were cancelled following the surprise Japanese surrender in the wake of atomic bomb attacks on Hiroshima and Nagasaki.

Extended-range P-47N

Despite its many qualities, the P-47 was never viewed in the same light as the P-51 by the USAAF brass, not least because it cost far more to produce. Although Republic managed to reduce the unit price of the aircraft from $US113,246 in 1941 to $US85,578 in 1944, it remained an extremely expensive fighter aircraft. By comparison, the P-51 Mustang cost $US50,985 and the P-40 came in at only $US44,892. Even as early as the end of 1943, the Thunderbolt's future looked in doubt, and the P-47 production plants were expected to cease production of the Thunderbolt during 1944, perhaps then switching to production of the rival P-51. Existing orders would keep the Farmingdale and Evansville plants busy for a while, but it was decided that no new orders would be placed for new P-47s.

In an effort to keep the aircraft in production, the Republic Aircraft Corporation eagerly accepted a USAAF order to design a prototype dedicated long-range escort fighter derivative of the Thunderbolt, with a new wing incorporating additional internal fuel tankage. The new model, it was hoped, would combine the performance of the P-47M with a combat radius of more than 1,000 miles (1610 km), enabling it to operate successfully over the very long distances common in the Pacific theatre. The new variant would at last allow the P-47 to rival the Mustang in the long-range escort fighter role.

It was decided that the new variant should be a minimum-change derivative of the P-47M, restricting such changes to the wing. This was seen as the best possible location for extra fuel because it would have the least impact on centre of gravity, and restricting changes to the wing would minimalise disruption to production. Rather than designing an entirely new wing, Republic modified the existing wing planform by adding a constant-chord 18-in (46-cm) section at each wingroot. These new sections each contained four small inter-connected tanks with a total capacity of 100 US gal (379 litres), raising total inter-

Above: After being flown from Bellows Field to an airfield adjacent to the Ford Island naval base, a dockside crane hoisted 318th Fighter Group P-47s aboard USS Natoma Bay, destination Saipan. Two escort carriers were used to convey a total of 72 aircraft to Saipan, leaving Hawaii on 5 June, refuelling at Eniwetok en route and arriving in the Marianas mid-month.

Above right: Despite attempts by Japanese Aichi D3A 'Val' dive-bombers to hinder their progress, all of the 318th's aircraft were successfully launched from the carriers on 22 and 23 June, landing at Aslito (later Isley) Air Field. Here P-47D-11 Dee-Icer of the 73rd Fighter Sqn has a catapult bridle attached in readiness for a launch from USS Manila Bay on the 23rd. Thirty-three aircraft were launched at two-minute intervals.

in Ceylon (now Sri Lanka). It had been intended to equip squadrons with either the Thunderbolt Mk I or the Thunderbolt Mk II, although, in the event, most units initially operated a mix of variants. This was partly the result of poor corrosion protection applied to the first P-47D-27-REs shipped to India, some of which had to be scrapped upon arrival.

The British viewed the Thunderbolt as an interim type, although additional aircraft were sought when a request for 180 Mustangs (intended to re-equip five remaining Hurricane squadrons) was turned down. Even with these extra aircraft, the RAF hoped to begin replacing Thunderbolts with Tempest IIs from the end of 1945. Interim aircraft it may have been, but the Thunderbolt was numerically important, and an extra conversion unit was established in Egypt in October 1944.

The RAF Thunderbolts flew standing 'cab rank' reactive close air support patrols and a variety of other fighter-

318th Fighter Group – by carrier to Saipan

P-47D-21, 19th Fighter Sqn
Having re-equipped with Thunderbolts at the end of 1943, the 318th Fighter Group was ordered to Saipan from Hawaii the following June to provide air support for beleaguered US Marines fighting on Saipan itself and nearby Tinian. Lt William Mathis, the pilot of this early D-21 (42-25343 *Joey*), went on to 'make ace' during June 1945, flying a P-47N-1.

325343

318th Fighter Group – summer on Saipan

The activities of the 318th Fighter Group during the period of its deployment to Saipan appear to have been the subject of considerable photographic coverage by the USAAF. During the summer of 1944 the group supported the invasions of Tinian and Guam, typically making use of such weapons such as napalm, dropped from low level, and pairs of 1,000-lb bombs. Clashes with Japanese fighters were few and far between; only eight kills were recorded by the group before it re-equipped with P-38s in late 1944.

nal fuel capacity to 570 US gal (2158 litres). Provision was made for two auxiliary underwing drop-tanks of up to 330 US gal (1250 litres) each. The new inboard sections increased wheel track and overall span, despite square-cropped tips that sliced 9 in (23 cm) – and Kartveli's graceful curved tip – from each wing. The new wing also increased the rate of roll from 79°/sec to 98°/sec, and improved turn performance.

An aircraft originally earmarked as a YP-47M (42-27387, first laid down as a P-47D-27-RE but not completed as such) was converted to serve as the official XP-47N prototype. Another YP-47M (42-27388) was used for tests of a refined version of the new Long Range Wing, with the addition of modified ailerons and flaps. The first aircraft to fly with the new wing was the XP-47K (42-8702), built as Farmingdale's last P-47D-5-RE and subsequently modified as the first Thunderbolt with a 'blown' bubble canopy. This aircraft, at the time, was the only available bubble-hooded Thunderbolt, and as such was thought to be most suitable as a prototype for the planned P-47N, since this would inevitably be fitted with the new canopy. Although it was known that the production P-47N would feature a new, more powerful version of the R-2800 engine, the XP-47K retained its standard 2,300-hp (1716-kW) B-series R-2800-59 engine with water injection. The aircraft lacked the power of the P-47M or the planned P-47N, but it was able to demonstrate the advantages offered by the new wing.

The true XP-47N (42-27387) was equipped with a C-Series R-2800-57 Double Wasp engine, rated at 2,800 hp (2089 kW) and fitted with a new Unilever Power Control System providing automatic adjustment and control of throttle, propeller pitch, rpm and supercharger boost. This was intended to ease pilot workload by effectively providing a 'single-lever' power control system, and it was hoped it would make the aircraft as easy to fly (over extended ranges) as a P-51D. Work on the prototype took only 56 days to complete, and the aircraft first flew on 22 July 1944.

Handling characteristics of the XP-47N proved to be similar to those of the P-47D when lightly loaded, and turn performance was rated as superior. Directional instability at high weight (with more than about 300 US gal/1136 litres of external fuel) necessitated the later addition of a dorsal fin fillet on the production version, although the prototype aircraft never received one. The Unilever Power Control System was not satisfactory, causing power surges and propeller over-speeds. It proved so erratic that the Wright Field test pilots (led by Major F.A. Borsodi and Captain R.B. Johnston) had the system disconnected before carrying out approach and landing trials.

Impressive reach

The XP-47N demonstrated a range of 2,170 miles (3492 km) with full internal fuel and 600 US gal (2272 litres) of external fuel, though this figure included 15 minutes at full combat power and five at War Emergency Power. This was extremely impressive by the standards of the day, and made possible 1,000-mile (1610-km) escort missions – a reach greater than that offered by even the P-51 Mustang.

In these views of 42-75351, a P-47D-11-RE retrofitted with wing pylons, it is pictured on 26 June 1944 (left), having been hit by flak while attacking Japanese forces on Saipan. Damage appears superficial after its wheels-up landing and a later view of the machine (right) sees it undergoing extensive repairs. Ground support missions from Aslito were only 15 minutes in duration and officially counted as 'half missions', to the dismay of many pilots!

The 318th FG reached Saipan seven days after the Marines had invaded. A handful of Japanese infantry remained, snipers making their presence felt from beyond the airfield's perimeter fence. On 26 June 42-75379 Hed-up N'locked was set alight and burnt out. Note that the armament in its starboard wing has been salvaged for reuse.

Above: On 16 July 1944, just over three weeks after arriving at Saipan, Dee-Icer (pictured opposite, aboard Manila Bay) is in the early stages of an engine change in the open air on Isley Field. Facilities on the island appear to have been fairly basic; tented accommodation for AAF personnel is visible in the background.

Left: Military policemen patrol a damp flightline at Isley Air Field while Apple Jack receives attention to its powerplant. The 'Bar Flies' emblem on the cowling of aircraft is a reference to the nickname of the 73rd FS, which further identified its aircraft with white banding on the cowling, fuselage and tail.

Piloted by Captain Harry McAfee of the 19th Fighter Sqn, Miss Mary Lou (an Evansville-built P-47D-20) was the first P-47 to land on Saipan on 22 June 1944, having been catapulted off the carrier Natoma Bay 60 miles (97 km) off the coast of Saipan. This later view shows the aircraft armed with triple-tube 4.5-in M10 'Bazooka' rocket launchers.

To build the new Long Range Wing, Republic set up a new manufacturing shop at Farmingdale (P-47D wings had been bought in from a sub-contractor). Initial output was very slow, with the first wingsets taking 8,600 man hours each, though this soon fell to 1,900 man hours.

These production difficulties with the new wing led to the construction of an interim batch of 130 aircraft, powered by the new R-2800-57 engine but retaining the basic airframe of the P-47D-30-RE. Designated P-47Ms, these aircraft had a brief and very limited service career in the European Theater of Operations, serving with the 56th Fighter Group of the Eighth Air Force.

As a result of the new wing difficulties and other problems, each of the first 50 P-47N-1-REs off the line took an average of 25,439 man hours to produce. Within six months this had fallen by over 70 per cent, to 7,236 man hours.

The first 550 production P-47N-1-REs (44-87784 to 88333) incorporated a prominent dorsal fin fillet to reduce longitudinal stability, but lacked the Unilever Power Control System that had caused such problems on the prototypes. The prototypes had new wing tanks containing 200 US gal (758 litres) of fuel, of which only 160 US gal (605 litres) could be used, whereas the P-47N-1-RE had new tanks containing 168 US gal (636 litres) of usable fuel. The type also had a new General Electric CH-5 turbo-supercharger, which gave greater carburettor air pressure and maintained a more constant air pressure. The engine also had an automatic water injection system that cut in if boost increased to a level where detonation would occur without water. There was automatic control of engine cowl flaps, intercooler radiator doors and cockpit heating.

The USAAF began taking delivery of large numbers of P-47Ns during January 1945; they went to the 318th Fighter Group already operational in the Pacific and equipped the newly formed 413th, 414th and 507th Fighter Groups. The 318th had been a long-serving P-47D unit within the 7th Air Force but had equipped with P-38s in mid-November 1944, although these aircraft did not serve long with the group. The experienced P-47D pilots of the 318th were flown to Hawaii in groups in March and April 1945, where they

Right: Ground personnel prepare a 19th FS aircraft for a bombing mission on a nearby island in the Marianas group. Its load comprises a pair of 500-lb bombs and a standard 75-US gal (284-litre) drop tank. During early July 1944 the 19th began stripping the OD paint from the cowling and tail assembly of its aircraft and applying light blue bands across the tailplane and fin. Note also the use of single letter codes by the 19th FS instead of numbers (in the '4xx' range by the 73rd FS, for example).

Thunderbolts equipped both the 5th and 6th Fighter Squadrons (Commando) of the 1st Air Commando Group from the late summer of 1944 until early 1945. This pair, in the process of being serviced and 'bombed up', displays the group's distinctive markings (five diagonal blue stripes on the aft fuselage). As well as strikes against Japanese front-line forces, the 1st ACG Thunderbolts hit supply lines, which were vital to the war effort in the dense jungle environment. There was virtually no aerial opposition, and the 1st ACG P-47 squadrons produced no aces for the history books.

China-Burma-India

Tenth Air Force was created in 1942, its deployment to the CBI being completed in May, when the last of its units arrived in India. Two P-47 groups were assigned – the 33rd and the 80th Fighter Groups – along with two Thunderbolt-equipped squadrons in the 1st Air Commando Group. Tenth Air Force operations were confined to Burma and India after the arrival of Fourteenth Air Force. From July 1945 the Tenth moved to China for the final assault on Japan.

Formed in China in March 1943 to support US and Chinese forces fighting the Japanese in China and northern Burma, the Fourteenth Air Force was commanded by former AVG ('Flying Tigers') commander Major General Claire Chennault. Two Thunderbolt-equipped fighter groups were assigned to the Fourteenth – the 81st FG (from mid-1944) and the 33rd FG (for a matter of a few months during the summer of 1944).

familiarised themselves with their newly assembled P-47Ns (which had been shipped to Hawaii by sea) before ferrying them the 4,100 miles (6598 km) to Ie Shima, a newly captured island only 325 miles (523 km) from the Japanese mainland.

The 413th, 414th and 507th FGs were formed specifically to support the final push against Japan by the USAAF's heavy bombers. They ferried their P-47Ns from Hawaii to their operational bases, the 413th moving to Ie Shima via Saipan, where they flew two ground attack missions against Japanese forces at Truk. The group was established alongside the 318th FG at Ie Shima by mid-May. The 507th went to Iwo Jima in June 1945, and was joined there about four weeks later by the 414th FG, which had flown some sorties against Truk while en route, operating from Guam. The Iwo Jima P-47Ns began flying operations in mid-July 1945.

By the time the P-47N entered service, much of the US bombing effort was being undertaken at night, reducing the need for long-range escort fighter sorties. The P-47N groups on Ie Shima and Iwo Jima were diverted to long-range fighter-bomber sorties against targets in Japan, occupied China, and Korea. With the Iwo Jima-based P-47N groups having to fly almost twice as far as those at Ie Shima, they were often led by B-29 pathfinders. Some aircraft were lost when they ran out of fuel on the long way home, their pilots usually bailing out close to pre-positioned US Navy ships. Another hazard came from the very high take-off weights of these aircraft, which were forced to use almost the full length of the islands' 1.5-mile (2.4-km) runways. When a Republic technical representative visited one of the islands, it is said that he felt that too much fuel was being used due to the application of excessive boost during take-off. Demonstrating the approved technique, this pilot failed to reach a safe flying speed and crashed off the end of the runway.

The P-47N-equipped Fighter Groups escorted US Navy PB4Ys and other reconnaissance and attack aircraft, but the 413th flew the P-47N's sole full-scale B-29 escort mission, on 8 August 1945. Some 151 P-47Ns escorted the 400 B-29s

*With ground attack as its primary role, the 80th Fighter Group arrived in India in May 1943, joining the 10th Air Force with P-38s and P-40s. Converting on to P-47s once again in 1944 (having been the second group to train on P-47Bs back in 1942), the group was based at Myitkyina, Burma, with regular detachments at various locations in Burma and India. The group's three squadrons are represented here: the appropriately named **Burma Yank** is 42-27480 of the 89th FS (above), 42-27419 '75' is a 90th FS machine pictured during 1944 and '27' **Carlotta** is an 88th FS aircraft, pictured at Barrackpore, India, and still in weathered OD paint in March 1945.*

to the steel-producing city of Yawata on the island of Kyushu. About 60 Japanese fighters rose to meet the bombers, and a massive air battle ensued. Thirteen Japanese fighters were claimed destroyed, at the cost of

Two 91st Fighter Squadron/ 81st Fighter Group P-47Ds stand idle outside a hangar on a Chinese air force field at Hsian during 1945; inside a 426th NFS P-61 Black Widow undergoes major work. Note that both the P-47s are equipped with a D/F loop, mounted behind the cockpit and considered an essential navigational aid over the rugged terrain of the CBI. Most of the action was ground attack, with little Japanese opposition in the air. The 81st was only credited with six kills. Many of its aircraft went on to serve with the Nationalist air force.

Right: The 91st FS identified its aircraft with a diagonal tailfin stripe and numbered its aircraft in the 900/930 range. '910' is an Evansville-built P-47D-30 with a retrofitted dorsal fin.

Below: My Better Half was the mount of Captain Hugh McLean of the 91st Fighter Squadron, pictured at Hangchow, possibly after VJ-Day. In the background is a long line-up of 75th FS/23rd FG P-51D Mustangs.

five Thunderbolts, four of whose pilots were recovered from the sea.

The P-47N was used only lightly in the air-to-air role, though the 318th Fighter Group did encounter enemy aircraft during a number of offensive patrols, and the score

of victories slowly began to mount. The group downed 102 enemy aircraft during June, and more were destroyed during August.

P-47N-5 and beyond

The basic P-47N was soon augmented by the first of 550 P-47N-5-REs (44-88334 to 88883), which introduced a number of improvements, including provision for five zero-length rocket launchers underwing, an AN/APS-13 tail warning radar, SCR-522 radio homer, better rudder pedal adjustment, and provision for a vacuum tap for a Berger 'g-suit'. The aircraft also had a catapult attachment, to allow P-47s to be delivered to the Far East aboard US Navy aircraft-carriers. Some aircraft had R-2800-73 engines in which a new General Electric ignition system replaced the usual Scintilla type. A number of P-47N-1-REs were brought up to the same standards as P-47N-2-REs.

There were no P-47N-10-REs, but Farmingdale built 200 -15-REs (44-88884 to 89083) with an improved gyro gunsight, armrests on the pilot's seat, revised instrumentation, and the improved S-1 bomb rack. There was provision for a General Electric C-1 autopilot (as fitted to earlier blocks) but this was not fitted due to supply shortages. Between July and September 1945 Evansville built 149 aircraft under the designation P-47N-20-RA (45-49975 to -50123), the final 72 (from 45-50051) having a new cockpit floor with a smooth rudder pedal track. These aircraft were otherwise similar to Farmingdale's -20-RE, 200 of which (44-89084 to 89283) were built with a new emergency fuel system, a revised water injection switch, and a new radio, and were powered by R-2800-73 or R-2800-77 engines (these differed mainly in the type of magneto used, the -77 re-introducing the Scintilla type).

The final 167 P-47Ns were -25-REs (44-89284 to 89450), few if any of which were delivered to front-line units before the war ended. These aircraft had the same new cockpit floor as the Evansville-built Ns, but also introduced aileron and flap modifications to withstand rocket efflux, and had a new automatic engine control system that finally replaced the Unilever Power Control System intended for the N. The new system was simpler, coupling the throttle and turbocharger regulator (boost), with the turbo activating only when the carburettor butterflies were fully open. This effectively gave the pilot 'single-lever' operation. The G-1 autopilot was finally reintroduced, and the last 147 aircraft had relocated wingtip navigation lights.

This brought P-47N production to 1,816, excluding the prototype. A total of 1,667 was built at Farmingdale. Some 233 P-47N-30-REs (44-89451 to 89683) were cancelled, along with planned Evansville-built P-47N-RAs (45-50124 to 55174).

XP prototypes

The P-47N was the most advanced Thunderbolt version to see front-line service, but there also was a series of XP-designated prototypes. Some were simple development mules whose task was to test new powerplants or other

The Thunderbolt was introduced by the RAF in the Far East to replace the aging Hawker Hurricane; here drop tank-carrying Thunderbolt Mk Is taxi past Hurricane Mk IICs on an unknown airstrip at the beginning of another sortie.

equipment items, but others were intended to form the basis of new production models.

Among the first category was the XP-47E (41-6065), a modified P-47B with a pressurised cockpit. It was later fitted with a new R-2800-59 engine in a new cowling, and a paddle-bladed Hamilton Standard propeller. The XP-47F (41-6065), another converted P-47B, was used to test a new, entirely redesigned laminar flow wing. The XP-47K (42-8702), converted from a P-47D-5-RE, tested the new bubble canopy used by late-block P-47Ds and later variants. (According to legend, this first bubble canopy was taken from a Hawker Typhoon.) The XP-47L (42-76614) was similar, but was based on the last P-47D-20-RE.

The Thunderbolt variant with the sleekest and most streamlined appearance was the XP-47H, two of which were produced through the conversion of P-47D-15-RAs (42-23297 and -23298). These aircraft were intended as test-beds for the new 2,500-hp (1865-kW), inverted V-16 Chrysler XI-2220-11 engine, and as such had a much extended nose, and a new liquid cooler in the belly, with a prominent ventral airscoop. They were converted by Chrysler at Evansville. The first XP-47H made its maiden flight on 26 July 1945, and the aircraft made 27 flights before it was grounded following a prop-shaft failure. Performance proved disappointing, the engine being prone to overheating unless the cooler regulator doors were open – but they provided so much drag that speed was severely limited. The second XP-47H flew in September but, with the advent of new jet fighters, the XI-2220 engine looked like a dead-end, and the aircraft was ferried straight to Wright Field for disposal.

The XP-47J, by contrast, was an experimental aircraft (and not a prototype as such) but was intended to form the basis of a new Thunderbolt production variant. The new version was to be designed as a lightweight Thunderbolt, maximising performance at the expense of a lighter structure, a simpler radio fit, six instead of eight guns (and only 267 rounds per gun), and removal of the rear gas tank. The aircraft used an improved 2,800-hp (2089-kW) R-2800 engine with water injection and a new General Electric CH-5 turbo-supercharger. The engine was closely cowled, with a separate staggered chin intake for the supercharger. The R-2800 originally was intended to drive a new contra-rotating propeller, but development problems led to its replacement by a conventional 13-ft (3.96-m) diameter Curtiss prop. The XP-47J prototype (42-46952), officially known as the Superbolt, first flew on 26 November 1943 and soon displayed the highest speeds ever achieved by a P-47. The USAAF demonstrated 493 mph (793 km/h) in the aircraft and the manufacturers claimed 504 mph (811 km/h), despite the fact that exhaust system limitations prevented the engine from being run at full power.

The production P-47J would have required a change to 70 per cent of production tooling and an unacceptable disruption to aircraft deliveries. Worse, the new XP-72 promised to be a better long-term option, so the P-47J programme was abandoned even before the planned second XP-47J could be completed.

The XP-72 was similarly conceived as a lightweight, improved Thunderbolt, but this time powered by a 3,000-hp (2238-kW) four-row, 28-cylinder R-4360 Wasp Major engine. Production aircraft were to feature the

Above: No. 30 Squadron spent almost two years, from March 1942, defending Trincomalee and Colombo in Ceylon against carrierborne attacks by the Japanese before taking on a more offensive role in Burma in early 1944. In May the unit re-equipped with Thunderbolts; these Mk IIs, pictured over East Bengal in December 1944, represent about three-quarters of the squadron's likely complement.

Above left: Trading its Spitfire Mk Vs for Thunderbolts in September 1944, No. 134 Sqn was active for nine months before being renumbered as No. 131 Sqn. Note that these aircraft have yet to have white recognition markings applied, dating the picture between September and November 1944.

Thunderbolt or Ki-44-II?

As Thunderbolt Mk IIs began operations over Burma in September 1944 concerns were raised regarding the similarity between the silhouettes of the Republic fighter and the Nakajima Ki-44-II Shoki (Allied codename 'Tojo'). White recognition markings were devised and applied, during October, to aircraft of Nos 146 and 261 Sqns. In the meantime, three USAAF fighter groups in the CBI theatre in the process of converting to the P-47D were adopting their own cowling and tail identification bands; in November those applied to aircraft of the 80th Fighter Group – white bands on the engine cowling, the wing, the fin and the tailplane – were adopted for the RAF's Thunderbolts, and the following January instructions were issued that these markings be applied to all ACSEA aircraft, with the exception of night-fighters and four-engined types.

Of the over 800 Thunderbolts delivered to the RAF, only a handful reached the UK, for evaluation by A&AEE Boscombe Down. As well as a couple of USAAF examples, A&AEE test flew four Thunderbolts, including Mk I FL844 (a P-47D-22-RE), pictured between test flights to evaluate the carriage and dropping of a variety of stores, including drop tanks.

42-27386 was the second of three YP-47Ms (hence the 'M2' marking), converted from P-47D-27-REs. The aircraft's cowling and empennage were sprayed yellow – a feature of experimental P-47s during 1944/45. The aircraft carries a standard 110-US gal (416-litre) drop tank on its centreline pylon.

Major Michael Jackson of the 62nd Fighter Sqn scored a total of eight victories in Thunderbolts, though all were claimed while flying P-47Ds. When the 56th FG re-equipped with P-47Ms, Jackson's assigned aircraft was 44-21117 Teddy. As well as his eight air-to-air kills, the tally below the cockpit of the aircraft includes 5.5 ground victories in white. Note also that this aircraft has yet to be fitted with a dorsal fin (as introduced in the P-47D-40 and retrofitted to many P-47D-30s), a modification made to most P-47Ms upon their arrival in England.

The 'M' – the fastest production 'Jug'

With a top speed in the vicinity of 470 mph (756 km/h) at 32,000 ft (9754 m), the P-47M-1-RE was the fastest of the production Thunderbolts, benefiting from an uprated R-2800-57 'C' series engine and a new turbo-supercharger design. Only 130 were built and only one unit – the Eighth Air Force's vaunted 56th Fighter Group – operated the type. First missions were flown in January 1945, however engine teething problems (low cylinder head temperatures, ignition failure and corrosion) meant that the P-47M did not properly enter service until April, just weeks before VE-Day.

3,450-hp (2574-kW) R-4360-19 engine with a mechanically-driven radial compressor instead of an exhaust-driven supercharger. Production aircraft were also intended to have a universal wing with provision for alternative armament options of either six 0.50-in machine-guns or four 37-mm cannon. The first of two prototypes (43-36598 and 36599) made its maiden flight on 2 February 1944, followed in July 1944 by the second, which had a contra-rotating six-bladed Aeroproducts airscrew. Neither prototype featured the planned shaft-driven compressor, but by the time they flew they effectively had been overtaken by jet fighters, and a planned pre-series aircraft was never completed. The 100 P-72-REs ordered by the USAAF were cancelled.

At the close of war

The three Thunderbolt groups of the 20th Air Force and the 7th Air Force's 318th FG ended the war with P-47Ns but, elsewhere in the Pacific, the P-47D remained dominant. The 7th Air Force in Hawaii included the P-47D-equipped second-line 508th FG, while the 15th FG had converted to Mustangs. In China, the 14th Air Force included a single Thunderbolt group (the 81st FG); the 10th Air Force (which had moved to China from India and Burma in July 1945) had one more (the 80th FG), its other P-47 units having converted to the Mustang (the 1st Air Commando Group) or the P-38 (the 33rd FG).

P-47Ms of the 56th Fighter Group

P-47M-1
Captain Witold Lanowski was one of a group of Polish volunteers that joined the 56th FG in 1944; Lanowski went on to score four kills. By May 1945 most of the 61st Fighter Sqn's aircraft were finished in variations on this unconventional scheme.

P-47M-1
Major George Bostwick, CO of the 63rd Fighter Sqn, finished the war with eight kills, one of which was the only Me 262 shot down by a Thunderbolt ace. His last two victories were a pair of Bf 109s downed on 7 April 1945 in the aircraft depicted here.

P-47Ds also served with the 5th Air Force as it supported the push from the southwest Pacific. Thunderbolt units remaining at war's end were the 35th, 58th and 348th FGs. The 58th Fighter Group was unusual in having a fourth squadron, this being the 201st Escuadrón Aéreo de Pelea of the Mexican Air Force. It operated alongside three USAAF squadrons, its aircraft wearing Mexican red, white and green stripes on their rudders, and Mexico's unusual triangular national insignia on the opposite wing to the USAAF star-and-bar, which was also carried. The Mexican P-47 pilots performed with great distinction from April 1945, notching up 791 combat sorties (in 59 combat missions, totalling 1,966 flying hours). General Kenney rated their combat record as "Splendid!".

The Mexican squadron returned home in November 1945, leaving behind its war-weary aircraft. They were replaced by 25 as-new P-47Ds in Mexico itself, under Lend-Lease provisions. The new aircraft equipped the 201st Escuadrón Aéreo de Pelea at Veracruz (and later Mexico City) and a training unit that moved from Mexico City to Guadalajara. A dwindling number remained in use until June 1958, when the last nine were retired.

Almost as numerous as the USAAF's Thunderbolts within the Pacific theatre were those operated by the Royal Air Force, and as the war ended they equipped 16 squadrons, based mainly in India and Burma. They were, however, destined to be disbanded or converted to British types very quickly after the end of hostilities.

Despite its very high price tag and its early replacement in the high-profile Eighth Air Force, the Thunderbolt was actually built in larger numbers than any other US wartime fighter, with an astonishing production total of 15,683 aircraft. About two thirds of the P-47s built were sent overseas, and 5,222 of these were lost (although only 824 fell in air combat). At the peak of its popularity, in 1944, P-47s equipped 31 front-line fighter groups. With these units, the Thunderbolt flew 545,575 combat missions (totalling 1,934,000 flying hours). During the course of these sorties, P-47 pilots claimed 12,000 enemy aircraft destroyed, dropped 132,482 tons of bombs, and fired 59,567 rockets against ground targets.

Reworked from a P-47D-27-RE, the XP-47N (left) was equipped with the new square-tipped long-range wing and R-2800-57 engine, though its fuselage resembled that of a P-47M and initially lacked the dorsal fin of production P-47Ns. Viewed from above (below left), the changes made to the wing are clear; four interconnected self-sealing fuel tanks mounted inboard provided a total of 200 US gal (757 litres) of much needed extra fuel capacity.

Below: Thunderbolt production ended with a batch of P-47N-25-REs from Farmingdale, the last of which was completed in October 1945, though not delivered to the USAAF until December. Four hundred aircraft were on order from Farmingdale, allocated the serials 44-89284/89683, though only 167 aircraft (to 44-89450) were completed. (Had the war continued, P-47N production would have been concentrated at Republic's Evansville, Indiana, plant.) P-47N-25 44-89444 was the last Thunderbolt delivered to the USAAF and, after ANG service, was preserved by the Cradle of Aviation Museum.

colonial control had been relatively light, so trouble in Malaya remained sporadic and controllable. The Thunderbolts of Nos 60 and 81 Squadrons did fly 'show of force' missions in the north when rioting followed the disbandment of the MPAJA on 31 December 1945.

The British were obliged to restore Dutch control in the Netherlands East Indies (now Indonesia), where there was a greater thirst for independence. The departing Japanese

Post-war unrest

The rapid and sudden defeat of the Japanese left a vacuum in some areas that the Japanese had occupied, and there was a struggle as new liberation movements (some of them the same groups who had resisted and fought the Japanese) jockeyed with former colonial powers to seize and control territory.

In Malaya (now Malaysia), resistance to the reimposition of British colonial rule was light, and largely restricted to an ethnic Chinese minority who had formed the backbone of the wartime Malayan People's Anti Japanese Army. The indigenous Malays were generally sanguine and British

Lt Oscar Perdomo – final USAAF fighter ace

P-47N-2 *Lil Meatie's Meat Chopper*
Having arrived at Ie Shima in June 1945, it seemed unlikely that Lt Oscar Perdomo of the 464th Fighter Sqn/507th Fighter Group would see Japanese aircraft in the air, let alone score any victories.
Yet, on 13 August 507th P-47Ns encountered 50 Japanese machines over Seoul, Korea; in the dogfighting that followed Perdomo downed five aircraft, becoming the last ace of the war.

Above: Protected from the elements and corrosive salt spray, new P-47Ns shipped from the US await final assembly on Guam during March 1945.

Right: Armourers clean and replenish the '50 calibers' of a 318th Fighter Group P-47N. By VJ-Day the group had been credited with 153 victories, all but five while flying the P-47N, and for the loss of just 10 of its own aircraft. Partly a reflection of the standard of Japanese fighter pilots by the end of the war, this nonetheless outstanding victory/loss ratio made the 318th one of the most successful P-47s groups of the war.

Far right: 1st Lieutenant John Dooling of the 318th FG surveys the damage caused to his Thunderbolt by AA fire over a Japanese island to the south of Kyushu. The majority of missions flown by the Twentieth Air Force's three P-47N groups were not bomber escort sorties, as originally intended, but raids on ground targets on the Japanese mainland, and shipping.

Thunderbolts again for the 318th

By far the the most experienced of the Thunderbolt-equipped fighter groups in the central Pacific, the 318th FG was the first group to receive P-47Ns and had the most success, in terms of aerial victories, with the variant. The 318th deployed to the Marianas in June 1944, famously making the journey aboard two aircraft-carriers, and flew its P-47Ds intensely in support of Marines fighting the Japanese before range limitations forced it to convert to the P-38 in November 1944, though these were replaced by P-47Ns just five months later. Group personnel flew these machines the 5,000 miles (8047 km) from Hawaii to Ie Shima – the longest ferry mission flown by a single-engined type in the USAAF's history.

had deliberately armed the nationalist forces of Dr Sukharno, who proclaimed Indonesian independence on 18 August 1945. Nos 60 and 81 Squadrons were sent to Batavia (now Jakarta) and, after covering a landing by British and Indian troops, went into action against nationalist forces on 1 November. Several aircraft were damaged, and two were lost. Military control reverted to the Dutch on 30 June. No. 81 Squadron disbanded, but No. 60 remained in the Dutch East Indies and was involved in further fighting until 28 November 1946, when the last operational front-line RAF Thunderbolts were struck off charge before No. 60 returned to Singapore to convert to the Spitfire Mk XVIII.

Although French Thunderbolt squadrons were involved in operations in French Indo-China, air and ground crews deployed to the area without their aircraft and flew other types (principally Spitfires, F6F Hellcats and F8F Bearcats) while in Vietnam, Laos and Cambodia.

Post-war use

With the rapid introduction of new jet-powered fighters in the air-to-air role with US forces, many observers expected the P-47 to form the backbone of those units that retained propeller-driven fighters and operated principally in the fighter-bomber role. The Thunderbolt had proved suited to the air-to-ground role, thanks to its better air-to-ground gun tracking and big radial engine, which was less vulnerable to ground fire. In its P-47N form, the Thunderbolt also enjoyed greater range than the P-51D, and lacked the Mustang's unpleasant and unpredictable departure characteristics, making it a better air-to-air fighter.

Although some P-47 units remained as part of the Occupation Forces in both Germany and Japan, they

With crew chiefs on their wings to ensure that they do not stray from the taxiway, 73rd FS/318th FG P-47Ns make their way from dispersals to the runway on Ie Shima at the beginning of another mission. Ie Shima was 3 miles (4.8 km) from Okinawa, in the Ryukyu Island chain, and 325 miles (523 km) from the Japanese mainland. Early P-47 missions from the island were ground support sorties for troops mopping up the last Japanese resistance on Okinawa.

wound down quite rapidly, and by July 1947 (when the USAAF became the USAF) only the 86th FG remained in Germany (at Nordholz) and only the 23rd FG remained in the Far East, at Guam.

Otherwise, the Thunderbolt was confined to three front-line USAF groups in the continental USA, including the 56th FG at Selfridge Field, Michigan, the 14th FG at Dow Field, Maine, and the African-American-manned 332nd FG (previously the 477th FG) at Lockbourne, Ohio. The 81st Fighter Group at Wheeler Field, Hawaii, was also equipped with P-47Ns. These units had all disbanded or converted to jet fighters by the end of 1949.

The Thunderbolt (redesignated as the F-47 on 11 June 1948) returned to front-line USAF service briefly during the winter of 1952-53, when Air Defense Command was rejuvenated in the face of a growing Soviet bomber threat. The 47th FIS at Niagara Falls, New York, and the 48th FIS at Grenier Field, New Hampshire, used P-47Ns as interim equipment, pending conversion to more modern fighters.

Thunderbolts (mainly P-47Ns, but including some P-47Ds) were used to equip a number of units after the Air National Guard was reactivated on 30 January 1946. The ANG's combat element of 12 wings included 20 fighter groups (with 62 squadrons), two light bombardment groups (with four squadrons), and five composite groups (with 12 fighter and six bomber squadrons). Each fighter squadron had 25 primary mission aircraft (Mustang or Thunderbolt) augmented by one or two C-47 or C-46 trans-

ports, two Stinson L-5s for liaison duties, four target-towing A-26s, and two T-6 trainers.

Fighter groups in the west and mid-west were equipped with Mustangs, while those in the east and south received P-47s. The squadrons built up slowly, the final 84 units (including five F-47 squadrons) being federally recognised in 1949. In the end, the Thunderbolt equipped 28 ANG fighter squadrons, and the ANG reached a peak holding of 500 F-47s in 1950, 350 of them 'in commission'.

When North Korea invaded South Korea on 25 June 1950, the USAF immediately took over 145 of the ANG's 764 F-51Ds for service in Korea. More were destined to follow, and several ANG F-51D squadrons were recalled to active duty. Quite why the Mustang, with its vulnerable liquid-cooled in-line Merlin engine, was chosen over the F-47 for service in Korea remains unclear, though the facts that the Mustang was already in service in-theatre, and that

The 507th and 413rd Fighter Groups, of the Twentieth Air Force, became operational with P-47Ns in May and June 1945, respectively. The 414th FG followed suit in July, based on Iwo Jima, though all three groups in fact operated alongside the 318th FG in the ground support role. On only one occasion did all three Twentieth Air Force groups escort B-29s on a daylight raid over Japan, on 8 August. Pictured is 414th Fighter Group P-47N-5 Detroit Miss II, the mount of Urban 'Ben' Drew, a P-51 ace in the ETO.

Wartime P-47s under foreign flags

Mexico
Other than the RAF, the Armée de l'Air and the Forca Aérea Brasiliera, two other air arms operated, or were at least supplied with Thunderbolts during World War II. After undergoing training in the US, Escuadrón Aereo 201 of the Fuerza Aérea Expedicionaria Mexicana (FAEM) briefly operated P-47Ds from May 1945, under the control of the USAAF's 58th FG in the Philippine Islands. This view of aircraft at Porac airfield includes a number of new P-47D-30-RAs supplied under Lend-Lease (including 44-33710, left) for assignment to Escuadrón 201. Mexican insignia is visible above the starboard wing of 44-33710 and would have been complemented by single stripes of green, white and red on the aircraft's rudder. Full USAAF insignia were retained, including black theatre markings.

Soviet Union
Two hundred and three Thunderbolts were allocated to the Soviet Union under Lend-Lease. These comprised three P-47D-10-REs (42-75201/75203), 100 P-47D-22-REs (42-25539/25638) and 100 P-47D-27-REs (42-27015/27064 and 42-27115/27164); seven aircraft were lost en route. This aircraft is a P-47D-22-RE of the 255th IAP (Istrebitel'nyi Avia Polk/Fighter Regiment) of the Northern Fleet Air Force during the unit's evaluation of the type during October 1944.

Thunderbolts in the post-war USAF

Above: Based at Straubing, Germany, the 368th Fighter Group was renumbered the 78th FG in August 1946. 'D3' codes were worn by aircraft of the 397th FS 'Jabo Angels'.

An 86th Fighter Group P-47D is pictured on a base in North Africa during live firing exercises, a regular destination for Europe-based fighter units. The 86th FG was one of six P-47 groups to remain in Germany as part of the Allied occupation force.

Far right: This P-47D-30-RA is a 527th FS/86th FG at Munich in March 1949, shortly before the last USAF Thunderbolts in Europe were replaced by jets.

Along with the P-51, the P-47 formed the backbone of the Air National Guard during the immediate post-war years. 44-89429 is a late-production P-47N-25.

large numbers of F-51Ds and F-51Ks were available, probably accounted for the decision.

Although the Thunderbolt was not used in the Korean War, F-47s wearing USAF markings did fire their guns in anger once more. Following a prison break-out by anti-US nationalists in Puerto Rico, the governor launched a 'show of force' operation against their strongholds in the towns of Jayuya and Utuado on 31 October 1950. Led by a B-26 pathfinder, and supported by tanks and infantry, two flights of four F-47s 'strafed' the townships: they fired after pulling out of their dives, so that people on the ground experienced the sound and fury of a full-scale strafing attack, with spent cartridges clattering noisily onto the corrugated steel roofs of the shacks and houses in the rebel area.

End of the Guard 'Jugs'

As first-generation jets were retired by regular USAF units, they became available to re-equip ANG squadrons. The result was that the F-47 disappeared from ANG service between 1952 and 1954, the Hawaiian National Guard's 199th FS being the last operator of the type in the fighter role, before converting to the F-86E in February 1954. The 198th FS of the Puerto Rico ANG kept its F-47s until July, and later became the last USAF unit to operate a Thunderbolt. One of its last airframes, relegated to ground instructional duties with a local vocational school, was restored to airworthy status in 1967 to celebrate the unit's

25th anniversary, flying a number of demonstration and training flights including a photo-hop with one of the squadron's F-104 Starfighters. The aircraft was painted in a colour scheme that echoed the markings of the 353rd and 354th Fighter Groups.

The Puerto Rico ANG F-47s saw action again after their retirement. When Guatemala elected Arbenz Guzman as its president, his nationalistic aims scared the US administration, and he was quickly smeared as a Communist by senior American figures, including CIA director Allan Dulles. He considered Guzman's drive for independence and capitalism (in place of a semi-feudal relationship with his old employer, the United Fruit Company) to be a threat to US economic interests, and he determined to mount a coup.

The F-47s were a key part of the CIA's ensuing Operation Success. Loaned to the CIA, the aircraft were then leased to the Nicaraguan government for a nominal $US1 each, and delivered to Puerto Cabezas. Flown by Nicaraguan and American mercenary pilots, the F-47s operated from airfields in Honduras and Nicaragua. On 18 June 1954 two strafed the National Palace and the port of San José. On 19 June another F-47 attacked the Fuerza Aérea Guatemalteca facility at La Aurora airport, damaging a Beech AT-11. F-47Ns mounted additional attacks over the next few days, until Guzman finally resigned on 27 June. The following days brought a few more attack sorties, including one in which the British freighter *Springfjord* (suspected of carrying arms and Avgas for the Guatemalan forces) was sunk in San José harbour by a Thunderbolt. Colonel Castillo Armas, who led the CIA-backed invasion force, was appointed president. One of the F-47s was retained (with US pilot Jerry DeLarm) as part of his body-

With the introduction of buzz numbers by the USAAF (initially by Eighth Air Force in occupied Germany in November 1945), the P-47 was allocated a 'PE' prefix (as shown on P-47N-15 44-88887, above right); this became 'FE' in 1948 when the 'P for Pursuit' designator was dropped in favour of 'F for Fighter'. Late production P-47N-25 44-89416 (right) displays its modified buzz code.

guard, while the others were transferred to the Nicaraguan air force.

Before being transferred to Nicaragua in exchange for an ex-Swedish Mustang, the Guatemalan Thunderbolt saw combat again, and may have made the type's last ever air-to-air kill. Another regime described by opponents as Communist was that of President Figueres of Costa Rica. The dictators who ruled Guatemala, Nicaragua and Venezuela banded together to assist the exiled former president, Rafael Calderon. On 12 January 1955 'rebel' aircraft (some of them Venezuelan aircraft on a 'goodwill visit' to Managua) attacked Costa Rican targets. F-47s were reported as being involved in these attacks several times, although these reports usually detailed a single Thunderbolt. This aircraft is believed to have been the single F-47 originally retained in Guatemala as part of Armas's bodyguard, and is understood to have been flown by the same US pilot, Jerry DeLarm. Costa Rica obtained four F-51Ds from US stocks to help it resist the invasion, one of which was lost while being flown by a local pilot. Acknowledged to have been lost to enemy action, the aircraft wreck was found to have been raked by heavy calibre machine-gun bullets and it is suspected that the aircraft was shot down by DeLarm and his F-47. Although the F-47N was successful, the invasion was not, and Figueres retained power. Nicaragua's F-47s remained in service until 1962, and although the type was preferred by Nicaraguan pilots, spares shortages and the increasing availability of F-51Ds led to their eventual replacement.

The Thunderbolt was exported to more countries than the US's wartime allies and Latin America, serving with a number of new operators as the US rushed to re-arm new allies when the Cold War began. In most cases, the new

Thunderbolt operators used the type only briefly, before acquiring first-generation jets like the F-84 Thunderjet. This was the pattern followed in Iran, Italy, Portugal, Turkey and Yugoslavia.

French Thunderbolts

Remarkably, though, the aircraft served longer in France, despite the rapid influx of new jet fighters and jet fighter-bombers. France was the only European Thunderbolt operator to use its P-47s in combat post-war. Although the 3ème Escadre de Chasse had been sent to Indo-China, it went without its aircraft, and its pilots flew other types until their return to France. Front-line P-47s were replaced by de Havilland Vampires from 1950, but the type served on in the advanced training role and with 'weekend warrior' reservist units, notably EC 10, divided between Villacoublay, Dijon and Rabat. This unit was redesignated as Escadre d'Entrainement à la Chasse 17 in 1954, centralised by moving together the scattered elements of EC 10. Thus EC 1/10 and 2/10 became EC 2/17, while EC 3/10 became EC 3/17, joining EC 1/17 at Creil.

Nationalist China acquired a large number of ex-10th and 14th Air Force Thunderbolts immediately post-war. They consisted of several variants, including P-47Ns. This pair of 11th Fighter Group aircraft are P-47Ds. Many RoCAF Thunderbolts retained evidence of their former users: the 'razorback' in the background still has the diagonal fuselage stripes it wore when assigned to the 1st Air Commando Group.

In the post-war era the P-47 became the standard fighter type to be handed out to Latin American nations eager to build an air arm around this still potent machine. Brazil (left) and Mexico (below) had, of course, famously taken Thunderbolts to war alongside the Americans, and in the post-war years built their P-47 fleets up with fresh machines as they were discarded by USAF units.

Above: All but one of the world's currently airworthy Thunderbolts reside in the USA. The exception is The Fighter Collection's P-47D No Guts, No Glory, based at Duxford, UK. Built at Republic's Evansville factory in 1945 as 45-49192, the aircraft served with Air Training Command before delivery to the Peruvian air force in 1953 as FAP119. In 1969 the aircraft was one of six recovered from that country by Ed Jurist. Purchased by TFC in 1984, the aircraft was restored to flying condition by Fighter Rebuilders, and shipped to Duxford in 1986. The aircraft is painted in the markings of Lt Col Ben Mayo – CO of the Duxford-based 82 FS, 78 FG.

Above right: Wartime 'Jug' and, at this time, Republic chief production test pilot Glenn Bach sits in the cockpit of a P-47G, specially restored to mark the 20th anniversary of the type's first flight. In stark contrast, it sits alongside a brand new Republic F-105D.

With the 1954 uprising in Algeria, the potential of the ageing P-47 in the COIN role, and its ability to operate with minimal support in the harsh conditions likely to be encountered in North Africa, meant that a new front-line unit, EC 20, was formed from EC 17 in 1956. The new escadre (wing) had two groupes (enlarged squadrons): GC I/20 'Ayres-Nementcha' and GC II/20 'Ouarsenis'. The P-47 saw extensive operational service in Algeria, before GC II/20 'Ouarsenis' finally converted to the Douglas Skyraider during 1960.

The other P-47 operator whose Thunderbolts saw combat was Nationalist China (Taiwan), which inherited many of the P-47s left behind by the USAAF's 10th and 14th Air Forces. These aircraft served until 1955, and had a number of encounters with People's Republic of China fighters, including MiG-15s. There are persistent reports that at least one MiG-15 was shot down by a Nationalist F-47 on 22 May 1954. The successful pilot is usually named as Chiang Tien-En, and his wingman, Pen, is sometimes given a share of the remarkable kill. One young Taiwanese F-47 pilot, Tang Fe (who graduated from the Air Academy in 1952, to fly Thunderbolts with the 5th Air Group), later became chief of staff of Taiwan's air force and, in 2000, became Taiwan's premier.

By 1962, only Nicaragua and Peru still operated the F-47, the other Latin American users having discarded the type over the previous 10 years. Venezuela's F-47s were the first to go, having ceased operations in 1952, though the aircraft were still nominally on charge until mid-1954 and were not put up for sale until 1956. Colombia grounded its Thunderbolts in December 1955; Brazil deactivated its surviving aircraft in October 1957, striking them off charge

in batches throughout 1957-58.

The Aviación Militar Dominicana declared its last F-47s surplus in November 1957, and 12 of Chile's surviving 13 aircraft were struck off charge on the last day of 1957. The last Mexican Thunderbolts were retired in June 1958 although they were not replaced by Vampires until December 1960. Cuba's last F-47D squadron was probably effectively grounded in 1958, but two aircraft survived in nominally airworthy condition until the Bay of Pigs adventure in April 1961, when one was destroyed on the ground at Havana. In Ecuador the last F-47s were relegated to target tugs and trainers from 1958, and the last Thunderbolt was finally grounded in July 1959, as a precondition for the supply of the last batch of F-80s. With the retirement of the last Nicaraguan aircraft, Peru was left as the last Thunderbolt user. Its aircraft had been used as trainers since 1958 and remained active well into the 1960s. The Peruvians suffered their last Thunderbolt loss in June 1963, and 12 aircraft were still nominally on charge in 1966.

Warbirds

These aircraft survived long enough for the P-47 to become interesting to the fledgling warbird movement. Whereas many P-51s had survived thanks to their suitability as air racers, the population of airworthy Thunderbolts in the US had dwindled to one or two aircraft. As early as 1961, when Republic wanted to celebrate the 20th anniversary of the P-47's first flight, it proved difficult to find an aircraft even capable of being returned to flight-worthy status. A P-47G-15-CU was eventually located and restored, and was sold to Republic. The Confederate Air Force's import of six Peruvian aircraft between 1967 and 1969 thereby provided an invaluable boost to the aircraft's presence in the warbird scene. Since then, some aircraft have been relegated to static display status and others have been returned to the air, so that about 10 P-47s are currently airworthy.

Jon Lake

The Thunderbolt still flies – warbird register

Model	Real serial	Flown as	Name	Civ reg.	Status	Location
P-47D	42-8066				Under rest. to fly	Sydney, Aus
P-47D	42-8205		Big Stud	N14519	Airworthy (on display)	AZ, USA
P-47G	42-25234	42-28487	Spirit of Atlantic City	N3395G	Airworthy	CA, USA
P-47D	42-26766				Under rest. to fly	CA, USA
P-47M	42-27385			N27385	Airworthy (not flown)	CA, USA
P-47D	42-27608				Under rest. to fly	Sydney, Aus
P-47D	44-32817			N767WJ		
P-47D	44-90368	44-33240	Tarheel Hal	N4747P	Airworthy	TX, USA
P-47D	44-90438			N647D	Airworthy (on display)	TN, USA
P-47N	44-90447		Jacky's Revenge	N1345B	Airworthy	NY, USA
P-47D	44-90460		Hun Hunter XVI	N9246B	Airworthy (on display)	TN, USA
P-47D	44-90471	42-26641	Hairless Joe	N47DA	Airworthy	IL, USA
P-47D	45-49181	42-26418		N444SU	Airworthy (not flown)	MI, USA
P-47D	45-49192	42-26671	No Guts, No Glory	N47DD	Airworthy	Duxford, UK
P-47D	45-49205	42-28473	Big Chief	N47RP	Airworthy	CA, USA
P-47D	45-49385			N47DF	Airworthy	CA, USA
P-47D	45-49406			N7159Z	Under rest. to fly	WA, USA
P-47N	45-53436			N47TB	Crashed 22/3/02	NM, USA

44-90447 is currently the world's only airworthy P-47N following the crash of the Commemorative Air Force's example in March 2002. It is operated by the American Airpower Museum, Farmingdale, New York.

UNITED STATES

US ARMY AIR FORCES/US AIR FORCE

P-47s were operational with the following groups/squadrons during World War II, during the periods shown. They are grouped by numbered Air Force. Note that most groups also included an HQ Flight to which a handful of aircraft was usually assigned:

Fifth Air Force (SW/Central Pacific)

8th Fighter Gp	36th FS	1943/44
35th Fighter Gp	39th FS	1943/44
	40th FS	
	41st FS	
49th Fighter Gp	9th FS	1943/44
58th Fighter Gp	69th FS	1944/45
	310th FS	
	311th FS	
348th Fighter Gp	340th FS	1943/45
	341st FS	
	342nd FS	
	460th FS	

Seventh Air Force (Central Pacific)

318th Fighter Gp	19th FS	1944
	73rd FS	
	333rd FS	

Also assigned to the Seventh Air Force was the 15th Fighter Group (45th, 47th and 78th Fighter Sqns) and the 508th Fighter Group (466th, 467th and 468th Fighter Sqns) which flew P-47s in defence of the Hawaii Islands during 1944 and 1945, respectively.

Eighth Air Force (ETO)

4th Fighter Gp	334th FS	1943/44
	335th FS	
	336th FS	
56th Fighter Gp	61st FS	1943/45
	62nd FS	
	63rd FS	
78th Fighter Gp	82nd FS	1943/44
	83rd FS	
	84th FS	
352nd Fighter Gp	328th FS	1943/44
	486th FS	
	487th FS	
353rd Fighter Gp	350th FS	1943/44
	351st FS	
	352nd FS	
355th Fighter Gp	354th FS	1943/44
	357th FS	
	358th FS	
356th Fighter Gp	359th FS	1943/44
	360th FS	
	361st FS	
358th Fighter Gp	365th FS	1943/45
	366th FS	
	367th FS	
transferred to the Ninth Air Force 1944		
359th Fighter Gp	368th FS	1943/44
	369th FS	
	370th FS	
361st Fighter Gp	374th FS	1944

	375th FS	
	376th FS	

Other Eighth Air Force units known to have flown Thunderbolts included:
5th Emergency Rescue Sqn
495th Fighter Training Gp
3rd Gunnery and Tow Target Flight
65th Fighter Wing HQ Flight
66th Fighter Wing HQ Flight

Ninth Air Force (ETO)

36th Fighter Gp	22nd FS	1944/45
	23rd FS	
	53rd FS	
48th Fighter Gp	492nd FS	1944/45
	493rd FS	
	494th FS	
50th Fighter Gp	10th FS	1944/45
	81st FS	
	313th FS	
354th Fighter Gp	353rd FS	1944/45
	355th FS	
	356th FS	

The 354th was re-equipped with P-51s before VE-Day

362nd Fighter Gp	377th FS	1944/45
	378th FS	
	379th FS	
365th Fighter Gp	386th FS	1944/45
	387th FS	
	388th FS	
366th Fighter Gp	389th FS	1944/45
	390th FS	
	398th FS	
367th Fighter Gp	392nd FS	1944/45
	393rd FS	
	394th FS	
368th Fighter Gp	395th FS	1944/45
	396th FS	
	397th FS	
371st Fighter Gp	404th FS	1944/45
	405th FS	
	406th FS	
373rd Fighter Gp	410th FS	1944/45
	411th FS	
	412th FS	
404th Fighter Gp	506th FS	1944/45
	507th FS	
	508th FS	
405th Fighter Gp	509th FS	1944/45
	510th FS	
	511th FS	
406th Fighter Gp	512th FS	1944/45
	513th FS	
	514th FS	

Tenth Air Force (China-Burma-India)

1st Air Commando Gp		1944/45
	5th FS(C)	
	6th FS(C)	

Both squadrons were re-equipped with P-51s before VJ-Day

33rd Fighter Gp	58th FS	1944
	59th FS	

80th Fighter Gp	60th FS	
	88th FS	1944/45
	89th FS	
	90th FS	

Twelfth Air Force (Mediterranean)

27th Fighter Gp	522nd FS	1944/45
	523rd FS	
	524th FS	
57th Fighter Gp	64th FS	1944/45
	65th FS	
	66th FS	
79th Fighter Gp	85th FS	1944/45
	86th FS	
	87th FS	
86th Fighter Gp	525th FS	1944/45
	526th FS	
	527th FS	
324th Fighter Gp	314th FS	1944/45
	315th FS	
	316th FS	
350th Fighter Gp	345th FS	1944/45
	346th FS	
	347th FS	

Fourteenth Air Force (China)

81st Fighter Gp	91st FS	1944/45
	92nd FS	
	93rd FS	

Fifteenth Air Force (Mediterranean)

325th Fighter Gp	317th FS	1943/44
	318th FS	
	319th FS	

The 332nd Fighter Gp is also known to have operated ex-325th FG P-47s for two months during 1945 before converting to P-51 Mustangs

Twentieth Air Force (Central Pacific)

413rd Fighter Gp	1st FS	1945
	21st FS	
	34th FS	
414th Fighter Gp	413rd FS	1945
	437th FS	
	456th FS	
507th Fighter Gp	463rd FS	1945
	464th FS	
	465th FS	

With the end of hostilities large numbers of Thunderbolts were made surplus overnight and orders for almost 6,000 P-47Ns were cancelled. In the ETO there were over 3,000 P-47s on stations in France, Germany, Italy and the UK. Aircraft recently arrived in Britain (mainly D-30s and D-40s, but including five P-47N-5-REs believed to have been destined for the 56th FG) were prepared for shipment to the US and the PTO. Some squadron aircraft were also shipped, while others were scrapped.

Six fighter groups (36th, 79th, 86th, 366th, 368th and 406th) remained in Germany as part of the Allied occupation force, though the 36th and 86th FGs were disbanded in February 1946. In August the 366th, 368th and 406th FGs were renumbered as the 27th, 78th and 86th FGs, respectively; by the following June only the 86th remained, based at Nordholz with the last USAAF Thunderbolts in Europe. These were exchanged for F-84 Thunderjets sometime after 1949.

In the PTO around 1,400 P-47s were on hand on VJ-Day, 1,000 on the islands surrounding Japan. Some aircraft were shipped back to the US and others scrapped, while the 58th and 413th FGs took up occupational duties in Japan; while one other P-47 group to remain in the region, the 414th FG, moved to the Philippines. All were inactivated during 1946.

In the US, Thunderbolts remained with a handful of regular groups during 1946/47, the last examples in the USAF's inventory being those of the 14th Fighter Group of Dow Field, Maine, which transitioned to the F-84B in 1947. A few units reformed to operate jet fighters were briefly equipped with piston-engined aircraft until their new mounts became available. An example is the 47th FIS which reformed as an Air Defense Command unit in late 1952 as the 47th Fighter-Interceptor Squadron. Based at Niagara Falls Municipal Airport, the unit was equipped with F-47D-30s for a year before receiving F-86 Sabres.

This immaculately turned out P-47N was assigned to the First Air Force in the US during 1946. Long-range Ns served for some time after the war in ANG units.

AIR NATIONAL GUARD

Thunderbolts (designated F-47 from 1948) were available in large numbers in 1945 and, with the P-51 Mustang, were the main equipment of the ANG post-war. F-47Ds and Ns equipped Guard units east of the Mississippi and in some southern states, plus those of Puerto Rico and Hawaii. Twenty-eight fighter squadrons flew the type as their main equipment, the first of these being Connecticut's 118th FS from the latter half of 1946. Thunderbolts were finally withdrawn by the ANG in 1954, the last user being Hawaii's 199th FS.

The following ANG fighter squadrons were equipped with Thunderbolts in the years shown:

101st FS	MA	F-47N	1946-50
104th FS	MD	F-47D	1946-51
105th FS	TN	F-47D	1947-52
118th FS	CT	F-47N	1946-52
119th FS	NJ	F-47D	1947-52
121st FS	DC	F-47D	1946-49
128th FS	GA	F-47N	1946-52
131st FS	MA	F-47D	1947-51
132nd FS	ME	F-47D	1947-48
133rd FS	NH	F-47D	1947-52
134th FS	VT	F-47D	1946-50
136th FS	NY	F-47D	1948-52
137th FS	NY	F-47D	1948-52
138th FS	NY	F-47D	1948-52
139th FS	NY	F-47D	1948-51
141st FS	NJ	F-47D	1949-52
142nd FS	DE	F-47N	1946-50
146th FS	PA	F-47N	1946-51
147th FS	PA	F-47N	1949-51
148th FS	PA	F-47D	1947-50
149th FS	VA	F-47D	1947-53
152nd FS	RI	F-47D	1948-52
153rd FS	MS	F-47N	1946-52
156th FS	NC	F-47D	1948-49
158th FS	GA	F-47N	1946-48
167th FS	WV	F-47D	1947-51
198th FS	PR	F-47N	1947-54
199th FS	HI	F-47N	1947-54

This P-47N-25-RE was one of the last Thunderbolts to be completed, and served with the 142nd FS until replacement by F-84Cs in 1950. Note the lack of national insignia.

161

BRAZIL

FORÇA AÉREA BRASILIERA

1° Grupo de Aviação de Caça (1° GAvCa), Forca Aérea Brasiliera operated P-47Ds under the control of the 350th FG, USAAF at Tarquinia, Italy, from October 1944 until VE-Day. Initially 67 aircraft were received under Lend-Lease, 26 of which were issued to the 1° GAvCa (to equip four flights with six aircraft each and provide two aircraft for an HQ flight) and the remainder kept in reserve at Naples. Twenty-two aircraft were lost to all

causes during the campaign and a further 19 stored machines were commandeered by US units in Italy when a shortage of P-47s occurred shortly after VE-Day.

With the end of hostilities in Europe the FAB was able to take a full complement of 26 Thunderbolts back to Brazil, where 1° GAvCa reformed at São Paulo. Nineteen near-new replacement aircraft (for those commandeered earlier) were supplied

from the US shortly afterwards.

An additional 25 P-47Ds were acquired in 1947 and a further batch of refurbished ex-Virginia ANG F-47D-40s (P-47D-30s brought up to D-40 standard by TEMCO) arrived in 1953, the latter under the Mutual Defense Assistance Program (MDAP). By then, however, the FAB was taking delivery of Gloster Meteor jet fighters and the Thunderbolt had lost its pre-eminent position in 1° GAvCa. Plans to acquire a further 25

Groundcrew work on 1° GAvCa P-47Ds in Italy, where the squadron fought while assigned to the USAAF's 350th Fighter Group, part of the 12th Air Force.

F-47Ds were cancelled and, although there were around 40 examples still 'on the books' at the beginning of 1958, the last airworthy FAB Thunderbolts were struck-off-charge at the end of 1958.

CHILE

FUERZA AÉREA DE CHILE

The Fuerza Aérea de Chile (FACh) received six P-47Ds in 1946 and, by the beginning of 1949, the multi-role Grupo de Aviación No. 2 at Quintero had 12

on strength alongside a number of other aircraft supplied by the US. Later in the year, fighter unit Grupo de Aviación No. 5 was established to operate the aircraft, though a reorganisation saw B-25 bombers join the unit briefly before a definitive fighter unit (Grupo de Aviación No. 11) was established for the P-47s before the end of the year.

Spares supply problems, which plagued a number of Latin American air forces equipped with P-47s, affected the FACh and, by 1952, only seven aircraft survived in an airworthy condition.

Seventeen further F-47D-40s (refurbished and upgraded D-30s) were delivered under MDAP in 1953 and a revitalised Grupo de Aviación No. 11

was able to field 21 airworthy aircraft in June 1954.

Serviceability subsequently suffered as experienced ground crews were transferred to newly arrived de Havilland Vampire jets and by the time F-80 Shooting Stars were delivered in 1958 the writing was on the wall for the last 12 FACh Thunderbolts; all were struck off on 31 December.

CHINA

KMT/CHINESE NATIONALIST AIR FORCE

A large (and probably unknown) number of ex-10th and 14th Air Force Thunderbolts were 'left behind' in China after the war. The exact number is unknown because, while some 40 aircraft were officially transferred to Chiang Kai Shek's nationalist Republic of China Air Force, others were recovered and repaired from the large numbers of abandoned and dumped aircraft. Many Chinese Thunderbolts retained the fuselage and fin stripes associated with their previous USAAF operators, and some even retained their wartime nose-art, to which Chinese insignia and unit markings were added.

A full-scale civil war between Communist and Nationalist forces in China resumed in 1946, following the departure of US forces. Peking fell to Communist forces in January 1949, and the People's Republic of China was

declared on the mainland on 1 October 1949. Some two million followers of the Nationalist cause (including the bulk of the air force and most of the P-47s) fled to Taiwan, where the Republic of China was re-established on 1 December 1949. Taiwan had been formally restored to Chinese control on 25 October 1945. Neither Chinese entity recognised the other, and there was continuing military tension and regular clashes.

From May 1951 the USA stepped up its military aid to Taiwan, which came to be seen as a staunchly anti-communist ally during the Cold War years, while mainland 'Red' China was an active enemy, backing Russia and

China's Thunderbolts were a mixed bag of variants. The black diagonal tailstripe on this 11th FG P-47D shows former assignment to the USAAF's 81st FG.

supporting the North Koreans during the Korean War. The Chinese Nationalists soon received further Thunderbolts, including an unknown number of F-47Ns, though probably not as many as the 'two squadrons' quoted in some sources. A mutual defence treaty between the USA and Taiwan was signed in 1954, and this led to the introduction of large numbers of jet

fighters which replaced Taiwan's ageing Thunderbolts and Mustangs.

The primary RoCAF operator of the Thunderbolt was the 11th Fighter Group, including the 8th, 41st, 42nd, 43rd and 44th Fighter Squadrons, though the type is also understood to have served with the 3rd, 4th and 5th Groups, and with the 17th, 26th, 27th and 28th Fighter Squadrons.

COLOMBIA

FUERZA AÉREA COLOMBIANA

The Fuerza Aérea Colombiana (FAC) purchased and took delivery of its first seven P-47D-30s in July 1947; these were augmented by another 12 refurbished examples (upgraded by TEMCO to D-40 standard) in 1949, the combined fleet being organised into the Escuadrón de Caza-Bombardero based at Madrid AB.

By 1 May 1950 only six of 15 remaining aircraft were still airworthy, four aircraft having been lost to various causes. FAC Thunderbolts were engaged in a series of clashes with guerrillas along the southern border and this activity influenced attempts to

acquire further aircraft. Thirteen aircraft remained in use by June 1952 and these were joined by a further group of F-47Ds, originally destined for Chile (but diverted as a gesture of thanks for Colombia's contribution of infantry and naval units to the UN police action in Korea), in May 1953.

At this time a life extension programme was initiated, but by the end of 1954 only 14 were 'combat ready'. All were now based at Palanquero with 1° Escuadrón de Caza. All were grounded 12 months later but were not replaced until 1957, when a batch of 16 F-80Cs was delivered.

CUBA

FUERZA AÉREA EJERCITO DE CUBA

Cuba's Fuerza Aérea Ejercito de Cuba (FAEC) received its first Thunderbolts on the 50th anniversary of its independence in mid-1952, five aircraft

being followed by a further 21 examples (including a two-seat 'TF-47') over the next few months. These aircraft equipped Escuadrón de

Persecución '10 de Marzo' based at Campo Colombia, near Havana.

Spares supply problems resulted in the fleet being reduced to 17 by the end of 1956 and ultimately led to the purchase of 15 Hawker Sea Fury FB.Mk 11s as replacements. Thunderbolts bombed Castro-led insurgents at the Cienfuegos naval base on 5 September 1957.

By June 1958 only 10 were still in use, though problems with turbocharger exhaust tubes kept most of these aircraft grounded. At least two Thunderbolts passed to Castro's Fuerza Aérea Revolucionaria, but saw little use. One of these aircraft is believed to have been destroyed by aircraft supporting the ill-fated Bay of Pigs invasion.

DOMINICAN REPUBLIC

AVIACION MILITAR DOMINICANA

After various attempts to purchase 17 ex-Mexican air force P-47Ds, a similar number of (later 32) US government surplus F-47Ns and 25 F-47Ds refurbished "for a NATO country" all came to nought, 14 F-47D-40s were finally delivered to the Aviación Militar Dominicana (AMD) in late 1952, with another 11 arriving the following year.

These aircraft equipped an Escadrón de Caza-Bombardero until mid-1955, when the (renamed) Fuerza Aérea Dominicana's first de Havilland Vampire jets arrived. The 19 remaining F-47s were declared surplus, largely due to the ever-present lack of spares for the aircraft, in 1957 and eventually scrapped.

FRANCE

ARMÉE DE L'AIR

Four hundred and seventy P-47Ds (mainly D-30s) were allocated to the Armée de l'Air from factory production, the intention being to re-equip Free French units originally raised in North Africa. This process began in March 1944, though only 446 new aircraft were delivered. Others were received from USAAF depots in Italy and France and by VE-Day 131 examples remained in use with six units in two Escadres de Chasse, namely Groupe de Chasse (GC) II/3 'Dauphiné', GC III/3 'Ardennes', GC I/4 'Navarre', GC I/5 'Champagne', GC II/5 'La Fayette' and GC III/5 'Roussillon'.

GC II/3 and GC II/5 were renumbered GC I/4 and GC II/4, respectively, in 1947 and remained in occupied Germany until 1949/50; personnel from GC I/4, GC I/5 and GC III/6 were sent to Indo-China where they operated Spitfires until 1948, when they returned to France and again flew Thunderbolts until converting to jet aircraft in 1950.

Other post-war units equipped with P-47s were GC I/1 'Provence', GC II/1 'Nice', GC I/2 'Cigognes', GC III/2 'Alsace', GC I/8 'Mahgreb', GC II/8

'Languedoc', GC I/6 'Corse', GC II/6 'Normandie-Niémen' and GC I/21 'Artois'.

French Thunderbolts saw further action in Algeria during the 1950s, two of three reserve training units formed in 1951 providing the basis of ground attack/close support units Escadre de Chasse (EC) 1/20 'Ayres-Nementcha' and EC 2/20 'Ouarsénis', formed in 1956. Based at Oran and Rabat, these units flew sorties against Algerian nationalists until 1960, when the more applicable Douglas AD-4N Skyraider entered service. In addition, EC 3/20 'Oranie' was formed as an operational training unit.

(Note that the 'Groupe' designation for a wing comprising between two and four 'escadrons', or squadrons, gave way to 'Escadre' in 1951.)

This P-47D served with EC 2/10, a reserve training wing which operated the Thunderbolt from Villacoublay and Dijon until 1954. Given its prowess in the attack role, it is surprising the Thunderbolt was not sent to Indo-China, where Spitfires, Bearcats and KingCobras were used. However, the type did see action on counter-insurgency strikes in Algeria in the late 1950s.

ECUADOR

FUERZA AÉREA ECUATORIANA

After wartime P-47 operators Brazil and Mexico, Ecuador was the next Latin American country to receive Thunderbolts, taking delivery of 12 P-47Ds (plus a Douglas C-47A and an Beech AT-7) in 1947 under the American Republic Projects (ARP) aid programme.

Equipping a Quito-based Fuerza Aérea Ecuatoriana (FAE) fighter unit – the Escuadrón de Caza – these aircraft were joined by another four attrition replacement aircraft in 1949, though by early 1953 serviceability levels had dropped and only 10 aircraft were airworthy.

The Escuadrón de Caza received a boost in mid-1953 with the arrival of 11 MDAP-funded, refurbished F-47D-40s. The Thunderbolt unit was now known as the 10° Escuadrón, 100° Ala de Caza-Bombardero and had a nominal strength of 19 aircraft, though nine of

these were in need of overhauls. Spares problems and a series of accidents (and subsequent groundings) after engine and turbo-exhaust failures (the latter were removed altogether from some aircraft) led to the Thunderbolt fleet dwindling to eight operational examples by mid-1954.

By early 1958 matters had improved slightly, 12 airworthy aircraft equipping the redesignated 108° Escuadrón de Caza; mid-year the squadron moved to General Ulpiano Paez Air Base near Salinas to make way for new Gloster Meteor FR.Mk 9s at Quito.

By this time the last of the FAE's F-47s were employed as target-tugs to support training for the air force's new Lockheed F-80Cs; all had been grounded by the end of the following year, a pre-condition of the delivery of F-80s being that they be scrapped without delay.

GREAT BRITAIN

ROYAL AIR FORCE

Under Lend-Lease arrangements, 826 Thunderbolts (of 919 ordered; four crashed in the US prior to delivery and 89 were cancelled) were supplied to the RAF and were employed exclusively by Southeast Asia Command in the Far East, replacing Hawker Hurricanes. Delivered between February and May 1944, the first 239 examples were designated Thunderbolt Mk I and were mainly P-47D-22-REs. Subsequent deliveries, made between May 1944 and June 1945, were a mixture of D-25s, D-27s, D-28s and D-30s, plus a single D-40. These later aircraft were known as Thunderbolt Mk IIs (which entered service in September 1944), the distinction being their 'bubble' canopies.

Sixteen squadrons flew the Thunderbolt at different times from late spring 1944; peak strength occurred in August 1945, when 12 units were operational with the type. Conversion training of crews was initially the responsibility of No. 1670 Conversion Unit at Yelahanka, India. Later in 1944 No. 73 Operational Training Unit at Fayid, Egypt, also began to train crews. The first operational RAF Thunderbolts were those of No. 261 Sqn at Yelahanka.

Squadron establishment was generally 16 aircraft, though units often had as many as 20 aircraft on strength; most flew a mixture of Mk Is and IIs, though most Mk Is had been withdrawn by mid-1945.

SEAC Thunderbolt squadrons were:

No. 5 Squadron	Jun 44-Mar 46
No. 30 Squadron	Jul 44-Dec 45
No. 34 Squadron	Mar 45-Oct 45
No. 60 Squadron	Jun 45-Oct 46
No. 79 Squadron	Jul 44-Dec 45
No. 113 Squadron	Apr 45-Oct 45
No. 123 Squadron	Sep 44-Jun 45
No. 134 Squadron	Sep 44-Jun 45
No. 135 Squadron	Jun 44-Jun 45
No. 146 Squadron	Jun 44-Jun 45
No. 258 Squadron	Sep 44-Dec 45
No. 261 Squadron	Jul 44-Sep 45

During June 1945, four squadrons were renumbered (an administrative exercise allowing certain squadrons to remain in existence), as follows:

No. 123 Sqn to No. 81 Sqn (finally disbanded June 1946)
No. 134 Sqn to No. 131 Sqn (disbanded December 1945)
No. 135 Sqn to No. 615 Sqn (disbanded September 1945);
No. 146 Sqn to No. 42 Sqn (disbanded December 1945)

By the end of 1945 most RAF Thunderbolt squadrons had been disbanded, though two (Nos 5 and 30 Sqns) remained in India until re-equipped with Tempest Mk IIs in 1946 and two others (Nos 60 and 81 Sqns) were transferred to Batavia in October 1945 as part of the British occupation force preparing to hand the Netherlands East Indies back to Dutch control. In late October both squadrons covered British landings on the island and, during November, attacked Indonesian guerrilla positions. No. 81 Sqn disbanded in June 1946, No. 60 finally re-equipping with Spitfire FR.Mk XVIIIs at the end of the year.

IRAN

IMPERIAL IRANIAN AIR FORCE

Little is known of the F-47Ds supplied to the Imperial Iranian Air Force under the MDAP programme. Fifty examples, probably D-30s, were supplied during 1948 and are likely to have remained in use until F-84s were delivered during the 1950s. They were used by both fighter and attack squadrons.

ITALY

AERONAUTICA MILITARE ITALIANA

A stop-gap measure pending the delivery of Republic F-84Gs delayed by the Korean War, the acquisition of 100 F-47Ds during December 1950 allowed the re-equipment of 101° and 102° Gruppi of 5° Stormo, and 20° and 21° Gruppi of 51° Stormo. Replacing Spitfire Mk IXs, the F-47D-30s were delivered from USAAF storage in Germany; 77 were made operational by the AMI, the remainder going to 1° Reparto Tecnico Aerea at Bresso for spares recovery.

The AMI's Thunderbolts appear to have been phased out during 1952/53, as F-84Gs began to enter service.

MEXICO

FUERZA AÉREA MEXICANA

Mexico initiated talks with the US regarding the provision of an operational unit for active service in 1943. These resulted in the formation of Escuadrón Aereo 201 the following year, under the auspices of the Fuerza Aérea Expedicionaria Mexicana (FAEM). Departing Mexico for the US in July

Wartime 'Jug' – this P-47D receives attention in the field during fighting in the Philippines in 1945. Escuadrón 201 operated as part of the USAAF's 58th FG.

1944, the squadron went through a screening programme before training on the P-47.

By the end of April 1945 the unit had been shipped to the Philippine Islands, where it would operate as an element of the 58th Fighter Group, Fifth Air Force. During its brief operational life fighter-bomber sorties were flown, mainly during June and early July, against the last pockets of Japanese resistance in the Philippines. Escuadrón 201's aircraft were a mixture of 18 'borrowed' and Lend-Lease P-47Ds; by mid-July the unit had a nominal strength of 30 aircraft.

After VJ-Day the squadron returned to Mexico via Texas, where it collected 25 near-new P-47Ds. These equipped a re-established Escuadrón de Pelea 201, FAM, based at Mexico City. In June 1947 eight aircraft were transferred to the Escuela de Aviación Militar.

During 1950 consideration was given to replacing the P-47s with F-51s, but it would be another 10 years before true replacement aircraft were delivered (de Havilland Vampire F.Mk 3s); by then the FAM's active P-47 fleet – down to nine aircraft by the end of June 1957 – had been retired for some two years.

NICARAGUA

FUERZA AÉREA DE LA GUARDIA NACIONAL DE NICARAGUA

At least eight ex-Puerto Rico ANG F-47Ns were 'delivered' to the Fuerza Aérea de la Guardia Nacional de Nicaragua in 1954, though they were operated by, or at least on behalf of, the CIA and were soon engaged in attacks to topple the government of neighbouring Guatemala (Operation Success). These aircraft remained in use into the early 1960s and were the last operational P-47s in Central America.

PERU

FUERZA AÉREA DEL PERU

After Brazil, Peru was the second largest operator of Thunderbolts in mainland South America, the Cuerpo de Aeronautica Peruana (Fuerza Aérea del Peru from July 1950) receiving its first 25 P-47D-40s during 1947.

After prolonged negotiations and a series of bureaucratic mix-ups, two further batches totalling 31 aircraft were delivered during 1952/53, the combined fleet equipping two Escuadrones de Caza (12 and 13), each with 21 aircraft.

By mid-1954 maintenance problems grounded a number of aircraft; each squadron had just 15 airworthy aircraft during June. By then the type was viewed by the FAP as a fighter-bomber and was soon relegated to a fighter-trainer role. It was not long before jet aircraft were on the horizon. Lockheed T-33As and North American F-86s arrived during 1955, though it was some time before the P-47s were finally retired. Retained as trainers, the last examples were grounded in the 1960s. The type was still in use in 1963

and in 1966 there were 12 examples nominally 'on the books', though by this time none is believed to have been airworthy.

Like many other Latin American air arms, the Peruvian air force was given P-47s to form the basis of its post-war fighter force. Following the delivery of jet fighters, the P-47s still had a part to play as fighter trainers. Six Peruvian aircraft returned to the US in 1967/69 as part of the warbird renaissance.

PORTUGAL

Portugal received 50 P-47Ds under MDAP, and these equipped Esquadra 10 and Esquadra 11 at BA2 Ota. They were replaced by Republic F-84 Thunderjets after about four years' service.

SOVIET UNION

VOENNO VOZDUSHNYE SILY

The Thunderbolt enjoyed an abbreviated career in Russia, where it proved unpopular and unfamiliar to a generation of fighter pilots used to much smaller and more agile aircraft, from the indigenous Yak-3 to the Lend-Lease Bell Airacobra. Nor did the Russian air forces have any requirement for an aircraft optimised for the long-range escort of heavy bombers, nor for a high-altitude fighter, while its own Shturmoviks (ground attack aircraft) out-classed even the Thunderbolt in the air-to-ground role. Despite this some 203 Thunderbolts (mainly P-47D-22-REs and P-47D-27-REs) were allocated to Russia under Lend-Lease provisions, and 196 of these reached Russia.

A single P-47D-10-RE was delivered to Russia in mid-1944, and was evaluated by the NII and by the LII. Engineers were most impressed by the aircraft, while Soviet pilots cordially hated it! The bulk of Russia's Thunderbolts were shipped by sea to Iran, and were then assembled at Abadan by Douglas, and handed over to the 6th PIAP (Peregonochnyi Istrebeitelnyi Aviatsionnyi Polk – Ferry Fighter Air Regiment) for ferrying via Teheran to Kirovobad. 112 of the 190 aircraft delivered during 1944 used this route.

The first of 111 Thunderbolts were handed over to the 11th ZBAP (Zapasnoy Bombardirovochnyi Aviatsionnyi Polk – Reserve Bomber Air Regiment) which served as the VVS's depot, training and conversion unit for the P-47. This unit removed the aircraft's incompatible radio and IFF equipment and used a handful for training pilots. Some 12 were trained during 1944, and 15 more between 1 January and 1 August 1945, but neither they nor any of the aircraft entered front-line service.

A handful of aircraft were sent to the PVO of the South-Western front, 11 equipping a training unit at Byelaya

Cervka near Kiev. However, most went to Stryi, near Lvov, ready to be handed back to the USA, though this proved uneconomic and during the winter of 1945-46 the aircraft were instead bulldozed flat by tanks.

The 78 aircraft delivered in 1944 additional to those using the Southern Lend-Lease route were probably those allocated to the Soviet Navy's 255th IAP (Istrebeitelnyi Aviatsionnyi Polk – Fighter Air Regiment) at Vayenga. This unit (which had previously flown the P-39) evaluated the Thunderbolt between 29 October and 5 November 1944, concluding that the type would be useful as a fighter-bomber.

The unit CO decided to press into service the 14 P-47s he had available to equip a single squadron, and was

surprised to be allocated 50 aircraft, allowing the whole regiment to convert to the big Republic fighter. These

aircraft were reportedly used 'for several years' in the attack role before being disposed of.

VENEZUELA

FUERZA AÉREA VENEZOLANA

Venezuela's first six P-47s arrived in 1947, with a further batch of 22 delivered in 1949. Most were referred to as 'TF-47D', indicating not a two-seater conversion, but simply a training role. Peak Thunderbolt strength in the Fuerza Aérea Venezolana (FAV), in which the aircraft equipped Grupo Aereo 9, was reached in April 1950 when 24 aircraft were in use, two aircraft having crashed and four reduced to spares. However, with the arrival of the FAV's first two jet fighters (de Havilland Vampires) in December 1949, the Thunderbolt's days in FAV service were numbered
Later reassigned to Escuadrón de

Caza 36 (which had 19 aircraft in June 1950, 10 of which were airworthy), the FAV's P-47s were finally withdrawn in 1952, though the type's last three months of service were to prove particularly eventful. Escuadrón de Caza 36 was put on full alert after a Colombian naval vessel violated Venezuelan coastal waters. Eight P-47s mounted patrols over the area and the confrontation passed off peacefully.

FAV Thunderbolt operations ended shortly afterwards, though eight aircraft were still nominally 'on the books' at the end of June 1954. Seven of these were announced as for sale in January 1956.

TURKEY

TÜRK HAVA KUVVETLERI

Although neutral during World War II, Turkey displayed an increasingly pro-Allied stance, and was able to replace most of its wartime fighters with new equipment from 1945. Thus, the survivors of 164 Hurricanes, 24 Curtiss 87A-2 (P-40D) Kittyhawk Mk Is, 42 Curtiss 81A-3 (P-40C) Tomahawk Mk IIBs and 72 Focke Wulf Fw 190A-3s were all phased out of service, though 175 new Spitfires were added to the survivors from earlier deliveries, and these remained in service into the jet age.

Turkey acquired increasingly powerful long-range fighter bombers in the shape of 26 Beaufighters, and 132 Mosquitoes and, for the first time, also received fighters from the USA. 180 P-47Ds were supplied, and these served between 1948 and 1954.

At least some of the Turkish Thunderbolts equipped 102 Fighter Bombing Squadron at Balikesir. This unit adopted the tiger in its badge in 1950, and was re-designated as 192 Squadron in 1952, re-equipping with F-84 Thunderjets the same year.

YUGOSLAVIA

JUGOSLOVENSKO RATNO VAZDUHOPLOVSTVO

Yugoslavia signed a Mutual Assistance Pact with the USA in 1951, and the first manifestation of this appeared in 1952, with the delivery of the first of 150 F-47Ds (or 126, according to some sources). These had been relegated to

training duties by 1957, when they were replaced by F-84 Thunderjets and F-86 Sabres. Some may have served on in the advanced and fighter training roles, but all are believed to have been retired by 1960.

Fiat G.91
NATO's lightweight striker

For many a minor adjunct to NATO's Cold War effort, the G.91 was, at its inception, a highly capable attack aircraft and, but for political infighting, would have sold in massive numbers. The later G.91Y (seen here) was less successful, but nevertheless continued in front-line service well into the 1990s.

Hardly pretty, although it was good-looking in a delicate way (at least in its original form), the G.91 played a vital role in NATO's Cold War defence for three decades. Affectionately known as the 'Gina', after the Italian actress Gina Lollobrigida, the G.91 is today a misunderstood and unappreciated aircraft, with few people aware of its real significance and impressive combat record.

Fiat's G.91, an Italian lightweight fighter-bomber, may not be considered by many to be in the top tier of post-World War II aircraft. And given that it was initially adopted by just two countries (plus a third, later), such a ranking seems justified.

However, the historical importance of an aircraft sometimes extends beyond numbers. The G.91 was the result of the first attempt to create a standard NATO force in Europe: it was the first aircraft developed and built under a NATO specification. It was also Italy's first successful attempt to recreate, post-war, a national aviation industry capable of bringing to fruition its own successful military aircraft. The G.91 was the first Italian combat jet to be produced in series, and the first all-Italian model to be adopted by the Aeronautica Militare Italiana (Italian Air Force). In addition, it played a major role in the rebirth of the German aviation industry, being built under licence in West Germany. From these perspec-

tives, it is possible to say that the G.91 was one of the most important projects in the foundation of a modern European aviation industry.

Unfortunately, the aircraft was also subject to the strong nationalism and the huge differences in policies of European countries. Its limited adoption by NATO-member air forces was due primarily to these factors.

Light fighter origins

In 1953 the Advisory Group for Aeronautical Research and Development (AGARD) headed by Theodore von Karman, under the Supreme Headquarters, Allied Powers in Europe (SHAPE) command, initiated a competition known as NATO Light Weight Strike Fighter (LWSF) NBMR-1. This originated in a NATO planning conference of 1952, in Lisbon, when it was decided to reorganise the air forces of the alliance. In the wake of World War II and Korean War operational experience, and subsequent studies, NATO set new criteria and stan-

Below: The very first prototype of the Fiat G.91, NC.1, is seen here framed by the aircraft that inspired much of the initial design work – the F-86K Sabre (this example belonging to 1ª Aerobrigata). NC.1 differed from later prototypes in the inclusion of a nose-mounted test probe, deeper wing pylons, a taller fin, a large single nosewheel door and in lacking an armament bay.

Left: The prototype was later fitted with underwing pylons, which were wrapped around the leading edge of the wing. The pylons were deeper than on the subsequent production aircraft and are seen here fitted with non-standard long and narrow auxiliary tanks.

Below: By September 1959 the second prototype G.91 (NC.1bis; MM.565) had been assigned to the RSV for further evaluation work and allocated the code RS-01. The NC-1bis replaced the original Bristol Orpheus B.Or.1 engine with the more powerful 4,850-lb st (21.57-kN) B.Or.3, and is seen here undergoing tests at Pratica di Mare alongside a Fiat G.82 trainer.

dards for the use of close air support (CAS) air power. A new aircraft was required that could be deployed to airstrips close to the forward edge of the battle area, so as to reduce to the minimum the time between missions.

A simple and economic aircraft was needed, which could be adopted in large numbers by the allied air forces. So, in 1953, the AGARD set the requirements for the LWSF-NBMR, and communicated them to the most important European aviation industries.

The required performance called for an empty weight of 4,850 lb (2200 kg); maximum take-off weight of 10,360 lb (4700 kg); range of 151 nm (280 km), with time over target of eight to 10 minutes; maximum speed of Mach 0.95; take-off run from grass strips or highways of no more than 3,609 ft (1100 m), over a 49-ft (15-m) obstacle. Other requirements included manoeuvrability, good instrumentation, a fixed armament of four 12.7-mm Browning machine-guns or two 20- to 30-mm cannon, external weapon load comprising bombs, rockets and various pods, armoured cockpit and fuel tanks, capability to attack tanks, troops, airfields and fuel depots, and an ability to undertake interdiction missions against ships and trains. The AGARD suggested the adoption of the Bristol Siddeley Orpheus turbojet engine.

According to operational requirements, the new aircraft also had to be easy to maintain, refuel and re-arm, without dedicated structures and in a short time.

In 1954 a dedicated commission received

proposals from 10 European industries, and on 3 June 1955 three designs were selected. Two were French – the Breguet 1001 Taon and the Dassault Etendard IV – and one Italian: Fiat's G.91. NATO ordered three prototypes for each project, which were funded mainly by the US government (under the Mutual Weapons Development Program, MWDP), with 25 per cent coming from the governments of both France and Italy.

From the beginning, the Italian project appeared to match NATO specifications most closely, and was the favoured front-runner. On 30 July 1955, NATO asked Fiat to prepare to build 27 pre-series aircraft, to be assembled after the final decision. Sure that it would be selected, Fiat acquired the rights to licence-production of

the Orpheus engine, which received the designation of Fiat 4023.02.

The father of the Italian project was Ing. Giuseppe Gabrielli, who opted for a general layout similar to that of the North American F-86K, which had been built under licence by Fiat. In this way, he was able to accelerate the design phase, and to anticipate the test and development activity.

The new aircraft was smaller than the F-86K, but maintained a low-mounted swept wing, under-nose air intake, and conventional fin surfaces. The landing gear was designed for grass strip operations, and the wing, equipped with just two pylons, was simpler, lacked slats, and had a sweep of 37°.

Above: The third prototype, NC.2, was adorned with this lightning flash cheat line during its initial period of test flights. It also featured a slightly lengthened nose, deeper cockpit canopy and a single strake beneath the rear fuselage to improve directional stability.

Left: A further prototype, NC.3/MM.567, was ordered to replace the lost first prototype. It is seen here framed by the nose of a pre-series aircraft, both having been fitted with the standard four 0.5-in (12.7-mm) machine-guns in forward fuselage weapons bays. The photograph has been retouched to show the aircraft with the shorter tail fin of production variants, although in reality it was fitted with the taller unit of the previous prototypes.

Based, like Fiat's car manufacturing business, in the city of Turin, all of the prototype, pre-series and production G.91s were constructed at Fiat's Caselle airfield facility. By early 1958 the pre-series aircraft were rolling off the final assembly line in the Caselle North workshop (right) and during the spring of that year unpainted examples were used for trials and evaluation by the RSV at Pratica di Mare (below right).

Below: Bristol Siddeley utilised a pair of G.91Rs in 1959 and 1961 for ongoing development of the Orpheus engine. The aircraft were operated from Bristol's Filton airfield with the UK Class B registrations G-45/4 (seen here) and G-45/5 respectively.

The first G.91 prototype (Numero di Costruzione 1 – NC.1, or construction number 1), in a silver finish, made its maiden flight on 9 August 1956 from the Fiat facility at Torino-Caselle airport. The test pilot was Riccardo Bignamini, and the aircraft had no military serial (Matricola Militare, or MM).

The prototype was powered by the Orpheus B.Or.801 engine, a turbojet delivering 4,049 lb (18 kN) of thrust. Initially, NC.1 was not equipped with the fixed armament, because the armament bays under the cockpit were used for housing test and registration equipment. A test probe was fitted on the nose. NC.1 was a good 20 per cent heavier than called for by the specifications.

During the 24th test flight, on 20 February 1957, NC.1 exceeded Mach 1 four times with no problem. However, only a few days later, the first accident occurred: on 26 February the prototype was lost at Cavour, near Turin, during high-speed, low-level trials. Flutter (high vibrations) in the tail section led to loss of control, and test pilot Bignamini was forced to eject.

Immediately, Fiat, with the help of NACA (the US National Advisory Committee for Aeronautics) and French engineers, started to consider necessary modifications to be introduced in the second prototype. They included a shorter fin, an enlarged stabilator, small air flux stabilisers at the fin root, and a new, small, ventral fin under the tail. In addition, the second prototype received a raised canopy (by 2.4 in/6 cm), the more powerful Orpheus B.Or.803 engine that delivered 4,848 lb (21.56 kN) of thrust, new landing gear, and four fixed machine-guns in the armament bays.

The second aircraft, designated NC.1bis (curiously, not NC.2), made its first flight on 26 July 1957. It carried the military serial of MM.565. Fiat then completed the third prototype (NC.3, MM.566) and the first pre-series aircraft (NC.4, MM.567), which were sent to the Centre d'Essais en Vol at Brétigny, France, to participate in the final selection trials of the LWSF competition. The second prototype continued its test activity in Italy.

At Brétigny, evaluations were conducted by a mixed international team of test pilots and engineers. Tests included live firing missions, operational configurations and support equipment assessments.

Finally, in January 1958 NATO announced that the winner of the contest was the Italian G.91, clearing Fiat to start production of the pre-series batch.

Italian service

At the end of NATO trials at Brétigny, NC.3 and 4 were assigned to the Reparto Sperimentale Volo (RSV, or Test Flying Wing) of the AMI, at the Pratica di Mare base, to continue operational evaluations and to prepare the operational manuals for the flying units. The first front-line squadron selected to re-equip with the new aircraft was 103° Gruppo Caccia Bombardieri (CB, or fighter-bomber) of 5ª Aerobrigata (5th Wing) based at Rimini-Miramare, which at the time flew F-84Fs. 103° Gruppo was deployed to Pratica di Mare in August 1958 to carry out the transition with the assistance of the Test Wing. The G.91s assigned were from the pre-series batch, easily identifiable by their pointed noses.

With the positive conclusion of the operational trial series, the 103° Gruppo returned to its final destination, the base at Treviso-San Angelo, where it was assigned on 11 April 1959. During this period, NATO kept a close eye on the capability shown by the G.91, and official observers remained positively impressed by its performance and reliability.

In February 1960 it was the turn of a second Italian squadron to re-equip with the G.91: 14° Gruppo of 2ª Aerobrigata, then operating the Canadair CL-13 (F-86E Sabre). On 1 May 1960,

Grass field operational trials – conquering the mud

Once re-equipment was complete, 103° Gruppo was deployed to the airfields at Frosinone, Guidonia and Furbara (in the Lazio region), to train on unpaved airstrips and on bases not equipped for jet operations. The aircraft's excellent bare-field performance was exemplified by operations after torrential rain at Frosinone. The G.91s sunk hub-deep in the mud and the refuelling trucks were bogged down, however, the aircraft managed to not only taxi themselves to the refuellers but also take-off and land on the sodden grass strip.

In April 1959 operational evaluations continued at the Rivolto air base, near Udine, where the Gruppo could take advantage of the firing range at Maniago. A total of five aircraft, eight pilots and 15 groundcrew lived and operated under field conditions on the grassy margins of the base, in what was the last major evaluation before the G.91's acceptance by NATO. Sorties were flown under operational conditions from dawn to dusk with each aircraft making up to 10 sorties per day, with turnaround times often less than 30 minutes (including re-arming, refuelling and replenishing nitrogen, oxygen and hydraulic oil supplies).

Finally, between 26 June and 3 July 1959, the 103° Gruppo moved to the airfield at Campoformido to carry out Exercises Guizzo Rosso and Roman Tree, designed to test the capability of the unit in real wartime operations from grass strips. In a final demonstration on 3 July, the G.91s gave a target practice demonstration on the Maniago artillery range, deploying 500-lb (227-kg) bombs and aerial rockets.

A pre-series G.91, belonging to 103° Gruppo, prepares to take-off from Rivolto air base during the grass field operational trials in June 1959. The armament panels and associated guns and ammunition were interchangeable between aircraft. This example has borrowed a panel from an unpainted aircraft.

Affixed with NATO-badges on their fins, 103° Gruppo, 5ª Aerobrigata pre-series G.91s complete a maximum-rate turnaround at Friuli in front of high-ranking NATO officials in July 1959. Once refuelled, the weapons packs in either side of the fuselage were removed and new ones installed within four minutes.

103° Gruppo was redesignated as a Gruppo Caccia Tattici Leggeri (CTL, or Light Tactical Fighter Squadron).

On 16 March 1961, 14° Gruppo re-equipped with the G.91R/1 model, which differed from the pre-series aircraft in having a new nose that housed three Vinten cameras, giving a reconnaissance capability. The AMI received a total of 22 G.91R/1s, plus four pre-series samples (MM.6251, 6253, 6257 and 6259) that were upgraded to R/1 standards. On 1 October 1962, 2ª Aerobrigata was disbanded and the two squadrons were provisionally assigned to 51ª Aerobrigata as the Reparto Volo G.91 (G.91 Flying Group).

The AMI completed the re-equipment of its front-line units with two other batches: 25 G.91R/1As and 50 G.91R/1Bs. The G.91R/1A featured more sophisticated instrumentation (similar to that selected by the Luftwaffe), including a Position Homing Indicator (PHI), IFF and a UHF radio. The G.91R/1B sported technical modifications, having a stronger airframe, tyres without inner tubes, shell case collectors, a more powerful braking system and additional instrument modifications.

Two-seat trainer

In addition, the single-seat G.91R, Fiat studied a two-seat model for use in the training role. The G.91T was quite similar to the G.91R/1B, distinguished by a second, slightly stepped cockpit in tandem, with a fuselage that was 55 in (140 cm) longer than the single-seater. The T also had a taller fin, and its fixed armament was reduced to two Colt-Browning 12.7-mm machine-guns in order to limit the weight increase. Maximum take-off weight was 13,337 lb (6050 kg), compared to the 12,125 lb (5500 kg) of the G.91R/1.

The first aircraft (NC.001, MM.6288) was a pre-series machine, and made its maiden flight from Torino-Caselle on 31 May 1960, piloted by Simeone Marsan. The first production batch was for the Luftwaffe, although the AMI ordered 46 aircraft (designated G.91T/1, first serial MM.6315), later increased to a final total of 101. The main Italian operator of the G.91T was to be the Scuola Volo Basico Avanzato Aviogetti (SVBAA), based at Amendola.

On 1 October 1964, as part of a general AMI reorganisation, 14° and 103° Gruppi at Treviso air base were assigned to the reactivated 2° Stormo Caccia Tattici Ricognitori Leggeri (CTRL, or 2nd Wing Light Tactical Reconnaissance Fighter). In March 1965 the AMI completed re-equipment with the new aircraft when 13° Gruppo, then flying the last CL-13 Sabre, relocated from Cameri to Treviso, to receive the G.91R/1B model. During that summer the Gruppo moved south to Brindisi, where on 1 September it became the flying unit of the reactivated 32° Stormo.

So, at the end of 1965, the G.91 fleet comprised two squadrons (14° and 103°) dedicated to close air support and tactical reconnaissance missions, plus one squadron (13°) that specialised in tactical support to maritime operations. This organisation remained stable for

Above: Having demonstrated its potential in the light attack role, it soon became apparent that the G.91 would also be useful in the tactical reconnaissance role. Four of the original pre-series aircraft (including NC.19/MM.6253; above) were redesigned with a camera nose to serve as G.91R prototypes in this bare metal finish. Having completed the trials work, three of the four (including NC.19) were converted to G.91PAN standard for the 'Frecce Tricolori'.

Below: At the height of the Cold War, the G.91 formed the backbone of light attack/tactical reconnaissance units in both Germany and Italy. The two 2° Stormo gruppi were distinguished by colour-coded air intake lips; yellow for 103° and white for 14°, as demonstrated by this formation of four G.91R/1Bs over the Alps.

Above: Pre-series 'pointed nose' G.91s line-up alongside early production G.91R/1s at Treviso-San Angelo in the early 1960s. The aircraft wore the codes of the re-activated 2° Stormo and the aircraft's individual constructor's number on the forward fuselage. The aircraft being loaded with HVAR rockets in the foreground was later converted to G.91PAN standard for service with the 'Frecce Tricolori' before being written-off in an accident in March 1967.

Right: The longest serving and most famous Italian operator of the original version of the G.91 was 2° Stormo. Based at Treviso-San Angelo, the constituent front-line 14° and 103° Gruppi did much to establish operational tactics and procedures, and regularly partook in NATO exercises and deployments. Although the aircraft's rough-field performance was rarely utilised, the G.91's ability to operate from dispersed locations with little support equipment was impressive. Here a 103° Gruppo 'R' is readied in its camouflaged hide during a NATO TacEval (Tactical Evaluation) in the mid-1970s.

Below: In 1987 one of 14° Gruppo's ex-'Frecce Tricolori' G.91PANs (MM6265) was painted in a bright red scheme adorned with the Stormo's black knight's head (below) to celebrate the unit's 200,000th flying hour and the G.91's 30th anniversary. Another 14° Gruppo aircraft was specially painted to mark the final flight of a G.91R in AMI service in April 1992, flown by Lt Col Luciano Monesi.

eight years, until 1 August 1973, when 13° Gruppo began to re-equip with the new G.91Y. The G.91R/1Bs that had been operated by 13° Gruppo were redistributed between 2° Stormo at Treviso and 313° Gruppo at Rivolto, to serve as interim attrition replacements.

In the same period, the AMI was trying to establish the requirements for the successor to the G.91. These specifications were published in 1977 under a project designated AM-X, a light fighter-bomber intended to replace all the G.91 types and the F-104Gs in the AMI fleet.

In 1989 the Italian air force began introducing into service the new AM-X, now usually referred to as AMX. The first squadron to re-equip was

103° Gruppo, which in January was moved from Treviso to nearby Istrana air base, home of the 51° Stormo. On 7 November 1989, the 103° Gruppo received its first AMX, completing the transition to the new aircraft in about one year.

In the 1980s, it became clear that the G.91R's operational capabilities were no longer suited to the modern battlefield. Lacking avionics, passive defences (RWR, chaff and flares), active defences (ECM), and modern navigation systems, and with a quite limited armament capability, the G.91R was an outdated combat aircraft. Nevertheless, it was still a good advanced trainer, capable of training competent fighter-bomber pilots thanks to its undemanding

313° Gruppo 'Frecce Tricolori'

In December 1963 a number of pre-series G.91s were retired from 103° Gruppo service and assigned to 313° Gruppo Addestramento Acrobatico (Aerobatic Training), also known as the 'Frecce Tricolori'. These early G.91s were modified to make them more suitable for aerobatic display, featuring upgraded engines, a smoke generator system, small tanks at the wing station to carry the special coloured oils, armament panels replaced by dedicated counterweights, a different stick control sensibility, a pitch damper, and a new blue colour scheme – receiving the designation G.91PAN (for Pattuglia Acrobatica Nazionale, or National Aerobatic Team). However, in consideration of 313° Gruppo's secondary role, the G.91PANs could be re-armed and used to maintain combat readiness in the fighter-bomber role.

After 15 years of excellent service the 'Frecce Tricolori' began the search for a replacement. This duly arrived on 6 January 1982 in the shape of the Aermacchi MB.339. On 27 April, during an official ceremony, the remaining G.91PANs were retired and reassigned to 2° Stormo.

flight characteristics, agility and speed. By 1988, the 2° Stormo was no longer being selected for the annual NATO Tactical Evaluations.

The last squadron to fly the type, 14° Gruppo of 2° Stormo, continued its activity on the G.91R until 1991, when the first AMX was assigned. In July 1991 the Gruppo moved to the base at Istrana (home of the SAS-AMX, the provisional OCU for the type), and received its first new jets. The last flight of an Italian G.91R occurred at Treviso on 7 April 1992, when G.91R/1B MM.6413 (coded 2-15), in a special red-white-green livery, landed for the last time.

Only the two-seaters then remained in service. On 3 July 1986, the SVBAA at Amendola had been redesignated as 60ª Brigata Aerea (Air Brigade), and therefore changed its prefix code from 'SA-' to '60-'. Under a minor reorganisation, on 1 July 1993 60ª Brigata Aerea was disbanded and its two squadrons, 201° and 204° Gruppi, passed to 32° Stormo. On 2 July that unit relocated from Brindisi to the Amendola air base. 32° Stormo – which operated three squadrons (13°, 201° and 204°) flying the pooled G.91T/1 fleet – was destined to re-equip with the AMX. The first examples of the new aircraft arrived on 11 November 1994, by which time only a few G.91T/1s were still operational.

The very last Italian G.91 flight came on 30 September 1995, when G.91T/1 MM.6363 (NC.93, coded 32-63), wearing a special colour scheme, made its final landing at Amendola. This brought to an end the 37-year Italian career of Fiat's fighter-bomber.

Projects cancelled

As winner of the NATO competition, the G.91 was expected to be selected by several air forces. However, it was opposed or ignored by many, overwhelmingly for political reasons. France, for instance – the big loser of the competition – initially placed an order for 48 G.91R/2s (and for a number of two-seat G.91T/2s), but the contract was soon cancelled. Norway also showed an interest in the aircraft,

The first of two G.91T prototypes made its maiden flight on 31 May 1960. Based on the G.91R/1, it conferred the same excellent vice-free handling, economy and maintainability as single-seat versions. Such was the ease with which pilots converted to the 'Gina' that the G.91T was actually more useful as an advanced trainer in its own right than as a conversion trainer.

and Fiat considered a dedicated version, the G.91R/5, which differed in having enlarged wing tanks that were designed to increase range to 810 nm (1500 km). This version never went beyond paper studies. Another study was for an upgraded version for the AMI, known as the G.91R/6. It would have featured stronger landing gear, enlarged wheels and airbrakes, four under-wing pylons, and a Doppler navigation system.

One more version was the G.91S, with extensive modifications to allow supersonic speed: it featured a 38° swept wing and the Orpheus B.Or.12, a turbojet delivering 7,053 lb (31.37 kN) with afterburner. None of these projects was developed.

Two other proposals progressed to prototype form. The G.91A was intended to improve short take-off and landing (STOL) capability, thanks to an enlarged wing (2 m²/21.5 sq ft bigger) and the introduction of slats on the leading edge; it also featured new internal fuel tanks to extend range by 10 per cent. Aircraft NC.31 was modified to test this model. The G.91N, on the other hand, was converted from a pre-series aircraft, and included a modified navigation system that comprised a Decca and a rho-theta system.

Fiat considered other variations, based on the G.91T airframe. The G.91BS/1 and BS/2 were, respectively, single- and two-seat models designed for air battle surveillance. From the two-seater, Fiat also proposed the G.91T/4, a lead-in fighter trainer for the F-104, equipped with a modified nose housing the NASARR radar; and the G.91TS, a dedicated trainer for the supersonic G.91S. None of these proposals made it to even prototype stage.

Luftwaffe operations

The only NATO country to adopt the NATO LWSF was West Germany, which showed interest in the project as early as 1958. This interest was largely prompted by the possibility of undertaking licence-production of the aircraft, as the German aviation industry tried to recover its expertise, following World War II.

The Luftwaffe selected a version of the G.91 that differed from the basic model adopted by Italy. Emerging as the G.91R/3, it featured two 30-mm DEFA guns, four underwing pylons, a stronger airframe, and more complete instru-

mentation, including Bendix Doppler and PHI systems. The first 50 single-seaters and the first 44 two-seaters (G.91T/3) were built by Fiat. The licence-building team was headed by Dornier, as prime contractor, with Messerschmitt, Heinkel, and others. It was initially planned to produce 100 aircraft in West Germany, a figure that rose to 232 in January 1960, and finally to 295.

In 1960, the first five G.91R/3s (coded from YA+011 to YA+015) were introduced into service with the Luftwaffe test unit, Erprobungsstelle 61 (EstBw 61) at Manching, for a series of trials. Later, the team was reinforced by the arrival of two G.91T/3s, coded YA+022 and YA+023.

The first 35 two-seaters were delivered to 1. Staffel of Waffenschule 50 (1./WSLw 50, or the 1st Squadron/Weapons School 50), which acted as OCU for the type from Erding air base.

The first operational unit to receive the new G.91R/3 was the newly activated Aufklärungsgeschwader 53 (AKG 53, or the 53rd Reconnaissance Wing). This unit, with two squadrons, started its conversion course at Erding and then moved to the base at Leipheim.

Italian training – SVBAA

The majority of AMI G.91Ts were assigned, from 15 November 1964, to the Scuola Volo Basico Avanzato Aviogetti (SVBAA; Basic Advanced Flight Jet School) at Amendola. The resident 201°, 204° and 205° Gruppi immediately began converting to the newly arrived G.91T/1s, relinquishing their Lockheed T-33As as more airframes became available. Conversion to the new type was completed in August 1965.

Finished with bright orange tail and nose sections, the G.91T/1s wore large 'SA-' codes beneath the front cockpit and the unit's badge on the tail. The G.91T had independent ejection seats, allowing the instructor to continue to fly the aircraft in the event of the student ejecting. The aircraft were later upgraded with Martin-Baker Mk 6 zero-zero ejection seats. Following the disbandment of 205° Gruppo in 1975, the remaining two SVBAA Gruppi continued in the training role until the unit was redesignated as 60ª Brigata in the mid-1980s.

Export prospects and foreign trials

Surprisingly, during the G.91's development phase, the United States' armed forces showed an interest in the aircraft. At that time, the US Army was trying to establish its own fixed-wing component so that it would not be dependent on the USAF for battlefield close air support, forward air control, and tactical reconnaissance missions. The USAF and the US Marines also decided to evaluate the Italian aircraft. Accordingly, in 1960-1961 four aircraft from Italy and Germany were shipped to the USA. The US Army received two G.91R/1s (NC.52, MM.6286, US Army 0042, and NC.53, MM.6287), one G.91R/3 (NC.65, EC+105, US Army 0065), and one G.91T (NC.2, US Army 0002). They were evaluated at the US Army base at Fort Rucker, Alabama, and later at Kirtland AFB, New Mexico. NC.53 was also evaluated in the McKinley Climatic Laboratory at Eglin AFB, Florida. On 1 February 1961, during a test with JATO-assisted take-off from Fort Rucker, NC.52 crashed, killing Fiat test pilot Riccardo Bignamini. This accident, coupled with the US Army's decision to abandon plans for a fixed-wing component due to USAF pressure, led to the end of the US evaluation programme. The remaining aircraft returned to Europe some months later.

G.91 exports appeared to gain momentum when, under orders from the air forces of Greece and Turkey, 50 G.91R/4s were produced by Fiat. However, these aircraft never served with the two countries. Some six aircraft were painted in Greek Air Force colours, and one (NC.109, serial 10109) was sent to Larissa on 6 September 1961 as the Greek Air Force was forming the G.91 Sminos (Flight) within the 110th Pterix (Wing). However, Greece decided to adopt the Northrop F-5A Freedom Fighter, as did Turkey. The 50 G.91R/4s eventually entered service with West Germany.

Austria, a non-NATO country, submitted an order for 14 aircraft (12 single-seaters and two twin-seaters), but cancelled it soon after. Even the Imperial Iranian Air Force showed some interest, but nothing came of it.

Above left: It is reported that six G.91Rs were painted in Greek colours, however, 10109 was the only example to reach Greece, conducting trials at Larissa before the US-funded deal collapsed. The aircraft was later operated by the Luftwaffe (as BD+235) and the Portuguese air force (as 5401).

Above and below: Competing against the Douglas A-4 Skyhawk and Northrop F-5 for a US Army order that never materialised, the G.91's evaluation period in the USA was particularly unhappy. Of the three aircraft originally loaned, the G.91R/1 (above) crashed fatally in February 1961, followed by the loss of the G.91R/3 (below) in another accident in July of the same year.

On 20 July 1961 at Oberpfaffenhofen, the maiden flight was made of the first G.91R/3 produced by Dornier, serialled WN.301 (Werkenummer, or construction number 301). Shortly after, a second front-line wing (AKG 54) was established at Oldenburg. It was at this time that Greece and Turkey decided to abandon the G.91 in favour of the F-5A, and the 50 G.91R/4s already produced were assigned to WSLw 50 for training activity. They were coded BD+231 to BD+255, and BD+361 to BD+385.

Further front-line units to form in the early 1960s included Jagdbombergeschwader 35 (JBG 35) based at Husum, (later redesignated JBG 41), JBG 42 at Pferdsfeld and JBG 43 at Oldenburg.

Tests and trials

In May 1962 the Luftwaffe arranged a series of operational tests from unpaved strips at Bad Tölz. In September 1963 a similar exercise was held in Italy, when six aircraft from AKG 53 were deployed to Rivolto for one week. Two G.91R/3s were sent to the base at Colomb-Bechar, in the Algerian Sahara, for two-month trials that included Nord SS.11 missile tests.

Under the evaluations and tests designed to exploit to the utmost the CAS role, the Luftwaffe undertook the Short Airfield for Tactical Support (SATS) programme, which examined the use of very short runways. Accordingly, some G.91R/3s were modified with hooks (two under the wing leading edge, and two near the nose landing gear) to enable launch with a catapult, plus an arrester hook under the tail. The concept was eventually abandoned, mainly because of high costs.

Other tests included flying activity that used normal motorways as runways. In order to develop non-conventional flight operations, a special training unit was established in 1965, the Lehr und Versuchsschwarm G.91 (LVS G.91, part of WSLw 50), which received six aircraft, coded from XB+101 to XB+106. One aircraft was later modified by Dornier into a drone. In 1964 WSLw 50 moved to its long term home – the air base at Fürstenfeldbruck.

The last G.91R/3 produced by Dornier (WN.594) was delivered on 26 May 1966. In the same year, the G.91R/4s assigned to WSLw 50 were retired from service; some 40 were handed to the Portuguese Air Force, and the rest were relegated to ground training, static display or recruiting promotion duties, and were recoded in the BR+xxx range.

In 1967 the Luftwaffe reorganised its units. The front-line wings equipped with the 'Gina' were redesignated Leichtenkampfgeschwadern (LKG, or Light Attack Wings), each having two Staffel: one dedicated to the attack role, the other tasked with reconnaissance. Staffel were equipped with some 20 single-seaters, plus a few two-seaters attached to the wing. The 1967 order of battle of the G.91 fleet was as follows: LKG 41 (formerly JBG 41) based at Husum; LKG 42 (formerly JBG 42) at Pferdsfeld; LKG 43 (formerly AKG 54 and JBG 43) at Oldenburg; and LKG 44 (formerly AKG 53) at Leipheim.

Above: Having formed at Manching in 1956, the Luftwaffe's test and trials unit, Erprobungsstelle 61 (ESt 61), was naturally the first unit to German unit to receive the G.91. Two of the initial 44 Fiat-built two-seat G.91T/3s were allocated to the unit, of which YA+023 was fitted with this nose-mounted instrumentation boom.

Left: Leipheim-based Aufklärungsgeschwader 53 was the first front-line Luftwaffe unit to equip with the G.91. The first digit of the three numeral code suffix referred to the squadron – this example is therefore a 1. Staffel aircraft.

Above: The first five G.91R/3s received by the Luftwaffe from Fiat's Turin factory went to ESt 61 (YA+011 to YA+015) at Manching. The aircraft were not only used for evaluation and testing, but also flew clearance trials for all ordnance to be carried. YA+012 (serial 3002) is seen here at Dijon in 1965 during testing of the Nord SS.11 missile.

Below: The sole operator of the G.91R/4 in Luftwaffe service was Waffenschule 50. This variant combined the four nose-mounted Colt-Browning M3 0.5-in (12.7-mm) machine-gun armament of the G.91R/1 with the four-pylon wing of the G.91R/3. The aircraft were originally intended to fulfil Greek and Turkish orders, and were the first to be disposed of by the Luftwaffe when they were transferred to Portugal in 1965-66.

Schiessplatzstaffel G.91R/3 target tugs

Lufthansa's cargo and charter arm, Condor Flugdienst, had operated F-86 Sabres and OV-10B Broncos in the target-towing role (with the Schiessplatzstaffel at Westmorland/Sylt) when it took on the first of 22 G.91R/3s (see table) and two T/3s. Initially operated with civil D-9*** series registrations, the aircraft were subsequently given military serials in the 99+** series.

They were given high-conspicuity Dayglo markings on the nose, wingtips, rudder and fuel tank(s), the same markings as had been used by G.91R/3s assigned to Waffenschule 50. The aircraft were equipped with a Del Mar SK-460 target-towing system, with a turbine-driven winch on the port inboard station and a target launcher outboard. This towed a bomb-shaped FK-460 target (equipped with IR or tracking flares) from a fitting above and behind the trailing edge. When carried, the target-towing gear was balanced by a fuel tank to starboard. A target-towing control panel was fitted in place of the gunsight, and the aircraft was given military and civil radios (AM 21.5-75 MHz and FM 118-135 MHz). The latter required the addition of new blade antennas on the spine. Once in service, the aircraft had their cannon barrels removed, and the ports were faired over.

C/n	Reg.	Serial
0059		99+42
0060	D-9597	99+01
0068	D-9598	99+02
328	D-9599	99+03
343	D-9600	99+04
378	D-9601	99+05
397		99+37
437		99+38
453		99+43
459	D-9602	99+06
460	D-9603	99+07
467	D-9604	99+08
468	D-9605	99+09
482	D-9606	99+10
484		99+48
486		99+44
499		99+45
511		99+46
513		99+47
515		99+39
518	D-9607	99+11
554	D-9608	99+12

On 1 January 1968, the Luftwaffe adopted a new serialling system, which abandoned the letters and grouped aircraft according to their types and functions. The G.91R/3s were given serials between 3001 and 3323, while the G.91T/3s ranged from 3401 to 3440. In 1971, the Luftwaffe placed a new order for 22 more G.91T/3s, and Dornier reopened its production line. The aircraft (WN.601 to 622, coded from 3441 to 3462) were assigned to various units; 10 went to WSLw 50, and were used to train the new F-4F Phantom II weapons system officers. Aircraft production closed in 1973.

A lesser-known German unit operating the G.91 was the Schiessplatzstaffel, based at Westerland/Sylt and fulfilling the target-towing role. This unit, also equipped with F-86 Sabres and OV-10 Broncos, had a detachment at the NATO base at Decimomannu in Sardinia. From 1985, these aircraft were operated by Condor Flug, a contractor of the West German Ministry of Defence.

In the mid-1970s the Luftwaffe started to retire its first 'Ginas'. On 1 April 1975, LKG 42 at Pferdsfeld received its first F-4Fs and was redesignated JBG 35. Three months later, on 1 July 1975, LKG 44 at Leipheim turned in its G.91, following the decision to deactivate the wing.

The successor to the G.91 in Luftwaffe service was the Franco-German Alpha Jet, a light twin-engined trainer designed by a consortium formed by Dassault-Breguet and Dornier and first flying on 26 October 1973. The version selected by Germany was the Alpha Jet A, a dedicated air-to-ground model. Initial Luftwaffe evaluations were carried out by ESt 61 at Manching, and in 1979 the first aircraft were

Left: In 1968 the letter codes were abandoned in favour of the aircraft's serial appearing either side of the fuselage cross. LKG 41's aircraft inherited the eagle's head badge from its previous identity – JBG 41.

Below left: WS 50 was the main operator of the two-seat G.91T/3. During their service life the T/3s received a number of modifications, including the addition of a second landing light on the nosewheel door and a more bulged rear cockpit canopy following the installation of Martin-Baker Mk 6 zero-zero ejection seats.

Below: A quartet of Oldenburg-based LKG 43 G.91R/3s performs an immaculate flypast at RAF Upper Heyford in 1971. The high-visibility Dayglo drop tanks would have been repainted grey/green for combat missions.

To mark the retirement of the G.91 from Portuguese service, G.91R/3 5445 was prepared in this special colour scheme, including the badges of all five units that operated the type. The aircraft flew for the last time on 17 June 1993.

Above: For a significant proportion of the time Portuguese G.91R/4s were committed to combat operations in Africa the aircraft wore this very pale grey scheme, as illustrated by 5433 belonging to Esquadra 502 'Jaguares'.

Top: To supplement and eventually replace the G.91R/4s in FAP service, a total of 70 cannon-armed ex-Luftwaffe G.91R/3s was delivered between 1976 and 1982. Of these only 34 saw operational service, the remainder being used for spares recovery and ground instruction. This example is seen off Portugal's Atlantic coast in May 1979.

delivered to the Lehr und Versuchsgeschwader Alpha Jet at Fürstenfeldbruck, for operational trials. This unit became the OCU for the type, and in 1980 was designated JBG 49, following the deactivation of WSLw 50.

In January 1981 LKG 43 started to receive the new Alpha Jets, changing its designation to JBG 43. In 1982 it was the turn of LKG 41, which became JBG 41. The last flight of a Luftwaffe G.91 occurred at Fürstenfeldbruck, and was carried out by G.91T/3 serial 3402 of JBG 49, in March 1982.

Condor Flug continued to operate its G.91s well into the 1990s, and the last flight of a German 'Gina' was registered on 28 January 1993, performed by G.91T/3 serial 9940.

Portuguese 'Ginas'

By 1961 the Força Aérea Portuguesa (FAP, or Portuguese Air Force) had shown an interest in the Fiat G.91. At that time, the country was facing a crisis in its African colonies of Angola and Mozambique, which highlighted FAP need for a new, light fighter-bomber. However, Portugal was subject to United Nations sanctions and any purchase of armaments was prohibited. In 1965, when West Germany retired its G.91R/4s, the two countries formed an agreement: the Luftwaffe was allowed to build and use a new base at Beja, including a firing range, and in exchange the FAP received 40 G.91R/4s.

The first aircraft arrived in Portugal on 4 December 1965 and were assigned to Esquadra 51 'Falcoes' at Monte Real, a unit previously flying the F-86F Sabre. The aircraft received FAP serials between 5401 and 5440. In April 1966, eight G.91s were sent to Bissalanca base in Portuguese Guinea (now Guinea-Bissau) to form the first African squadron, Esq. 121

'Tigres'. In 1968-1969 two other squadrons were formed: Esq. 502 'Jaguares' based at Nacala, and Esq. 702 'Escorpioes' at Tete, both in Mozambique. During combat operations in the northern part of that country, the G.91 operated from six secondary airfields: Port Amelia, Mueda, Nampula, Nova Freixa, Vila Cabral and Beira.

In 1974 the April Revolution swept Portugal, and following the change of government the armed forces withdrew from Mozambique. Esq. 702 was merged with Esq. 502, and was assigned to the base at Luanda, in Angola, replacing the F-84Fs of Esq. 93 'Magnificos' and flying reconnaissance missions for some months, until the final withdrawal of Portugal from its colonies.

In August 1974 the remaining G.91R/4s were collected by Esq. 62 at Montijo, and the squadron was placed under AFSOUTH command to perform CAS and tactical reconnaissance for the southern NATO flank.

In 1975 the Luftwaffe disbanded two wings, and more 'Ginas' became available for the FAP. This second batch arrived in Portugal in 1976, comprising 20 G.91R/3s (which received serials from 5441 to 5460) and six G.91T/3s (serials 1801 to 1806). They were assigned to Esq. 62, which under a reorganisation in 1978 became Esq. 301.

In August 1980 the FAP established a G.91R/4 detachment at Lajes, in the Azores Islands. The number of aircraft was increased to 25, and on 31 January 1981 the detachment was upgraded to squadron status, becoming Esq. 303 'Tigres'.

More 'Ginas' came to Portugal after the final retirement of the type from Luftwaffe service.

At the beginning of the 1990s, the G.91's career was also coming to its end in Portugal. In January 1990 the FAP disbanded Esq. 303, marking the end of G.91R/4 operations. The

The 'Tigers' of Esquadra 301

Although their official unit name was 'Jaguares', Esq. 301 were regular participants in NATO Tiger Meets between 1977-93. The unit won the Silver Tiger trophy at the 20th Tiger Meet at Cameri in 1980, and repeated this success at Kleine Brogel in 1985. The aircraft received a variety of 'Tiger' schemes, including modestly-painted G.91T/3 1810 (top) and two more lavishly painted G.91R/3s, which competed in the 1991 (middle) and 1987 (bottom) competitions.

Above: Gun armament on the G.91 could be tailored to meet customer needs. Italy's G.91Rs and R/1s were fitted with four nose-mounted Colt-Browning M3 0.5-in (12.7-mm) machine-guns, whereas the Luftwaffe opted for a pair of 30-mm DEFA 552 cannon (each with 125 rounds per gun), mounted either side of the chin air intake (illustrated). This cannon-armament arrangement was subsequently chosen by the AMI for the G.91Y.

Above: AMI G.91Rs were handicapped by having only two fixed wing weapons pylons compared with the G.91R/3's four, although this was later rectified when removeable outer wing pylons were introduced. Most NATO-standard rockets (2.75-in, 3-in or 5-in) could be carried along with rocket pods (seen here), 500-lb (227-kg) bombs or napalm.

G.91 weapon options

The G.91 was designed as a lightweight strike aircraft, blessed with quick turnaround times and the ability to operate close to the front-line. The aircraft therefore was never endowed with the same load-carrying capabilities as the majority of its contemporaries. The Italian G.91R/1 was routinely fitted with only two wing pylons, each capable of carrying two 250-lb (113-kg) or 500-lb (227-kg) HE bombs, napalm tanks, a range of rocket projectiles, honeycomb packs of 31 folding-fin rockets. In addition, the aircraft was cleared to fire Nord 5103 air-to-air missiles, although these were not carried operationally. When auxiliary fuel tanks were carried a pair of additional pylons could be added to the outer wing with a usual armament of rocket packs. In later years, AMI G.91Rs also carried NATO-standard cluster bombs.

Luftwaffe G.91R/3s replaced the four Colt-Browning 0.5-in (12.7-mm) machine-guns with two 30-mm DEFA cannon. Other standard weapons carried by front-line Luftwaffe units included 2.75-in or 70-mm rockets in various pods (including LAU-3As and LAU-51As), MATRA SAMP Type 25ED 250-kg bombs, M61 A1 500-lb bombs, M116 napalm tanks or BL755 CBUs. When transferred to Portugal the R/3s were modified to carry Snakeye retarded bombs and Rockeye CBUs as well as Sidewinder air-to-air missiles on the outboard pylons.

The G.91Y incorporated the same DEFA cannon as Luftwaffe G.91R/s and could carry up to 4,000 lb (1814 kg) of assorted weapons on its four wing pylons. 500- and 1,000-lb bombs were regularly carried, as were most NATO-standard rocket pods (see above). In addition, the G.91Y was wired to carry B57 nuclear weapons.

Various air-to-surface missiles were cleared for carriage on the G.91R. These included Nord SS.11, AS.20 and AS.30L missiles mounted on the underwing pylons. Here the second G.91R/3 built for the Luftwaffe, and assigned to the test wing ESt 61, is seen during initial evaluation of the AS.20 during 1967. Extensive trials of both the AS.20 and AS.30 in that year included the launch of some 65 missiles.

Along with NATO-standard bombs and rockets, the Luftwaffe trialled a number of unique weapons programmes. This ESt 61 example carries an unusual store beneath its port wing, which must have added considerable drag during flight.

Above: The first prototype G.91Y (NC.2001) streams its brake parachute on landing at Fiat's Turin-Caselle headquarters in 1967. Wearing overall yellow primer with large photo-calibration markings, the aircraft had already received a number of modifications by this time, including the two small intakes for the afterburner on the upper rear fuselage replacing the original single fin root-mounted intake.

Below: Belonging to the Reparto Sperimentale di Volo (RSV) and wearing the unit's silver angel on a black shield badge, G.91Y RS-11 poses for the camera over the Alps during initial trials work. The RSV did much of the clearance work associated with the carriage of US-owned B57 nuclear bombs, which were held on a 'dual key' basis.

Above: Seen with the original fin root afterburner intake, NC.2001 (MM.579) poses for the camera on an early test flight in the winter of 1966-67. The brake parachute is contained in the fairing above the twin jetpipes.

very last flight of a Portuguese G.91 occurred on 17 June 1993, when a special 16-aircraft formation from Esq. 301 overflew the base at Montijo, led by G.91R/3 serial 5445, wearing an all-over silver scheme.

From 'Gina' to 'Yankee'

The use of a light fighter-bomber for close air support was still considered appropriate in the 1960s, but following the experiences of wars in the Middle East and Vietnam, and with NATO operational training activity, it became clear that the G.91R was no longer suited to this role. A need was identified for an aircraft capable of increased range, warload and instrumentation. Fiat decided to develop a new combat aircraft, using the G.91 as a base. From the very beginning, close links were established with the Italian Air Force, and this co-operation between industry and armed forces represented a first in Italy. The Air Force wanted a light fighter-bomber optimised for low-level flight, but also able to carry out medium-range interdiction missions. The new aircraft had to have the same good characteristics of the G.91R in terms of speed, agility, field maintenance and unpaved surface flight operations, but with

A quartet of G.91Ys belonging to 13° Gruppo and wearing the diving eagle badge of 32° Stormo formates over the Puglia region of southern Italy. The unit, famous for its shark mouth intake artwork, operated the 'Y' from 1975 until 1992 and was mainly tasked in the reconnaissance and TASMO roles.

clear improvements in range, armament, navigation systems and survivability.

The new project adopted the general layout of the G.91T design, but the aircraft was very different to either the 'R' or the 'T' models. The decision to maintain the G.91 designation (which was intended to mark continuity with the first successful project) was later recognised as counter-productive, because the new project was considered by many to be just a new version of the original aircraft.

The main feature of the new programme, designated G.91Y, was its engine configuration: it now called for two turbojets with afterburner, fed by a single Y-shaped air intake (this explained the 'Y' designation of the model). The twin-engined solution, suggested by war experiences, would bestow higher survivability

on the battlefield and greater flight safety. The G.91Y featured an increased empty weight (+24 per cent), payload (+73 per cent) and onboard fuel (+100 per cent), improvements made possible by the presence of two engines (delivering 63 per cent more take-off thrust) coupled with an enlarged wing surface. The selected engine was the General Electric J85-GE-13A, delivering 4,077 lb (18.14 kN) with afterburner.

The wing, initially tested on a G.91T, had a 37.4° sweep, a dihedral of 1.3° and a chord/thickness ratio of 10. It was nearly 21.5 sq ft (2 m²) larger, and equipped with automatic slats. The new wing gave 30 per cent more lift than the G.91R wing, offering better control at lower speeds and during take-off and landing. The Fowler flaps at the internal trailing edge had a maximum deflection of 40°. All the tail surfaces were enlarged and modified, and two ventral fins were added, as was a ventral hook, to be used in case of emergency landing. The landing gear complex was similar to the previous aircraft, but was strengthened and

Rocket-assisted take-off

In an attempt to improve the G.91's already impressive short-field performance, a series of trials with belly-mounted auxiliary rockets were conducted on a pre-series G.91, a G.91R and later on the second-prototype G.91Y. RATO allowed operations from semi-prepared strips as short as 2,000 ft (610 m), even with a full warload.

The deviated nozzle Aerojet-General 14-DS-1000-MB rockets could be attached to the aircraft in groups of two or four on a ventrally-mounted steel plate at the aircraft's centre of gravity.

Above: The second prototype G.91Y blasts skywards during a demonstration of the RATO-equipment at Caselle, Turin. Although successful in reducing the G.91Y's take-off run from 3,800 ft (1158 m) to around 2,000 ft (610 m), it was never routinely used operationally by front-line units.

Left: Early tests of the four-rocket system were conducted by Fiat on G.91R/1 MM.6292 (seen here), and also by the West German air force. The whole assembly could be jettisoned in flight once the rocket motors had exhausted their fuel.

Larger wing
The G.91Y's wing benefitted from a 10 per cent increase in area and the addition of wing-leading edge slats that, in unaccelerated flight, opened automatically at 230 kt (265 mph; 426 km/h) providing between 30-40 per cent more lift. The slats also reduced the wing loading on landing, bestowing a landing run similar to the lighter G.91R.

Cockpit
A number of new features were introduced to the AMI by the G.91Y, including a head-up display (HUD). Designed by Specto and built under licence by OMI in Italy, the HUD was linked to the flight computer and provided rudimentary flight and targeting data. In case of emergency the pilot sat on a Martin-Baker zero-zero rocket ejection seat.

Aeritalia G.91Y

13° Gruppo, 32° Stormo, Brindisi, 1985

G.91Y specification

Powerplant: two General Electric J85-GE-13A turbojet engines, each rated at 2,725 lb (12.12 kN) dry and 4,080 lb (18.15 kN) with afterburning
Maximum speed: Mach 0.95 at 30,000 ft; 715 mph (1150 km/h) at sea level
Max. rate of climb at sea level: 17,000 ft (5180 m) per minute
Operational ceiling: 45,930 ft (14000 m)
Take-off run: 2,820 ft (860 m) at 17,150 lb (7780 kg)
Landing run: 1,970 ft (600 m) from 50 ft (15 m)
Combat radius: 372 miles (600 km) on lo-lo-lo mission with 2,910 lb (1320 kg) load
Ferry range: 2,175 miles (3500 km)
Weights: basic empty 8,117 lb (3682 kg); normal take-off 17,196 lb (7800 kg); maximum take-off 19,180 lb (8700 kg)
Fuel load: internal 704 Imp gal (3200 litres); max. with drop tanks 1,056 Imp gal (4800 litres)
Weapons load: up to 4,000 lb (1814 kg)
Wing span: 29 ft 6.5 in (9.01 m)
Wing area: 195.15 sq ft (18.13 m²)
Length: 38 ft 3.5 in (11.67 m)
Height: 14 ft 6.5 in (4.43 m)

Nav/attack system
A key advance over the G.91R was the addition of an attack and navigation aid. The system was built under licence and consisted of a Computing Devices of Canada Model 5C-15 position and homing indicator (PHI), a Sperry SYP twin-axis gyroscopic platform and a Bendix RDA-12 Doppler radar. Up to 12 pre-selected fixed points or targets could be set-up in the PHI computer which, thanks to the Doppler, was fed ground-speed data allowing continuous update of the aircraft's position in relation to any of the pre-selected points.

Weapons and fire control
The larger wing and twin engines allowed the G.91Y to significantly increase the type's payload (by some 73 percent). All four wing pylons were stressed to carry loads of up to 1,000 lb (454 kg) with the inner pylons plumbed for the carriage of auxiliary fuel tanks. In addition to bombs, rockets, napalm and nuclear devices, the G.91Y had a fixed armament of two 30-mm DEFA cannon aimed using a Ferranti ISIS B gyro gun sight.

Powerplant
The most significant difference between the G.91Y and other models was the incorporation of two General Electric J85-GE-13A turbojets, increasing total thrust to 8,160 lb (36.3 kN). Although offering twin-engined safety and improvements in performance, the twin-engine layout added unnecessary complexity and raised operating and maintenance costs.

Undercarriage
To cope with the additional weight, the main landing gear was 'beefed-up' with the addition of a larger shock strut, additional lateral bracing, more powerful hydraulic brakes and larger tyres pressurised to 59-76 psi, depending on weight.

AMI G.91Ys became a favourite canvas of budding artists within both 8° and 32° Stormi. For Christmas 1983 101° Gruppo, 8° Stormo painted MM.6488 in a dark blue scheme, on which was superimposed a yellow comet stretching back to the fin, and the word 'Auguri' (greetings) painted on the belly (above). Other special schemes included a 13° Gruppo, 32° Stormo G.91Y (MM.6494), with a large black shark running down the fuselage sides (above right) and, to commemorate the G.91's retirement from Italian service, MM.6244 received a graduated grey-black scheme with a large red lightning flash on the fuselage onto which the gruppo's identity of 101° CBR was superimposed (right).

improved by adopting bigger, low-pressure tyres and stronger speed brakes. A parachute-brake was also installed.

The fuel system was formed by six tanks in the fuselage, for a total of 572 Imp gal (2600 litres), plus two tanks in the wings for another 66 Imp gal (300 litres) each. Two external fuel tanks could be carried on the inner wing pylons, capable of holding 57, 114 or 176 Imp gal (260, 520 or 800 litres), according to the type; the 114-Imp gal (520-litre) version was most commonly used.

Armament was composed of two 30-mm DEFA 552 guns (with 125 rounds each) and a maximum external load of 4,000 lb (1815 kg), divided between four underwing pylons. G.91Y weaponry included 500-lb Mk 82 and 1,000-lb Mk 83 slick and retarded bombs, BL755 cluster bombs, 500-lb napalm bombs, Orione, LAU-3A and LAU-18A rocket pods.

In accordance with its reconnaissance capability, the G.91Y was equipped with a range of photo-cameras in the nose. It received three de Oude Delft TA-7M2 and one vertical Fairchild KA-60C camera.

The first prototype, NC.2001 (military serial MM.579), took off for the first time from Torino-Caselle on 12 December 1966, piloted by Vittorio Sanseverino. In September 1967 the second prototype (NC.2002, MM.580) flew. This aircraft introduced modifications that had come from the first test activity, such as new air intakes for the afterburner sections, a new windscreen

and canopy, and a complete armament suite.

G.91Y operations

The Italian Ministry of Defence ordered an initial batch of 20 series aircraft, the first making its maiden flight in June 1968. Military serials assigned were in the range MM.6441 to 6460. The first unit to fly the type was the Reparto Sperimentale Volo, which received some aircraft from 1968 for operational test and evaluation.

In 1969, Italy underwent sweeping industrial reorganisation. In November, Fiat Divisione Aviazione (Aviation Division, including airframe and engines production) merged with Aerfer and Salmoiraghi (the latter part of IRI-Finmeccanica, a holding controlled by the Italian government) to become the new Aeritalia. From this entity, the MoD ordered a second batch of G.91Ys – 55 aircraft, later reduced to 45, with assigned serials from MM.6461 to 6495 and from MM.6951 to 6960. Eight other MMs (from 6961 to 6968) were allocated but never used. The second prototype was later introduced into service, changing its serial from MM.580 to MM.6440.

Along with the G.91YS, a second version studied by Aeritalia was the two-seat G.91YT, for use in pilot conversion, but – already having a numerous fleet of excellent G.91Ts – the Italian Air Force decided not to buy this model. The most advanced development was the G291, a fighter featuring a larger wing (with increased chord and flap section) and two additional pylons under the fuselage, near the wing leading

edge. It never went beyond the design board.

The first series production G.91Ys were assigned to 101° Gruppo of 8° Stormo, a unit still flying the old F-84F at Cervia air base. In October 1969 a first group of pilots and ground crew was detached to the Test Wing to convert to the new aircraft; from 1 December 1969 the squadron, together with the RSV personnel, started flying intensive trials of the weapon system. On 1 April 1970 the first 'Yankee' (as the aircraft was nicknamed) arrived at Cervia, and the squadron was redesignated Gruppo Caccia Bombardieri Ricognitori (CBR). On 23 January 1971, 101° Gruppo, with its complete fleet of 18 aircraft, was declared operational on the type.

Problems emerge

From the start, the G.91Y suffered several teething problems, a typical occurrence with many new aircraft. One in particular caused great concern: in certain flight conditions or during harsh use of the throttles, the peculiar air intake design led to an irregular air influx to one of the engines, causing a compressor stall. It was probable that the other engine would face the same emergency, because they shared a common duct. At least one aircraft was lost due to this problem.

The pilot of an 8° Stormo G.91Y, wearing the unit's famous red lightning strike, prepares for take-off during a NATO exercise in 1991. Note the guard surrounding the cannon muzzle to prevent damage on the ground, and the forward and lateral camera ports.

G.91YS for Switzerland

Designed to meet a Swiss requirement, the G.91YS incorporated a third wing pylon for the carriage of AIM-9L air-to-air missiles. The single demonstrator made its maiden flight in October 1970 and, although it was judged only half as effective as the A-7G, it would have cost some 50 per cent of its rival, allowing many more to be procured.

In 1969, Fiat (later Aeritalia) started its marketing efforts, trying to sell its new aircraft to foreign customers. In that year, the G.91Y participated in the Swiss Air Force contest to find a new fighter-bomber to replace its old de Havilland Venoms. To meet the Swiss specifications, the aircraft received a new Saab fire control system, a laser rangefinder in a modified nose, and two additional wing pylons, designed to carry AIM-9 Sidewinder air-to-air missiles. The prototype of the so-called G.91YS was NC.2023 (MM.6461), which was also equipped with two more ventral fins. This version had a maximum take-off weight increased to 19,841 lb (9000 kg), a warload increased to 4,410 lb (2000 kg), and a take-off run 20 per cent longer than the standard G.91Y; general performance remained unchanged.

The two shortlisted finalists of the Swiss contest were the American LTV A-7G Corsair II and the Italian G.91YS. The Swiss delayed making a final decision, possibly in an attempt to refuse both aircraft. On 4 May 1971 the Corsair was declared the winner, but additional refurbished Hawker Hunters were acquired instead. In 1976 the Swiss Air Force decided to buy the Northrop F-5E/F as a successor to the Venom.

In the meantime, MM.6461 was converted by Aeritalia back to its original standard and delivered to the Italian air force. It was assigned first to the Test Wing (code RS-12), and later to operational squadrons. Aside from the serial number, the aircraft remained recognisable only by its different nose.

G.91Y export campaign

With Fiat's marketing efforts lacking official support it was hard to gain orders from foreign air forces. From the outset, the G.91Y's prospects in the export market were far less positive than those of its predecessor. In addition, the Italian aircraft fared badly compared to the aggressive marketing policy of other industries, especially American ones, which were clearly supported by their own governments. In September 1975 Aeritalia made a promotional tour of Egypt, introducing the G222 transport, the G.91T trainer, and the G.91Y (aircraft NC.2048, MM.6486, for the occasion wearing civil registration I-MARH) at Almaza airport (Cairo), but collected no orders.

A solution was found by introducing a separating wall in the air duct, and by changing procedures in both normal and emergency operations.

In July 1973 the G.91Y had an opportunity to show off its abilities. The 5th Allied Tactical Air Force (ATAF) organised the Best Hit exercise, which was based at Istrana to take advantage of the Maniago firing range. Units from Italian, Greek, Turkish, USAFE and US Navy forces participated in this exercise/competition. 101° Gruppo team gained a respectable second place, and one of its pilots, Lt Col Paolo Francini, won the Top Gun Trophy.

The re-equipment of a second 'Yankee' squadron was delayed. First aircraft was assigned to 13° Gruppo of 32° Stormo only on 1 August 1973, with conversion completed by September 1974. With the new aircraft, 13° Gruppo maintained its original roles of maritime attack and reconnaissance, as already performed with the G.91R.

The Italian Air Force was never fully satisfied with the G.91Y, and, adding budget problems to this fact, it is easy to understand why the 'Y' never completely replaced the G.91R. Another 40 aircraft would have been necessary to replace the 'R' in the two squadrons of the 2° Stormo, but they were never ordered. For the same reasons, no upgrading projects were carried out on the 'Yankee', as the AMI preferred to concentrate its efforts on the new Panavia MRCA programme.

The situation of the 'Yankee' in Italian service

Right: In the early 1990s the remaining G.91Ys adopted low-visibility code markings, as demonstrated by this line-up of 101° Gruppo, 8° Stormo examples at their home base of Cervia in June 1993.

remained stable until 1993, when it became possible to begin re-equipping the two squadrons with the new AMX fighter-bomber. In late June 1993, 13° Gruppo passed all its remaining operational aircraft to 101° Gruppo, and moved to the air base at Amendola, flying the G.91T/1 until the arrival of the first AMX on 30 November 1994. 101° Gruppo continued its operations with the G.91Y through 1994, marking the end of the aircraft's career on 26 November 1994 when the official last flight took place.

Conclusions

According to many observers, Fiat's G.91R and 'T' were quite successful aircraft when introduced into service, and continued so until the mid-1970s. The former was well thought of in the close air support role, and the latter in the training mission. The G.91T/1 was used by the AMI for advanced training, prior to pilot assignment to the F-104 or Tornado OCU.

It is difficult to summon such positive considerations when talking about the G.91Y. This type was much more expensive, heavier and complex than the 'R' model, yet offered offensive capabilities not much much better than the German G.91R/3. Its J85 engines were too sophisticated and delicate for a tactical aircraft that was designed to be able to operate from unpaved strips. In addition, the Y-shaped single air intake that fed two engines presented too many problems. The aerodynamic design, good for a simple and light aircraft of the 1950s, was outdated by the 1960s.

Due to these shortcomings, and the total lack of support from the Italian government, Aeritalia was never able to sign an export contract. Nevertheless, the G.91Y served for 24 years in the AMI's front-line fleet, and within NATO's 5th ATAF, playing an important role, especially in the tactical reconnaissance mission, until the end of its career.

Dr Riccardo Niccoli

In the final days of German, Italian and Portuguese front-line service, G.91s were painted in a range of gaudy schemes to celebrate their retirement. The last Luftwaffe G.91 to fly, in March 1982, was this specially-painted T/3 (32+02) which had passed, at the end of its career, from Waffenschule 50 to JBG 49, during that unit's conversion to the Alpha Jet. Shortly after the last flight the aircraft was placed on display at Fürstenfeldbruck. The other surviving G.91T/3s were either scrapped, used as ground trainers, passed onto museums or delivered for use by the Portuguese air force.

G.91 production data

Model	Total	C/n	Serials	Manufacturer
G.91 prototypes	4	1-3	MM.565-567	Fiat (1)
G.91 pre-series	27	4-30	MM.6238-6264	Fiat
G.91A	1	31	MM.6265	Fiat (2)
G.91R/1	22	32-53	MM.6266-6287	Fiat
G.91R/1A	25	154-178	MM.6290-6314	Fiat
G.91R/1B	50	179-228	MM.6375-6424	Fiat
G.91R/3	50	54-89, 91-97, 102-108	3001-3043	Fiat (3)
G.91R/4	50	90, 98-101, 109-153	-	Fiat (4)
G.91R/3	295	301-595	3044-3323	Dornier (3)
G.91T	2	1-2	MM.6288-6289	Fiat
G.91T/1	101	45-144	MM.6315-6374,	Fiat
			MM.6425-6439,	Fiat
			MM.54392-54417	Fiat
G.91T/3	44	1-44	3401-3440	Fiat (5)
G.91T/3	22	601-622	3441-3462	Dornier (5)
G.91Y prototypes	2	2001-2002	MM.579-580	Fiat (6)
G.91Y	55	2003-2057	MM.6441-6495	Fiat (7)
G.91Y	10	2058-2067	MM.6951-6960	Aeritalia

(1) First prototype NC.1 had no MM, second was NC.1bis, MM.565
(2) Later converted to G.91R/1
(3) German serial system introduced on 1 January 1968. Twenty-one G.91R/3s and four G.91T/3s were written-off before this date and were not allocated serials
(4) No serials issued for the G.91R/4 (unit codes: BD+231-255, BD+361-385)
(5) Sometimes referred to as G.91T/1
(6) MM.580 later re-serialled as MM.6440
(7) MM.6461 became G.91YS prototype, later re-converted to G.91Y standard

ITALY

The Italian government had been a staunch supporter of Fiat's bid to win the NATO LWSF competition, and therefore unsurprisingly ordered the aircraft for the Aeronautica Militare Italiana (AMI) even before the results of the competition had been announced. In total Fiat delivered four prototypes, 225 pre-production and production G.91s/G.91Rs, two G.91T two-seat trainer prototypes and 101 production G.91T/1s. These were later supplemented by two prototype and 65 production G.91Ys – taking the total production for Italy to an impressive 399 aircraft.

At the time of the G.91's introduction into service the AMI was in the process of a major re-organisation of its command and structure. The Aerobrigata (Aerobrigades) were transformed into smaller stormi (equivalent to wings) each with two or three gruppi (equivalent to squadrons) which in turn each had two or three squadriglie (equivalent to flights). At the same time the unit names changed from role-based titles to an association with famous Italian airmen.

Prototype and early pre-series aircraft were delivered for test, trials and early evaluation work in a bare metal finish. In front-line service the aircraft wore the standard NATO grey/green camouflage with cerulean blue undersides, large white codes and large AMI roundels on the fuselage sides and outer sections of the wings. Unit insignia were usually carried on the tail fin. Later the undersides were changed to pale grey/silver and, from the mid-1980s, large stencil-type outline codes were introduced. In 1989 toned-down and smaller national insignia were introduced along with pale blue/dark grey codes, and all colourful unit insignia was removed.

With the force structure in a state of flux, gruppi moved between stormi and often acted as autonomous

units for short periods. Both Aerobrigata/Stormo and individual autonomous Gruppi are presented below.

Aerobrigata/Stormo/Scuola

Unit	Variants	Gruppi	Period
2ª Aerobrigata	G.91	14°	4/60-3/61
2° Stormo	G.91, R, R/1 R/1A, R/1B, PAN, T/1	14°, 103°, 602ª Sq.[1]	10/64-4/92
5ª Aerobrigata	G.91, R	103°	6/58-9/62
8° Stormo	G.91Y	101°, 608ª Sq.[1]	4/70-12/94
32° Stormo	G.91R/1 R/1B, Y, T/1	13°, 201° 632ª Sq.[1]	9/67-9/74(R) 8/73-11/92(Y) 11/92/9/95(T)
51ª Aerobrigata	G.91, R	Reparto Volo CTRL	9/62-10/64
60ª Brigata	G.91T/1	201°, 204°	6/85-3/93
SVBAA	G.91T/1	201°, 204°, 205°	11/64-6/85

Automonous units

Unit	Variants	Role	Period
13° Gruppo	G.91R, R/1	attack/recce	3/65-9/67
14° Gruppo	G.91, R	conversion and training	9/62-10/64
101° Gruppo	G.91Y	conversion to G.91Y	6/69-4/70
303° Gruppo	G.91T/1	liaison/support	3/78-?
311° Gruppo (RSV)	G.91, R, R/1 R/1A, R/1B, A, T, T/1, Y	test and evaluation	1956-92
313° Gruppo (Freece Tri.)	G.91PAN	aerobatic display team	12/63-4/82
512ª Sq.[1]	G.91T/1	liaison	3/78-?

[1] Squadriglia Collegamento – liaison flight

Above: A yellow tail band was worn by 103° Gruppo, 2° Stormo aircraft in the mid-1980s, as seen on this G.91R/1.

Above: In January 1998 101° Gruppo of 8° Stormo applied these special markings to one of its G.91Ys in celebration of reaching 50,000 hours on the type.

Above: 311° Gruppo G.91T/1s of the Reparto Sperimentale Volo (RSV) during evaluation at Pratica di Mare in 1960.

In September 1962 a Reparto Volo was formed as part of 51ª Aerobrigata to evaluate the G.91R. The aircraft wore 51ª codes and the unit's distinctive black cat badge.

In 1986 the SVBAA was redesignated as 60ª Brigata Aerea and the unit's G.91Ts correspondingly changed their codes to match their new identity.

WEST GERMANY

After receiving 50 Fiat-built G.91Rs, a German consortium, led by Dornier, built a further 294 examples. Added to these were the 50 Fiat-built G.91R/4s, originally intended for Greece and Turkey. In addition, 66 G.91T/3 two-seat trainers were acquired (22 built locally), bringing total G.91 procurement to 460 – making Germany the largest operator of the type.

Originally the front-line units consisted of two Aufklürungsgeschwadern (AKGs; reconnaissance wings) and two Jagdbombergeschwadern (JBGs; fighter-bomber wings) each with three staffeln (squadrons). In 1966-67 these were reorganised into four Leichtenkampfgeschwadern (LKGs; light combat wings), each with one attack and one recce staffel.

Training and conversion was the responsibility of Waffenschule 50 (WS 50) which, for the majority of its

existence, was based at Fürstenfeldbruck operating the majority of T/3s and the machine-gun-equipped R/4s, as well as standard R/3s.

Luftwaffe G.91s wore standard grey/green camouflage, at first with cerulean and later with light grey undersides. Codes were originally all-black and encompassed a two-letter unit identifier, followed by a three-digit number (the first of which referred to the allocated staffeln). In 1967 a white outline was added before, in 1968, this code system was abandoned in favour of a four-digit numerical serial.

Unit badges were routinely applied to the tailfin and more 'exotic' markings, including 'shark's mouths' were occasionally added in later years.

During the 1970s many aircraft had the centre portion of their underwing fuel tanks painted Dayglo orange, while aircraft assigned to training units also had Dayglo noses, wingtips and rudders.

Above: JBG 41, based at Husum, was one of the original front-line recipients of the G.91 wearing 'DG+' codes.

Above: During the introduction of the Starfighter into Luftwaffe service the test unit employed this G.91T/3.

Left: LKG G.91s, such as this G.91R/3 from LKG 44, wore two-letter unit codes for a brief period between 1966-68.

Luftwaffe front-line units

Unit	Variant	Code	Base	Period
AKG 53	G.91R/3	EC+	Erding	1961-66
AKG 54	G.91R/3	ED+	Oldenburg	1964-66
JBG 35	G.91R/3	DE+	Husum/ Pferdsfeld	1961,75
JBG 41	G.91R/3	DG+	Husum	1961-66 1981-82
JBG 42	G.91R/3	DH+	Pferdsfeld	1964-67
JBG 43	G.91R/3	-		1981
JBG 49	G.91R/3, T/3	-	Fürstenfeld.	1980-82
JBG 72	G.91R/3	JB+	Oldenburg	1964
JG 73	G.91R/3	JC+	Pferdsfeld	1964
LKG 41	G.91R/3, T/3	MA+	Husum	1966-81
LKG 42	G.91R/3, T/3	MB+	Pferdsfeld	1967-75
LKG 43	G.91R/3, T/3	MC+	Oldenburg	1966-81
LKG 44	G.91R/3, T/3	MD+	Leipheim	1966-75

German support/training units

Unit	Variant	Code	Base	Period
AKG Res.	G.91R/3	ER+	various	1962-66
ESt 61	G.91R/3, R/4, T/3	YA+	Manching	1960-80
F-104 Test	G.91T/3	KC+	various	1961-64
G.91 Test	G.91R/3, T/3	KD+	various	1961-63
'Kom Deci'	G.91R/3	-	Decimomannu	1974-77
Condor Flugdienst	G.91R/3, T/3	D- or 99+	Westerland-Sylt	1978-93
LKG Res.	G.91R/3	MR+	various	1966-80
LVS G.91	G.91R/3, T/3	XB+	Erding	1960-64
WS 50	G.91R/3, R/4, T/3	MD+	Fürstenfeld.	1961-80

Above: WS 50 was responsible for all G.91 training and operated a mix of G.91R/3s (seen here), R/4s and T/3s.

Above: Luftwaffe target-towing needs were met by Schiessplatzstaffel G.91s operated by Condor Flugdienst.

Above: During Exercise Bulls Eye in 1979 LKG 41's G.91R/3s were decorated with this simple shark's mouth.

Above: In the 1980s this ex-LKG 43 G.91R/3 was used for ground instruction by the F-4 Werft (overhaul unit).

Above: As the Alpha Jet (behind) was introduced, remaining WS 50 G.91T/3s were reallocated to JBG 49.

Above: Among the non-flying units to regularly receive G.91s was the repair specialist Versorgungsregimenten 6.

PORTUGAL

Portugal's G.91s were the only examples to see combat operating in the former Portuguese colonies of Guinea-Bissau, Angola and Mozambique from 1966-74. The FAP initially received an initial batch of 40 G.91R/4s from West Germany in 1965-66. Originally intended for Greece and Turkey, these aircraft had different armament and avionics from the standard Luftwaffe G.91R/3s, and were the logical choice for disposal.

Between 1976-82 Portugal received a further 26 G.91T/3s and 70 R/3s as they were retired from Luftwaffe service, many of which were used for spares recovery. The G.91R/4s were phased-out in the 1980s, leaving Esq. 301 as the sole surviving unit with G.91R/3s and T/3s until the type was finally retired from service in June 1993.

Força Aérea Portuguesa units

Unit	Variant	Combat	Period
Esq. 51 'Falcoes'	G.91R/4	None	1966-75
Esq. 62 'Jaguares'	G.91R/3, R/4, T/3	None	1974-77
Esq. 93 'Escorpioes'	G.91R/4	Angola (1974)	1974-75
Esq. 121 'Tigres'	G.91R/4	Guinea-Bissau (1966-74)	1966-74
Esq. 301 'Jaguares'	G.91R/3, T/3	None	1977-93
Esq. 303 'Tigres'	G.91R/4, T/3	None	1981-89
Esq. 502 'Jaguares'	G.91R/4	Mozambique (1969-74)	1969-74
Esq. 702 'Escorpioes'	G.91R/4	Mozambique (1970-74)	1970-74

Right: One of the very first ex-Luftwaffe G.91R/4s delivered to the Portuguese air force was 5432. The aircraft is seen here in 1966 at Montijo air base shortly after delivery to Esquadra 51 'Falcoes'. The unit was responsible for much of the G.91 evaluation work in preparation for operational deployment in Africa.

Above: In 1982 the Portuguese air force changed its G.91 scheme from NATO grey/green to this green/tan camouflage more suited to the Mediterranean theatre. Here, at a NATO Tiger Meet at RAF Gütersloh, West Germany, in September 1982, an Esquadra 301 G.91T/3 and G.91R/3 display the old and new schemes respectively.

Surviving Portuguese single-seat G.91s in the 1990s were the cannon-armed G.91R/3s flown by the 'Jaguares' of Esq. 301. From 1980 Esq 301's 'Ginas' gained a secondary air defence commitment following a modification to carry the AIM-9 Sidewinder missile.

OTHER OPERATORS

Greece and Turkey were due to take half each of 50 G.91R/4s built by Fiat with US funding. Around six aircraft were painted in Greek colours and aircraft 109 (below) was evaluated at Larissa before both orders were cancelled. Additionally, two G.91Rs (G-45-4 and -5) were used by Bristol Siddeley for Orpheus development work at Filton, UK, and a total of three G.91s (an R/1 an R/3 and a T/1) was evaluated by the US Army to fulfil a 'Fast FAC' requirement. Later a further R/1 was evaluated by the USAF at Kirtland AFB. No export orders for the G.91Y ever materialised, despite the development of the G.91YS to meet a Swiss air force requirement.